Student's Soluti

Jerome E. Kaufmann

ALGEBRA FOR COLLEGE STUDENTS

fifth edition

prepared by

Karen L. Schwitters
Jesse R. Turner
Seminole Community College

PWS Publishing Company

I(T)P An International Thomson Publishing Company

Boston • Albany • Bonn • Cincinnati • Detroit • London • Madrid • Melbourne
Mexico City • New York • Paris • San Francisco • Singapore • Tokyo • Toronto • Washington

I(T)P™
International Thomson Publishing
The trademark ITP is used under license.

ISBN: 0-534-94911-8

Printed and bound in the United States of America by Malloy Lithographing
6 7 8 9 10 — 99

PREFACE

This Student's Solutions Manual is a supplement to accompany the text Algebra For College Students, 5th Edition by Jerome Kaufmann. The manual shows the solutions for all the odd problems in the problem sets. If the solution to the problem involves a graph as a solution, consult the answer portion of your textbook for the graph. This manual does not include solutions for Thoughts into Words, graphing calculator activities, and Further Investigations problems.

To use this manual effectively, it should be consulted only after the student has attempted the solution and has been unsuccessful in solving the problem. As the student uses this manual remember that the manual presents only one method of solution for a problem. The solutions presented here parallel the methods used in the textbook.

Karen L. Schwitters
Jesse R. Turner
Seminole Comunity College
Sanford, FL

CONTENTS

Chapter 1 Basic Concepts and Properties

Problem Set 1.1 Sets, Real Numbers, and Numerical Expressions

1. True
3. False
5. True
7. False
9. True
11. 0 and 14
13. 0, 14, $\frac{2}{3}$, $-\frac{11}{14}$, 2.34, $3.2\overline{1}$, $6\frac{7}{8}$, -19, and -2.6
15. 0 and 14
17. All of them
19. $R \nsubseteq N$
21. $I \subseteq Q$
23. $Q \nsubseteq H$
25. $N \subseteq W$
27. $I \nsubseteq N$
29. $\{1, 3, 5, 7, \ldots\} \subseteq I$
31. $\{0, 3, 6, 9, \ldots\} \nsubseteq N$
33. $\{1, 2\}$
35. $\{0, 1, 2, 3, 4, 5\}$
37. $\{\ldots, -1, 0, 1, 2\}$
39. $\emptyset$
41. $\{0, 1, 2, 3, 4\}$
43. -6
45. 2
47. $3x+1$
49. $5x$
51. $16+9-4-2+8-1$
    $25-4-2+8-1$
    $21-2+8-1$
    $19+8-1$
    $27-1$
    26
53. $9 \div 3 \bullet 4 \div 2 \bullet 14$
    $3 \bullet 4 \div 2 \bullet 14$
    $12 \div 2 \bullet 14$
    $6 \bullet 14$
    84
55. $7+8 \bullet 2$
    $7+16$
    23
57. $9 \bullet 7-4 \bullet 5-3 \bullet 2+4 \bullet 7$
    $63-20-6+28$
    $43-6+28$
    $37+28$
    65
59. $(17-12)(13-9)(7-4)$
    $(5)(4)(3)$
    $(20)(3)$
    60
61. $13+(7-2)(5-1)$
    $13+(5)(4)$
    $13+20$
    33
63. $(5 \bullet 9-3 \bullet 4)(6 \bullet 9-2 \bullet 7)$
    $(45-12)(54-14)$

$(33)(40)$

1320

65. $7[3(6-2)]-64$

$7[3(4)]-64$

$7(12)-64$

$84-64$

20

67. $[3+2(4\bullet 1-2)][18-(2\bullet 4-7\bullet 1)]$

$[3+2(4-2)][18-(8-7)]$

$[3+2(2)][18-1]$

$(3+4)(17)$

$(7)(17)$

119

69. $14+4\left(\frac{8-2}{12-9}\right)-2\left(\frac{9-1}{19-15}\right)$

$14+4\left(\frac{6}{3}\right)-2\left(\frac{8}{4}\right)$

$14+4(2)-2(2)$

$14+8-4$

$22-4$

18

71. $[7+2\bullet 3\bullet 5-5]\div 8$

$[7+6\bullet 5-5]\div 8$

$[7+30-5]\div 8$

$(37-5)\div 8$

$(32)\div 8$

4

73. $\frac{3\bullet 8-4\bullet 3}{5\bullet 7-34}+19$

$\frac{24-12}{35-34}+19$

$\frac{12}{1}+19$

$12+19$

31

Problem Set 1.2 Operations with Real Numbers

1. $8+(-15)$

-7

3. $(-12)+(-7)$

-19

5. $-8-14$

-22

7. $9-16$

-7

9. $(-9)(-12)$

108

11. $(5)(-14)$

-70

13. $(-56)\div(-4)$

14

15. $\frac{-112}{16}$

-7

17. $-24+52$

28

19. $19-(-14)$

$19+14$

33

21. $(-17)(12)$

-204

23. $78\div(-13)$

-6

25. $0\div(-14)$

0

27. $(-21)\div 0$

undefined

31. $-42-(-25)$

$-42+25$

-17

33. $(-4)(-3)(6)$

$(12)(6)$

72

35. $9(-8)(5)$

$(-72)(5)$

-360

37. $(-3)(-7)(-6)$

$(21)(-6)$

-126

39. $-\left(\frac{-56}{-7}\right)$

$-(8)$

-8

41. $9-12-8+5-6$

$-3-8+5-6$

$-11+5-6$

$-6-6$

-12

43. $-21+(-17)-11+15-(-10)$

$-38-11+15+(+10)$

$-49+15+(+10)$

$-34+(+10)$

-24

45. $7-(9-17)$

$7-(-8)$

$7+8$

15

47. $16-18+19-[14-22-(31-41)]$

$16-18+19-[14-22-(-10)]$

$16-18+19-(14-22+10)$

$16-18+19-(-8+10)$

$16-18+19-(2)$

$16-18+19-2$

$-2+19-2$

$17-2$

15

49. $[14-(16-18)]-[32-(8-9)]$

$[14-(-2)]-[32-(-1)]$

$(14+2)-(32+1)$

$16-33$

-17

51. $6-5(-3)$

$6+15$

21

53. $-5+(-2)(7)-(-3)(8)$

$-5+(-14)-(-24)$

$-5+(-14)+24$

$-19+24$

5

55. $5(-6)-(-4)(11)$

$-30-(-44)$

$-30+44$

14

57. $(-6)(-9)+(-7)(4)$

$54-28$

26

59. $3(5-9)-3(-6)$

$3(-4)+18$

$-12+18$

6

61. $(6-11)(4-9)$

$(-5)(-5)$

25

63. $-6(-3-9-1)$
$-6(-12-1)$
$-6(-13)$
78

65. $56 \div (-8) - (-6) \div (-2)$
$-7-(3)$
$-7-3$
-10

67. $-3[5-(-2)]-2(-4-9)$
$-3(5+2)-2(-13)$
$-3(7)+26$
$-21+26$
5

69. $\frac{-6+24}{-3}+\frac{-7}{-6-1}$
$\frac{18}{-3}+\frac{-7}{-7}$
$-6+1$
-5

71. $[(-3)(4)-(-2)(1)][(-2)(-7)-(-8)(6)]$
$[-12-(-2)][14-(-48)]$
$(-12+2)(14+48)$
$(-10)(62)$
-620

73. $(7-11)[-3-(6-7)]-12$
$(-4)[-3-(-1)]-12$
$(-4)(-3+1)-12$
$(-4)(-2)-12$
$8-12$
-4

75. $(-5.4)+(-7.2)$
-12.6

77. $(12.2)+(-5.6)$
6.6

79. $-17.3+12.5$
-4.8

81. $(-2.1)(-3.5)$
7.35

83. $(5.4)(-7.2)$
-38.88

85. $\frac{-1.2}{-6}$
$.2$

87. $(16.8) \div (-1.2)$
-14

89. $-21.4-(-14.9)$
$-21.4+14.9$
-6.5

91. $1.42-7.29$
-5.87

93. $-2.11-4.67$
-6.78

Problem Set 1.3 Properties of Real Numbers and the Use of Exponents

1. Associative property of addition

3. Commutative property of addition

5. Additive inverse property

7. Multiplicative property of negative one

9. Commutative property of multiplication

11. Distributive property

13. Associative property of multiplication

15. $36+(-14)+(-12)+21+(-9)-4$
$36+21+(-14)+(-12)+(-9)+(-4)$
$57+(-39)$
18

17. $[83+(-99)]+18$

$-16+18$

2

19. $25(-13)(4)$

$25(4)(-13)$

$100(-13)$

-1300

21. $17(97)+17(3)$

$17(97+3)$

$17(100)$

1700

23. $14-12-21-14+17-18+19-32$

$14-14-21-12-18-32+17+19$

$0-83+36$

-47

25. $(-50)(15)(-2)-(-4)(17)(25)$

$15(-50)(-2)-(17)(-4)(25)$

$15(100)-17(-100)$

$1500+1700$

3200

27. 2^3-3^3

$8-27$

-19

29. -5^2-4^2

$-25-16$

-41

31. $(-2)^3-3^2$

$-8-9$

-17

33. $3(-1)^3-4(3)^2$

$3(-1)-4(9)$

$-3-36$

-39

35. $7(2)^3+4(-2)^3$

$7(8)+4(-8)$

$56-32$

24

37. $-3(-2)^3+4(-1)^5$

$-3(-8)+4(-1)$

$24-4$

20

39. $(-3)^2-3(-2)(5)+4^2$

$9-3(-2)(5)+16$

$9+6(5)+16$

$9+30+16$

$39+16$

55

41. $2^3+3(-1)^3(-2)^2-5(-1)(2)^2$

$8+3(-1)(4)-5(-1)(4)$

$8-3(4)+5(4)$

$8-12+20$

$-4+20$

16

43. $[4(-3)+5(-2)^2]^2$

$[4(-3)+5(4)]^2$

$(-12+20)^2$

$(8)^2$

64

45. $[3(-2)^2-2(-3)^2]^3$

$[3(4)-2(9)]^3$

$(12-18)^3$

$(-6)^3$

-216

47. $2(-1)^3-3(-1)^2+4(-1)-5$

$2(-1)-3(1)+4(-1)-5$

$-2-3-4-5$

$-5-4-5$

$-9-5$

-14

49. $2^4-2(2)^3-3(2)^2+7(2)-10$

$16-2(8)-3(4)+7(2)-10$

$16-16-12+14-10$

$0-12+14-10$

$-12+14-10$

$2-10$

-8

51. $5(-1)^4-4(-1)^3-3(-1)^2+8(-1)-14$

$5(1)-4(-1)-3(1)+8(-1)-14$

$5+4-3-8-14$

$9-25$

-16

53. $-3^2-2^3+4(-1)^4-(-2)^3+11$

$-9-8+4(1)-(-8)+11$

$-9-8+4+8+11$

$-17+4+8+11$

$-13+8+11$

$-5+11$

6

55. Calculator problems

57. 3^7

2187

59. $(-2)^{11}$

-2048

61. -5^6

$-1 \bullet 5^6$

$-1 \bullet 15625$

$-15,625$

63. $(1.41)^4$

3.95254161

Problem Set 1.4 Algebraic Expressions

1. $-7x+11x$

$(-7+11)x$

$4x$

3. $5a^2-6a^2$

$(5-6)a^2$

$-a^2$

5. $4n-9n-n$

$(4-9-1)n$

$-6n$

7. $4x-9x+2y$

$(4-9)x+2y$

$-5x+2y$

9. $-3a^2+7b^2+9a^2-2b^2$

$-3a^2+9a^2+7b^2-2b^2$

$(-3+9)a^2+(7-2)b^2$

$6a^2+5b^2$

11. $15x-4+6x-9$

$15x+6x-4-9$

$(15+6)x-13$

$21x-13$

13. $5a^2b-ab^2-7a^2b$

$5a^2b-7a^2b-ab^2$

$(5-7)a^2b-ab^2$

$-2a^2b-ab^2$

15. $3(x+2)+5(x+3)$

$3x+6+5x+15$

$3x+5x+6+15$

$8x+21$

17. $-2(a-4)-3(a+2)$

$-2a+8-3a-6$

$-2a-3a+8-6$

$-5a+2$

19. $3(n^2+1)-8(n^2-1)$
$3n^2+3-8n^2+8$
$3n^2-8n^2+3+8$
$-5n^2+11$

21. $-6(x^2-5)-(x^2-2)$
$-6x^2+30-x^2+2$
$-6x^2-x^2+30+2$
$-7x^2+32$

23. $5(2x+1)+4(3x-2)$
$10x+5+12x-8$
$10x+12x+5-8$
$22x-3$

25. $3(2x-5)-4(5x-2)$
$6x-15-20x+8$
$6x-20x-15+8$
$-14x-7$

27. $-2(n^2-4)-4(2n^2+1)$
$-2n^2+8-8n^2-4$
$-2n^2-8n^2+8-4$
$-10n^2+4$

29. $3(2x-4y)-2(x+9y)$
$6x-12y-2x-18y$
$6x-2x-12y-18y$
$4x-30y$

31. $3(2x-1)-4(x+2)-5(3x+4)$
$6x-3-4x-8-15x-20$
$6x-4x-15x-3-8-20$
$-13x-31$

33. $-(3x-1)-2(5x-1)+4(-2x-3)$
$-3x+1-10x+2-8x-12$
$-3x-10x-8x+1+2-12$
$-21x-9$

35. $3x+7y$
$3(-1)+7(-2)$
$-3-14$
-17

37. $4x^2-y^2$
$4(2)^2-(-2)^2$
$4(4)-(4)$
$16-4$
12

39. $2a^2-ab+b^2$
$2(-1)^2-(-1)(-2)+(-2)^2$
$2(1)-(-1)(-2)+4$
$2-2+4$
4

41. $2x^2-4xy-3y^2$
$2(1)^2-4(1)(-1)-3(-1)^2$
$2(1)-4(1)(-1)-3(1)$
$2+4-3$
$6-3$
3

43. $3xy-x^2y^2+2y^2$
$3(5)(-1)-(5)^2(-1)^2+2(-1)^2$
$3(5)(-1)-(25)(1)+2(1)$
$-15-25+2$
$-40+2$
-38

45. $7a-2b-9a+3b$
$7a-9a-2b+3b$
$-2a+b$
$-2(4)+(-6)$
$-8-6$
-14

47. $-9a^2+6+7a^2-14$
$-9a^2+7a^2+6-14$
$-2a^2-8$
$-2(-5)^2-8$

$-2(25)-8$

$-50-8$

-58

49. $-2a-3a+7b-b$

$-5a+6b$

$-5(-10)+6(9)$

$50+54$

104

51. $-2(x+4)-(2x-1)$

$-2x-8-2x+1$

$-2x-2x-8+1$

$-4x-7$

$-4(-3)-7$

$12-7$

5

53. $2(x-1)-(x+2)-3(2x-1)$

$2x-2-x-2-6x+3$

$2x-x-6x-2-2+3$

$-5x-1$

$-5(-1)-1$

$5-1$

4

55. $3(x^2-1)-4(x^2+1)-(2x^2-1)$

$3x^2-3-4x^2-4-2x^2+1$

$3x^2-4x^2-2x^2-3-4+1$

$-3x^2-6$

$-3(-4)^2-6$

$-3(16)-6$

$-48-6$

-54

57. $5(x-2y)-3(2x+y)-2(x-y)$

$5x-10y-6x-3y-2x+2y$

$5x-6x-2x-10y-3y+2y$

$-3x-11y$

$-3(-1)-11(-2)$

$3+22$

25

59. πr^2

$(3.14)(8.4)^2$

$(3.14)(70.56)$

221.5584

221.6

61. $\pi r^2 h$

$(3.14)(4.8)^2(15.1)$

$(3.14)(23.04)(15.1)$

$(72.3456)(15.1)$

1092.41856

1092.4

63. $2\pi r^2+2\pi rh$

$2(3.14)(7.8)^2+2(3.14)(7.8)(21.2)$

$2(3.14)(60.84)+(6.28)(7.8)(21.2)$

$(6.28)(60.84)+(48.984)(21.2)$

$382.0752+1038.4608$

1420.536

1420.5

65. $n+12$

67. $n-5$

69. $n(50)$

$50n$

71. $\frac{1}{2}n-4$

73. $\frac{n}{8}$

75. $2n-9$

77. $10(n-6)$

79. $n+20$

81. $2t-3$

83. $n+47$

85. $8y$

87. $25m$

89. $\frac{c}{25}$

91. $n+2$

93. $\frac{c}{5}$

95. $12d$

97. $3y+f$

99. $5280m$

Chapter 1 Review

1a. 67

b. $0, -8$, and 67

c. 0 and 67

d. $0, \frac{3}{4}, -\frac{5}{6}, 8\frac{1}{3}, -8, .34, .2\overline{3}, 67$, and $\frac{9}{7}$

e. $\sqrt{2}$ and $-\sqrt{3}$

3. Substitution property for equality

5. Distributive property

7. Commutative property for addition

9. Multiplicative inverse property

11. $-8+(-4)-(-6)$
$-8+(-4)+(+6)$
$-12+(+6)$
-6

13. $-8-7+4+6-10+8-3$
$-15+4+6-10+8-3$
$-11+6-10+8-3$
$-5-10+8-3$
$-15+8-3$
$-7-3$
-10

15. $-3(2-4)-4(7-9)+6$
$-3(-2)-4(-2)+6$
$6+8+6$
$14+6$
20

17. $[5(-2)-3(-1)][-2(-1)+3(2)]$
$(-10+3)(2+6)$
$(-7)(8)$
-56

19. $(-2)^4+(-1)^3-3^2$
$16+(-1)-9$
$15-9$
6

21. $[4(-1)-2(3)]^3$
$(-4-6)^3$
$(-10)^3$
-1000

23. $3a^2-2b^2-7a^2-3b^2$
$3a^2-7a^2-2b^2-3b^2$
$-4a^2-5b^2$

25. $ab^2-3a^2b+4ab^2+6a^2b$
$ab^2+4ab^2-3a^2b+6a^2b$
$5ab^2+3a^2b$

27. $3(2n^2+1)+4(n^2-5)$
$6n^2+3+4n^2-20$
$6n^2+4n^2+3-20$
$10n^2-17$

29. $-(n-1)-(n+2)+3$
$-n+1-n-2+3$
$-n-n+1-2+3$
$-2n+2$

31. $4(a-6)-(3a-1)-2(4a-7)$
$4a-24-3a+1-8a+14$

$4a - 3a - 8a - 24 + 1 + 14$

$-7a - 9$

33. $-5x + 4y$

$-5(3) + 4(-4)$

$-15 - 16$

-31

35. $-5(2x - 3y)$

$-5[2(1) - 3(-3)]$

$-5(2 + 9)$

$-5(11)$

-55

37. $a^2 + 3ab - 2b^2$

$(2)^2 + 3(2)(-2) - 2(-2)^2$

$4 + 3(2)(-2) - 2(4)$

$4 - 12 - 8$

$-8 - 8$

-16

39. $3(2x - 1) + 2(3x + 4)$

$6x - 3 + 6x + 8$

$6x + 6x - 3 + 8$

$12x + 5$

$12(4) + 5$

$48 + 5$

53

41. $2(n^2 + 3) - 3(n^2 + 1) + 4(n^2 - 6)$

$2n^2 + 6 - 3n^2 - 3 + 4n^2 - 24$

$2n^2 - 3n^2 + 4n^2 + 6 - 3 - 24$

$3n^2 - 21$

$3(3)^2 - 21$

$3(9) - 21$

$27 - 21$

6

43. $4 + 2n$

45. $\frac{2}{3}n - 6$

47. $5n + 8$

49. $5(n + 2) - 3$

51. $37 - n$

53. $2y - 7$

55. $1p + 5n + 25q$

57. length = y yards = 36y inches
width = f feet = 12f inches

$P = 2l + 2w$

$P = 2(36y) + 2(12f)$

$P = (72y + 24f)$ inches

59. $12f + i$

Chapter 1 Test

1. Symmetric property

3. $-4 - (-3) + (-5) - 7 + 10$

$-4 + (+3) + (-5) - 7 + 10$

$-1 + (-5) - 7 + 10$

$-6 - 7 + 10$

$-13 + 10$

-3

5. $5(-2) - 3(-4) + 6(-3) + 1$

$-10 + 12 - 18 + 1$

$2 - 18 + 1$

$-16 + 1$

-15

7. $-4(3 - 6) - 5(2 - 9)$

$-4(-3) - 5(-7)$

$12 + 35$

47

9. $3(-2)^3 + 4(-2)^2 - 9(-2) - 14$

$3(-8) + 4(4) - 9(-2) - 14$

$-24 + 16 + 18 - 14$

$-8+18-14$

$10-14$

-4

11. $[-2(-3)-4(2)]^5$

$(6-8)^5$

$(-2)^5$

-32

13. $3(3n-1)-4(2n+3)+5(-4n-1)$

$9n-3-8n-12-20n-5$

$9n-8n-20n-3-12-5$

$-19n-20$

15. $3a^2-4b^2$

$3(-1)^2-4(3)^2$

$3(1)-4(9)$

$3-36$

-33

17. $-5n^2-6n+7n^2+5n-1$

$2n^2-n-1$

$2(-6)^2-(-6)-1$

$2(36)-(-6)-1$

$72+6-1$

$78-1$

77

19. $-2xy-x+4y$

$-2(3)(9)-(3)+4(9)$

$-54-3+36$

$-57+36$

-21

21. $6n-30$

23. $\frac{72}{n}$

25. length $=$ x yards $=$ 3x feet
width $=$ y feet

Perimeter $= 2l+2w$
Perimeter $= 2(3x)+2(y)$
Perimeter $= (6x+2y)$ feet

Chapter 2 Equations and Inequalities

Problem Set 2.1 Solving First-Degree Equations

1. $3x+4=16$

   $3x+4-4=16-4$

   $3x=12$

   $\frac{1}{3}(3x)=\frac{1}{3}(12)$

   $x=4$

   The solution set is $\{4\}$.

3. $5x+1=-14$

   $5x+1-1=-14-1$

   $5x=-15$

   $\frac{1}{5}(5x)=\frac{1}{5}(-15)$

   $x=-3$

   The solution set is $\{-3\}$.

5. $-x-6=8$

   $-x-6+6=8+6$

   $-x=14$

   $-1(-x)=-1(14)$

   $x=-14$

   The solution set is $\{-14\}$.

7. $4y-3=21$

   $4y-3+3=21+3$

   $4y=24$

   $\frac{1}{4}(4y)=\frac{1}{4}(24)$

   $y=6$

   The solution set is $\{6\}$.

9. $3x-4=15$

   $3x-4+4=15+4$

   $3x=19$

   $\frac{1}{3}(3x)=\frac{1}{3}(19)$

   $x=\frac{19}{3}$

   The solution set is $\left\{\frac{19}{3}\right\}$.

11. $-4=2x-6$

    $-4+6=2x-6+6$

    $2=2x$

    $\frac{1}{2}(2)=\frac{1}{2}(2x)$

    $1=x$

    The solution set is $\{1\}$.

13. $-6y-4=16$

    $-6y-4+4=16+4$

    $-6y=20$

    $-\frac{1}{6}(-6y)=-\frac{1}{6}(20)$

    $y=-\frac{20}{6}$

    $y=-\frac{10}{3}$

    The solution set is $\left\{-\frac{10}{3}\right\}$.

15. $4x-1=2x+7$

    $4x-1+1=2x+7+1$

    $4x=2x+8$

    $4x-2x=2x-2x+8$

    $2x=8$

    $\frac{1}{2}(2x)=\frac{1}{2}(8)$

    $x=4$

    The solution set is $\{4\}$.

17. $5y+2=2y-11$

    $5y+2-2=2y-11-2$

    $5y=2y-13$

    $5y-2y=2y-2y-13$

    $3y=-13$

    $\frac{1}{3}(3y)=\frac{1}{3}(-13)$

    $y=-\frac{13}{3}$

    The solution set is $\left\{-\frac{13}{3}\right\}$.

19. $3x+4=5x-2$

$3x+4-4=5x-2-4$

$3x=5x-6$

$3x-5x=5x-5x-6$

$-2x=-6$

$-\frac{1}{2}(-2x)=-\frac{1}{2}(-6)$

$x=3$

The solution set is $\{3\}$.

21. $-7a+6=-8a+14$

$-7a+6-6=-8a+14-6$

$-7a=-8a+8$

$-7a+8a=-8a+8a+8$

$a=8$

The solution set is $\{8\}$.

23. $5x+3-2x=x-15$

$3x+3=x-15$

$3x+3-3=x-15-3$

$3x=x-18$

$3x-x=x-x-18$

$2x=-18$

$\frac{1}{2}(2x)=\frac{1}{2}(-18)$

$x=-9$

The solution set is $\{-9\}$.

25. $6y+18+y=2y+3$

$7y+18=2y+3$

$7y+18-18=2y+3-18$

$7y=2y-15$

$7y-2y=2y-2y-15$

$5y=-15$

$\frac{1}{5}(5y)=\frac{1}{5}(-15)$

$y=-3$

The solution set is $\{-3\}$.

27. $4x-3+2x=8x-3-x$

$6x-3=7x-3$

$6x-3+3=7x-3+3$

$6x=7x$

$6x-6x=7x-6x$

$0=x$

The solution set is $\{0\}$.

29. $6n-4-3n=3n+10+4n$

$3n-4=7n+10$

$3n-4+4=7n+10+4$

$3n=7n+14$

$3n-7n=7n-7n+14$

$-4n=14$

$-\frac{1}{4}(-4n)=-\frac{1}{4}(14)$

$n=-\frac{14}{4}$

$n=-\frac{7}{2}$

The solution set is $\left\{-\frac{7}{2}\right\}$.

31. $4(x-3)=-20$

$4x-12=-20$

$4x-12+12=-20+12$

$4x=-8$

$\frac{1}{4}(4x)=\frac{1}{4}(-8)$

$x=-2$

The solution set is $\{-2\}$.

33. $-3(x-2)=11$

$-3x+6=11$

$-3x+6-6=11-6$

$-3x=5$

$-\frac{1}{3}(-3x)=-\frac{1}{3}(5)$

$x=-\frac{5}{3}$

The solution set is $\left\{-\frac{5}{3}\right\}$.

35. $5(2x+1)=4(3x-7)$

$10x+5=12x-28$

$10x+5-5=12x-28-5$

$10x=12x-33$

$10x - 12x = 12x - 12x - 33$

$-2x = -33$

$-\frac{1}{2}(-2x) = -\frac{1}{2}(-33)$

$x = \frac{33}{2}$

The solution set is $\left\{\frac{33}{2}\right\}$.

37. $5x - 4(x - 6) = -11$

$5x - 4x + 24 = -11$

$x + 24 = -11$

$x + 24 - 24 = -11 - 24$

$x = -35$

The solution set is $\{-35\}$.

39. $-2(3x - 1) - 3 = -4$

$-6x + 2 - 3 = -4$

$-6x - 1 = -4$

$-6x - 1 + 1 = -4 + 1$

$-6x = -3$

$-\frac{1}{6}(-6x) = -\frac{1}{6}(-3)$

$x = \frac{3}{6}$

$x = \frac{1}{2}$

The solution set is $\left\{\frac{1}{2}\right\}$.

41. $-2(3x + 5) = -3(4x + 3)$

$-6x - 10 = -12x - 9$

$-6x - 10 + 10 = -12x - 9 + 10$

$-6x = -12x + 1$

$-6x + 12x = -12x + 12x + 1$

$6x = 1$

$\frac{1}{6}(6x) = \frac{1}{6}(1)$

$x = \frac{1}{6}$

The solution set is $\left\{\frac{1}{6}\right\}$.

43. $3(x - 4) - 7(x + 2) = -2(x + 18)$

$3x - 12 - 7x - 14 = -2x - 36$

$-4x - 26 = -2x - 36$

$-4x - 26 + 26 = -2x - 36 + 26$

$-4x = -2x - 10$

$-4x + 2x = -2x + 2x - 10$

$-2x = -10$

$-\frac{1}{2}(-2x) = -\frac{1}{2}(-10)$

$x = 5$

The solution set is $\{5\}$.

45. $-2(3n - 1) + 3(n + 5) = -4(n - 4)$

$-6n + 2 + 3n + 15 = -4n + 16$

$-3n + 17 = -4n + 16$

$-3n + 17 - 17 = -4n + 16 - 17$

$-3n = -4n - 1$

$-3n + 4n = -4n + 4n - 1$

$n = -1$

The solution set is $\{-1\}$.

47. $3(2a - 1) - 2(5a + 1) = 4(3a + 4)$

$6a - 3 - 10a - 2 = 12a + 16$

$-4a - 5 = 12a + 16$

$-4a - 5 + 5 = 12a + 16 + 5$

$-4a = 12a + 21$

$-4a - 12a = 12a - 12a + 21$

$-16a = 21$

$-\frac{1}{16}(-16a) = -\frac{1}{16}(21)$

$a = -\frac{21}{16}$

The solution set is $\left\{-\frac{21}{16}\right\}$.

49. $-2(n - 4) - (3n - 1) = -2 + (2n - 1)$

$-2n + 8 - 3n + 1 = -2 + 2n - 1$

$-5n + 9 = 2n - 3$

$-5n + 9 - 9 = 2n - 3 - 9$

$-5n = 2n - 12$

$-5n - 2n = 2n - 2n - 12$

$-7n = -12$

$-\frac{1}{7}(-7n) = -\frac{1}{7}(-12)$

$n = \frac{12}{7}$

The solution set is $\left\{\frac{12}{7}\right\}$.

51. Let $x =$ number.

$3x - 15 = 27$

$3x - 15 + 15 = 27 + 15$

$3x = 42$

$\frac{1}{3}(3x) = \frac{1}{3}(42)$

$x = 14$

The number is 14.

53. Let $n =$ 1st integer.
$n + 1 =$ 2nd integer.
$n + 2 =$ 3rd integer.

$n + n + 1 + n + 2 = 42$

$3n + 3 = 42$

$3n + 3 - 3 = 42 - 3$

$3n = 39$

$\frac{1}{3}(3n) = \frac{1}{3}(39)$

$n = 13$

The integers are 13, 14, and 15.

55. Let $n =$ 1st odd integer.
$n + 2 =$ 2nd odd integer.
$n + 4 =$ 3rd odd integer.

$3(n + 2) - (n + 4) = n + 11$

$3n + 6 - n - 4 = n + 11$

$2n + 2 = n + 11$

$2n + 2 - 2 = n + 11 - 2$

$2n = n + 9$

$2n - n = n - n + 9$

$n = 9$

The integers are 9, 11, and 13.

57. Let $x =$ smaller number.
$6x - 3 =$ larger number.

$6x - 3 - x = 67$

$5x - 3 = 67$

$5x - 3 + 3 = 67 + 3$

$5x = 70$

$\frac{1}{5}(5x) = \frac{1}{5}(70)$

$x = 14$

smaller number $= 14$
larger number $= 6(14) - 3 = 84 - 3 = 81$
The numbers are 14 and 81.

59. Let $x =$ normal rate of pay.
Dan worked 40 hours at regular pay and 6 hours at double pay.

$40x + 6(2x) = 572$

$40x + 12x = 572$

$52x = 572$

$\frac{1}{52}(52x) = \frac{1}{52}(572)$

$x = \$11.00$

Dan's normal hourly rate is $11.00 per hour.

61. Let $x =$ number of pennies.
$2x - 10 =$ number of nickels.
$3x - 20 =$ number of dimes.

$x + 2x - 10 + 3x - 20 = 150$

$6x - 30 = 150$

$6x - 30 + 30 = 150 + 30$

$6x = 180$

$\frac{1}{6}(6x) = \frac{1}{6}(180)$

$x = 30$

number of pennies $= 30$
number of nickels $= 2(30) - 10 = 60 - 10 = 50$
number of dimes $= 3(30) - 20 = 90 - 20 = 70$

There are 30 pennies, 50 nickels, and 70 dimes.

63. Let $x =$ cost of the ring.

$3x - 150 = 750$

$3x - 150 + 150 = 750 + 150$

$3x = 900$

$\frac{1}{3}x(3x) = \frac{1}{3}(900)$

$x = 300$

The cost of the ring is $300.

65. Let $x =$ number of 3-bedroom apartments.
$3x + 10 =$ number of 2-bedroom apartments.
$2(3x + 10) =$ number of 1-bedroom apartments.

$x + 3x + 10 + 2(3x + 10) = 230$

$x + 3x + 10 + 6x + 20 = 230$

$10x + 30 = 230$

$10x + 30 - 30 = 230 - 30$

$10x = 200$

$\frac{1}{10}(10x) = \frac{1}{10}(200)$

$x = 20$

number of 3-bedroom apartments $= 20$
number of 2-bedroom apartments $= 3(20) + 10 = 20 + 10 = 70$
number of 1-bedroom apartments $= 2(70) = 140$

There are 20 three-bedroom apartments, 70 two-bedroom apartments, and 140 one-bedroom apartments.

Problem Set 2.2 Equations Involving Fractional Forms

1. $\frac{3}{4}x = 9$

$\frac{4}{3}\left(\frac{3}{4}x\right) = \frac{4}{3}(9)$

$x = 12$

The solution set is $\{12\}$.

3. $-\frac{2x}{3} = \frac{2}{5}$

$-\frac{3}{2}\left(-\frac{2x}{3}\right) = -\frac{3}{2}\left(\frac{2}{5}\right)$

$x = -\frac{3}{5}$

The solution set is $\left\{-\frac{3}{5}\right\}$.

5. $\frac{n}{2} - \frac{2}{3} = \frac{5}{6}$

$6\left(\frac{n}{2} - \frac{2}{3}\right) = 6\left(\frac{5}{6}\right)$

$6\left(\frac{n}{2}\right) + 6\left(-\frac{2}{3}\right) = 6\left(\frac{5}{6}\right)$

$3n - 4 = 5$

$3n - 4 + 4 = 5 + 4$

$3n = 9$

$\frac{1}{3}(3n) = \frac{1}{3}(9)$

$n = 3$

The solution set is $\{3\}$.

7. $\frac{5n}{6} - \frac{n}{8} = -\frac{17}{12}$

$24\left(\frac{5n}{6} - \frac{n}{8}\right) = 24\left(-\frac{17}{12}\right)$

$24\left(\frac{5n}{6}\right) + 24\left(-\frac{n}{8}\right) = 24\left(-\frac{17}{12}\right)$

$20n - 3n = -34$

$\frac{1}{17}(17n) = \frac{1}{17}(-34)$

$n = -2$

The solution set is $\{-2\}$.

9. $\frac{a}{4} - 1 = \frac{a}{3} + 2$

$12\left(\frac{a}{4} - 1\right) = 12\left(\frac{a}{3} + 2\right)$

$12\left(\frac{a}{4}\right) + 12(-1) = 12\left(\frac{a}{3}\right) + 12(2)$

$3a - 12 = 4a + 24$

$3a - 12 + 12 = 4a + 24 + 12$

$3a = 4a + 36$

$3a - 4a = 4a - 4a + 36$

$-a = 36$

$-1(-a) = -1(36)$

$a = -36$

The solution set is $\{-36\}$.

11. $\frac{h}{4} + \frac{h}{5} = 1$

$20\left(\frac{h}{4} + \frac{h}{5}\right) = 20(1)$

$20\left(\frac{h}{4}\right) + 20\left(\frac{h}{5}\right) = 20(1)$

$5h + 4h = 20$

$9h = 20$

$\frac{1}{9}(9h) = \frac{1}{9}(20)$

$h = \frac{20}{9}$

The solution set is $\left\{\frac{20}{9}\right\}$.

13. $\frac{h}{2}-\frac{h}{3}+\frac{h}{6}=1$

$6\left(\frac{h}{2}-\frac{h}{3}+\frac{h}{6}\right)=6(1)$

$6\left(\frac{h}{2}\right)+6\left(-\frac{h}{3}\right)+6\left(\frac{h}{6}\right)=6(1)$

$3h-2h+h=6$

$2h=6$

$\frac{1}{2}(2h)=\frac{1}{2}(6)$

$h=3$

The solution set is $\{3\}$.

15. $\frac{x-2}{3}+\frac{x+3}{4}=\frac{11}{6}$

$12\left(\frac{x-2}{3}+\frac{x+3}{4}\right)=12\left(\frac{11}{6}\right)$

$12\left(\frac{x-2}{3}\right)+12\left(\frac{x+3}{4}\right)=12\left(\frac{11}{6}\right)$

$4(x-2)+3(x+3)=2(11)$

$4x-8+3x+9=22$

$7x+1=22$

$7x+1-1=22-1$

$7x=21$

$\frac{1}{7}(7x)=\frac{1}{7}(21)$

$x=3$

The solution set is $\{3\}$.

17. $\frac{x+2}{2}-\frac{x-1}{5}=\frac{3}{5}$

$10\left(\frac{x+2}{2}-\frac{x-1}{5}\right)=10\left(\frac{3}{5}\right)$

$10\left(\frac{x+2}{2}\right)-10\left(\frac{x-1}{5}\right)=10\left(\frac{3}{5}\right)$

$5(x+2)-2(x-1)=2(3)$

$5x+10-2x+2=6$

$3x+12=6$

$3x+12-12=6-12$

$3x=-6$

$\frac{1}{3}(3x)=\frac{1}{3}(-6)$

$x=-2$

The solution set is $\{-2\}$.

19. $\frac{n+2}{4}-\frac{2n-1}{3}=\frac{1}{6}$

$12\left(\frac{n+2}{4}-\frac{2n-1}{3}\right)=12\left(\frac{1}{6}\right)$

$12\left(\frac{n+2}{4}\right)-12\left(\frac{2n-1}{3}\right)=12\left(\frac{1}{6}\right)$

$3(n+2)-4(2n-1)=2$

$3n+6-8n+4=2$

$-5n+10=2$

$-5n+10-10=2-10$

$-5n=-8$

$-\frac{1}{5}(-5n)=-\frac{1}{5}(-8)$

$n=\frac{8}{5}$

The solution set is $\left\{\frac{8}{5}\right\}$.

21. $\frac{y}{3}+\frac{y-5}{10}=\frac{4y+3}{5}$

$30\left(\frac{y}{3}+\frac{y-5}{10}\right)=30\left(\frac{4y+3}{5}\right)$

$30\left(\frac{y}{3}\right)+30\left(\frac{y-5}{10}\right)=30\left(\frac{4y+3}{5}\right)$

$10y+3(y-5)=6(4y+3)$

$10y+3y-15=24y+18$

$13y-15=24y+18$

$13y-15+15=24y+18+15$

$13y=24y+33$

$13y-24y=24y-24y+33$

$-11y=33$

$-\frac{1}{11}(-11y)=-\frac{1}{11}(33)$

$y=-3$

The solution set is $\{-3\}$.

23. $\frac{4x-1}{10}-\frac{5x+2}{4}=-3$

$20\left(\frac{4x-1}{10}-\frac{5x+2}{4}\right)=20(-3)$

$20\left(\frac{4x-1}{10}\right)-20\left(\frac{5x+2}{4}\right)=20(-3)$

$2(4x-1)-5(5x+2)=20(-3)$

$8x-2-25x-10=-60$

$-17x-12=-60$

$-17x-12+12=-60+12$

$-17x = -48$

$-\frac{1}{17}(-17x) = -\frac{1}{17}(-48)$

$x = \frac{48}{17}$

The solution set is $\left\{\frac{48}{17}\right\}$.

25. $\frac{2x-1}{8} - 1 = \frac{x+5}{7}$

$56\left(\frac{2x-1}{8} - 1\right) = 56\left(\frac{x+5}{7}\right)$

$56\left(\frac{2x-1}{8}\right) - 56(1) = 56\left(\frac{x+5}{7}\right)$

$7(2x-1) - 56 = 8(x+5)$

$14x - 7 - 56 = 8x + 40$

$14x - 63 = 8x + 40$

$14x - 63 + 63 = 8x + 40 + 63$

$14x = 8x + 103$

$14x - 8x = 8x - 8x + 103$

$6x = 103$

$\frac{1}{6}(6x) = \frac{1}{6}(103)$

$x = \frac{103}{6}$

The solution set is $\left\{\frac{103}{6}\right\}$.

27. $\frac{2a-3}{6} + \frac{3a-2}{4} + \frac{5a+6}{12} = 4$

$12\left(\frac{2a-3}{6} + \frac{3a-2}{4} + \frac{5a+6}{12}\right) = 12(4)$

$12\left(\frac{2a-3}{6}\right) + 12\left(\frac{3a-2}{4}\right) + 12\left(\frac{5a+6}{12}\right) = 12(4)$

$2(2a-3) + 3(3a-2) + 1(5a+6) = 12(4)$

$4a - 6 + 9a - 6 + 5a + 6 = 48$

$18a - 6 = 48$

$18a - 6 + 6 = 48 + 6$

$18a = 54$

$\frac{1}{18}(18a) = \frac{1}{18}(54)$

$a = 3$

The solution set is $\{3\}$.

29. $x + \frac{3x-1}{9} - 4 = \frac{3x+1}{3}$

$9\left(x + \frac{3x-1}{9} - 4\right) = 9\left(\frac{3x+1}{3}\right)$

$9(x) + 9\left(\frac{3x-1}{9}\right) - 9(4) = 3(3x+1)$

$9x + 3x - 1 - 36 = 9x + 3$

$12x - 37 = 9x + 3$

$12x - 37 + 37 = 9x + 3 + 37$

$12x = 9x + 40$

$12x - 9x = 9x - 9x + 40$

$3x = 40$

$\frac{1}{3}(3x) = \frac{1}{3}(40)$

$x = \frac{40}{3}$

The solution set is $\left\{\frac{40}{3}\right\}$.

31. $\frac{x+3}{2} + \frac{x+4}{5} = \frac{3}{10}$

$10\left(\frac{x+3}{2} + \frac{x+4}{5}\right) = 10\left(\frac{3}{10}\right)$

$10\left(\frac{x+3}{2}\right) + 10\left(\frac{x+4}{5}\right) = 3$

$5(x+3) + 2(x+4) = 3$

$5x + 15 + 2x + 8 = 3$

$7x + 23 = 3$

$7x + 23 - 23 = 3 - 23$

$7x = -20$

$\frac{1}{7}(7x) = \frac{1}{7}(-20)$

$x = -\frac{20}{7}$

The solution set is $\left\{-\frac{20}{7}\right\}$.

33. $n + \frac{2n-3}{9} - 2 = \frac{2n+1}{3}$

$9\left(n + \frac{2n-3}{9} - 2\right) = 9\left(\frac{2n+1}{3}\right)$

$9n + 9\left(\frac{2n-3}{9}\right) - 9(2) = 3(2n+1)$

$9n + 2n - 3 - 18 = 6n + 3$

$11n - 21 = 6n + 3$

$11n - 21 + 21 = 6n + 3 + 21$

$11n = 6n + 24$

$11n - 6n = 6n - 6n + 24$

$5n = 24$

$\frac{1}{5}(5n) = \frac{1}{5}(24)$

$n = \frac{24}{5}$

The solution set is $\left\{\frac{24}{5}\right\}$.

35. $\frac{3}{4}(t-2) - \frac{2}{5}(2t-3) = \frac{1}{5}$

$20\left[\frac{3}{4}(t-2) - \frac{2}{5}(2t-3)\right] = 20\left(\frac{1}{5}\right)$

$20\left[\frac{3}{4}(t-2)\right] - 20\left[\frac{2}{5}(2t-3)\right] = 4$

$15(t-2) - 8(2t-3) = 4$

$15t - 30 - 16t + 24 = 4$

$-t - 6 = 4$

$-t - 6 + 6 = 4 + 6$

$-t = 10$

$-1(-t) = -1(10)$

$t = -10$

The solution set is $\{-10\}$.

37. $\frac{1}{2}(2x-1) - \frac{1}{3}(5x+2) = 3$

$6\left[\frac{1}{2}(2x-1) - \frac{1}{3}(5x+2)\right] = 3(6)$

$6\left[\frac{1}{2}(2x-1)\right] - 6\left[\frac{1}{3}(5x+2)\right] = 18$

$3(2x-1) - 2(5x+2) = 18$

$6x - 3 - 10x - 4 = 18$

$-4x - 7 = 18$

$-4x - 7 + 7 = 18 + 7$

$-4x = 25$

$-\frac{1}{4}(-4x) = -\frac{1}{4}(25)$

$x = -\frac{25}{4}$

The solution set is $\left\{-\frac{25}{4}\right\}$.

39. $3x - 1 + \frac{2}{7}(7x-2) = -\frac{11}{7}$

$7\left[3x - 1 + \frac{2}{7}(7x-2)\right] = 7\left(-\frac{11}{7}\right)$

$7(3x) + 7(-1) + 7\left[\frac{2}{7}(7x-2)\right] = -11$

$21x - 7 + 2(7x-2) = -11$

$21x - 7 + 14x - 4 = -11$

$35x - 11 = -11$

$35x - 11 + 11 = -11 + 11$

$35x = 0$

$\frac{1}{35}(35x) = \frac{1}{35}(0)$

$x = 0$

The solution set is $\{0\}$.

41. Let $x =$ number.

$\frac{1}{2}x = \frac{2}{3}x - 3$

$6\left(\frac{1}{2}x\right) = 6\left(\frac{2}{3}x - 3\right)$

$3x = 6\left(\frac{2}{3}x\right) - 6(3)$

$3x = 4x - 18$

$3x - 4x = 4x - 4x - 18$

$-x = -18$

$x = 18$

The number is 18.

43. Let $x =$ length

$\frac{1}{4}x + 1 =$ width

$P = 2l + 2w$

$42 = 2(x) + 2\left(\frac{1}{4}x + 1\right)$

$42 = 2x + \frac{2}{4}x + 2$

$42 = 2x + \frac{1}{2}x + 2$

$2(42) = 2\left(2x + \frac{1}{2}x + 2\right)$

$84 = 2(2x) + 2\left(\frac{1}{2}x\right) + 2(2)$

$84 = 4x + x + 4$

$84 = 5x + 4$

$84 - 4 = 5x + 4 - 4$

$80 = 5x$

$\frac{1}{5}(80) = \frac{1}{5}(5x)$

$16 = x$

length = 16 inches

width $= \frac{1}{4}(16) + 1 = 4 + 1 = 5$ inches

The length is 16 inches and the width is 5 inches.

45. Let $x = 1\text{st integer}$.
$x + 1 = 2\text{nd integer}$.
$x + 2 = 3\text{rd integer}$.

$x + \frac{1}{3}(x+1) + \frac{3}{8}(x+2) = 25$

$24\left[x + \frac{1}{3}(x+1) + \frac{3}{8}(x+2)\right] = 24(25)$

$24(x) + 24\left[\frac{1}{3}(x+1)\right] + 24\left[\frac{3}{8}(x+2)\right] = 600$

$24x + 8(x+1) + 9(x+2) = 600$

$24x + 8x + 8 + 9x + 18 = 600$

$41x + 26 = 600$

$41x + 26 - 26 = 600 - 26$

$41x = 574$

$\frac{1}{41}(41x) = \frac{1}{41}(574)$

$x = 14.$

The integers are 14, 15, and 16.

47. Let $x = \text{one piece}$.

$\frac{2}{3}x = \text{other piece}$.

$x + \frac{2}{3}x = 20$

$3\left(x + \frac{2}{3}x\right) = 3(20)$

$3(x) + 3\left(\frac{2}{3}x\right) = 60$

$3x + 2x = 60$

$5x = 60$

$\frac{1}{5}(5x) = \frac{1}{5}(60)$

$x = 12$

one piece $= 12$ ft.
other piece $= \frac{2}{3}(12) = 8$ ft.

The shorter piece is 8 ft.

49. Let $x = \text{present age of Angie}$.
$64 - x = \text{present age of mother}$.
$x + 8 = \text{Angie's age in eight years}$.
$64 - x + 8 = 72 - x = \text{mother's age in eight years}$.

$\frac{3}{5}(72 - x) = x + 8$

$5\left[\frac{3}{5}(72 - x)\right] = 5(x + 8)$

$3(72 - x) = 5x + 40$

$216 - 3x = 5x + 40$

$216 - 3x + 3x = 5x + 3x + 40$

$216 = 8x + 40$

$216 - 40 = 8x + 40 - 40$

$176 = 8x$

$\frac{1}{8}(176) = \frac{1}{8}(8x)$

$22 = x$

Angie's present age $= 22$
Mother's present age $= 64 - 22 = 42$

At the present time Angie is 22 years old and her mother is 42 years old.

51. Let $x = 1\text{st exam score}$.
$x + 10 = 2\text{nd exam score}$.
$x + 10 + 4 = 3\text{rd exam score}$.

$\frac{x + x + 10 + x + 10 + 4}{3} = 88$

$\frac{3x + 24}{3} = 88$

$3\left(\frac{3x + 24}{3}\right) = 3(88)$

$3x + 24 = 264$

$3x + 24 - 24 = 264 - 24$

$3x = 240$

$\frac{1}{3}(3x) = \frac{1}{3}(240)$

$x = 80$

1st exam score $= 80$
2nd exam score $= 80 + 10 = 90$
3rd exam score $= 90 + 4 = 94$

The exam scores are 80, 90, and 94.

53. Let $x = \text{one angle}$.

$\frac{1}{3}x + 4 = \text{other angle}$.

$x + \frac{1}{3}x + 4 = 180$

$3\left(x + \frac{1}{3}x + 4\right) = 3(180)$

$3x + 3\left(\frac{1}{3}x\right) + 3(4) = 540$

$3x + x + 12 = 540$

$4x + 12 = 540$

$4x + 12 - 12 = 540 - 12$

$4x = 528$

$\frac{1}{4}(4x) = \frac{1}{4}(528)$

$x = 132$

One angle is 132°

Other angle $= \frac{1}{3}(132) + 4 = 44 + 4 = 48°$

The angles are 132° and 48°.

55. Let $x =$ angle.
$90 - x =$ complement of the angle.
$180 - x =$ supplement of the angle.

$90 - x = \frac{1}{6}(180 - x) - 5$

$6(90 - x) = 6\left[\frac{1}{6}(180 - x) - 5\right]$

$6(90) - 6x = 6\left[\frac{1}{6}(180 - x)\right] - 6(5)$

$540 - 6x = 180 - x - 30$

$540 - 6x = 150 - x$

$540 - 6x + 6x = 150 - x + 6x$

$540 = 150 + 5x$

$540 - 150 = 150 - 150 + 5x$

$390 = 5x$

$\frac{1}{5}(390) = \frac{1}{5}(5x)$

$78 = x$

The angle is 78°.

Problem Set 2.3 Equations Involving Decimals

1. $.14x = 2.8$

$100(.14x) = 100(2.8)$

$14x = 280$

$x = 20$

The solution set is $\{20\}$.

3. $.09y = 4.5$

$100(.09y) = 100(4.5)$

$9y = 450$

$y = 50$

The solution set is $\{50\}$.

5. $n + .4n = 56$

$10(n + .4n) = 10(56)$

$10n + 4n = 560$

$14n = 560$

$n = 40$

The solution set is $\{40\}$.

7. $s = 9 + .25s$

$100(s) = 100(9 + .25s)$

$100s = 900 + 25s$

$100s - 25s = 900 + 25s - 25s$

$75s = 900$

$s = 12$

The solution set is $\{12\}$.

9. $s = 3.3 + .45s$

$100(s) = 100(3.3 + .45s)$

$100s = 330 + 45s$

$100s - 45s = 330 + 45s - 45s$

$55s = 330$

$s = 6$

The solution set is $\{6\}$.

11. $.11x + .12(900 - x) = 104$

$100[.11x + .12(900 - x)] = 100(104)$

$100(.11x) + 100[.12(900 - x)] = 10400$

$11x + 12(900 - x) = 10400$

$11x + 10800 - 12x = 10400$

$-x + 10800 = 10400$

$-x + 10800 - 10800 = 10400 - 10800$

$-x = -400$

$x = 400$

The solution set is $\{400\}$.

13. $.08(x + 200) = .07x + 20$

$100[.08(x + 200)] = 100(.07x + 20)$

$8(x + 200) = 100(.07x) + 100(20)$

$8x + 1600 = 7x + 2000$

$x + 1600 = 2000$

$x = 400$

The solution set is {400}.

15. $.12t - 2.1 = .07t - .2$

$100(.12t - 2.1) = 100(.07t - .2)$

$12t - 210 = 7t - 20$

$12t = 7t + 190$

$5t = 190$

$t = 38$

The solution set is {38}.

17. $.92 + .9(x - .3) = 2x - 5.95$

$100[.92 + .9(x - .3)] = 100(2x - 5.95)$

$92 + 90(x - .3) = 200x - 595$

$92 + 90x - 27 = 200x - 595$

$90x + 65 = 200x - 595$

$90x = 200x - 660$

$-110x = -660$

$x = 6$

The solution set is {6}.

19. $.1d + .11(d + 1500) = 795$

$100[.1d + .11(d + 1500)] = 100(795)$

$10d + 11(d + 1500) = 79500$

$10d + 11d + 16500 = 79500$

$21d + 16500 = 79500$

$21d = 63000$

$d = 3000$

The solution set is {3000}.

21. $.12x + .1(5000 - x) = 560$

$100[.12x + .1(5000 - x)] = 100(560)$

$12x + 10(5000 - x) = 56000$

$12x + 50000 - 10x = 56000$

$2x + 50000 = 56000$

$2x = 6000$

$x = 3000$

The solution set is {3000}.

23. $.09(x + 200) = .08x + 22$

$100[.09(x + 200)] = 100(.08x + 22)$

$9(x + 200) = 8x + 2200$

$9x + 1800 = 8x + 2200$

$9x = 8x + 400$

$x = 400$

The solution set is {400}.

25. $.3(2t + .1) = 8.43$

$100[.3(2t + .1)] = 100(8.43)$

$30(2t + .1) = 843$

$60t + 3 = 843$

$60t = 840$

$t = 14$

The solution set is {14}.

27. $.1(x - .1) - .4(x + 2) = -5.31$

$100[.1(x - .1) - .4(x + 2)] = 100(-5.31)$

$10(x - .1) - 40(x + 2) = -531$

$10x - 1 - 40x - 80 = -531$

$-30x - 81 = -531$

$-30x = -450$

$x = 15$

The solution set is {15}.

29. Let $p =$ original price.

$(100\%)p - (20\%)p = 72$

$1.00p - .20p = 72$

$.80p = 72$

$100(.80p) = 100(72)$

$80p = 7200$

$p = 90$

The original price is $90.

31. Let $s =$ sale price.

$s = 64 - (15\%)(64)$

$s = 64 - .15(64)$

$s = 64 - 9.60$

$s = 54.40$

The sales price is \$54.40.

33. Let $s =$ selling price.

$s = 30 + (60\%)(30)$

$s = 30 + .60(30)$

$s = 30 + 18$

$s = 48$

The selling price should be \$48.

35. Let $s =$ selling price.

$s = 200 + 50\%(s)$

$s = 200 + .50s$

$100(s) = 100(200 + .50s)$

$100s = 20000 + 50s$

$50s = 20000$

$s = 400$

The selling price is \$400.

37. Amount of profit $= 39.60 - 24 = 15.60$

Rate of profit $= \frac{15.60}{24.00} = 0.65 = 65\%$

39. Let $x =$ Robin's present salary.

$34775 = x + (7\%)(x)$

$34775 = x + .07x$

$34775 = 1.07x$

$100(34775) = 100(1.07x)$

$3477500 = 107x$

$32,500 = x$

Robin's present salary is \$32,500.

41. Let $x =$ amount invested at 10%.
$x + 1500 =$ amount invested at 11%.

$(10\%)(x) + (11\%)(x + 1500) = 795$

$.10x + .11(x + 1500) = 795$

$100[.10x + .11(x + 1500)] = 100(795)$

$10x + 11(x + 1500) = 79500$

$10x + 11(x + 1500) = 79500$

$10x + 11x + 16500 = 79500$

$21x + 16500 = 79500$

$21x = 63000$

$x = 3000$

amount invested at 10%
$= \$3000$
amount invested at 11%
$= 3000 + 1500 = \$4500$

\$3000 is invested at 10% and \$4500 is invested at 11%.

43. Let $x =$ additional amount invested at 9%.

$(6\%)(500) + (9\%)(x) = (8\%)(x + 500)$

$.06(500) + .09x = .08(x + 500)$

$100[.06(500) + .09x] = 100[.08(x + 500)]$

$6(500) + 9x = 8(x + 500)$

$3000 + 9x = 8x + 4000$

$9x = 8x + 1000$

$x = 1000$

\$1000 additional must be invested at 9%.

45. Let $x =$ number of pennies.
$2x - 1 =$ number of nickels.
$2x - 1 + 3 = 2x + 2 =$ number of dimes.

$.01x + .05(2x - 1) + .10(2x + 2) = 2.63$

$100[.01x + .05(2x - 1) + .10(2x + 2)] = 100(2.63)$

$1x + 5(2x - 1) + 10(2x + 2) = 263$

$1x + 10x - 5 + 20x + 20 = 263$

$31x + 15 = 263$

$31x = 248$

$x = 8$

number of pennies $= 8$
number of nickels $= 2(8) - 1 = 16 - 1 = 15$
number of dimes $= 2(8) + 2 = 16 + 2 = 18$

There are 8 pennies, 15 nickels, and 18 dimes.

47. Let $x =$ number of dimes.
$3x =$ number of quarters.
$70 - (x + 3x) = 70 - 4x$
$=$ number of half-dollars.

$.10x + .25(3x) + .50(70 - 4x) = 17.75$

$100[.10x+.25(3x)+.50(70-4x)] = 100(17.75)$

$10x + 25(3x) + 50(70-4x) = 1775$

$10x + 75x + 3500 - 200x = 1775$

$-115x + 3500 = 1775$

$-115x = -1725$

$x = 15$

numbers of dimes
$= 15$
number of quarters
$= 3(15) = 45$
number of half-dollars
$= 70 - 4(15) = 70 - 60 = 10$

There are 15 dimes, 45 quarters, and 10 half-dollars.

Problem Set 2.4 Formulas

1. $i = Prt$

   $i = 300(8\%)(5)$

   $i = 300(.08)(5)$

   $i = 120$

   The interest is $120.

3. $i = Prt$

   $132 = 400(11\%)t$

   $132 = 400(.11)t$

   $132 = 44t$

   $3 = t$

   The time is 3 years.

5. $i = Prt$

   $90 = 600(r)\left(2\frac{1}{2}\right)$

   $90 = 600(r)(2.5)$

   $90 = 1500r$

   $.06 = r$

   $6\% = r$

   The rate is 6% per year.

7. $i = Prt$

   $216 = P(9\%)(3)$

   $216 = P(.09)(3)$

   $216 = P(.27)$

   $800 = P$

   The principal is $800.

9. $A = P + Prt$

   $A = 1000 + 1000(12\%)(5)$

   $A = 1000 + 1000(.12)(5)$

   $A = 1000 + 600$

   $A = 1600$

   The amount is $1600.

11. $A = P + Prt$

    $1372 = 700 + 700(r)(12)$

    $1372 = 700 + 8400r$

    $672 = 8400r$

    $.08 = r$

    $8\% = r$

    The rate is 8% per year.

13. $A = P + Prt$

    $326 = P + P(7\%)(9)$

    $326 = P + P(.07)(9)$

    $326 = P + .63P$

    $326 = 1.63P$

    $200 = P$

    The principal is $200.

15. $A = \frac{1}{2}h(b_1 + b_2)$

    $2(A) = 2\left[\frac{1}{2}h(b_1 + b_2)\right]$

    $2A = h(b_1 + b_2)$

    $\frac{2A}{h} = b_1 + b_2$

    $\frac{2A}{h} - b_1 = b_2$

    $b_2 = \frac{2(98)}{14} - 8$ $\quad b_2 = \frac{2(104)}{8} - 12$ $\quad b_2 = \frac{2(49)}{7} - 4$

    $b_2 = 14 - 8$ $\quad b_2 = 26 - 12$ $\quad b_2 = 14 - 4$

    $b_2 = 6$ $\quad b_2 = 14$ $\quad b_2 = 10$

$b_2 = \frac{2(162)}{9} - 16$ $\quad b_2 = \frac{2\left(16\frac{1}{2}\right)}{3} - 4$ $\quad b_2 = \frac{2\left(38\frac{1}{2}\right)}{11} - 5$

$b_2 = 36 - 16$ $\quad b_2 = 11 - 4$ $\quad b_2 = 7 - 5$

$b_2 = 20$ $\quad b_2 = 7$ $\quad b_2 = 2$

Area(A)	98	104	49	162	$16\frac{1}{2}$	$38\frac{1}{2}$
Height(h)	14	8	7	9	3	11
One base(b_1)	8	12	4	16	4	5
Other base(b_2)	6	14	10	20	7	2

17. $V = Bh$ for h

$\frac{1}{B}(V) = \frac{1}{B}(Bh)$

$\frac{V}{B} = h$

19. $V = \pi r^2 h$ for h

$\frac{1}{\pi r^2}(V) = \frac{1}{\pi r^2}(\pi r^2 h)$

$\frac{V}{\pi r^2} = h$

21. $C = 2\pi r$ for r

$\frac{1}{2\pi}(C) = \frac{1}{2\pi}(2\pi r)$

$\frac{C}{2\pi} = r$

23. $I = \frac{100M}{C}$

$C(I) = C\left(\frac{100M}{C}\right)$

$CI = 100M$

$\frac{1}{I}(CI) = \frac{1}{I}(100M)$

$C = \frac{100M}{I}$

25. $F = \frac{9}{5}C + 32$ for C

$F - 32 = \frac{9}{5}C$

$\frac{5}{9}(F - 32) = \frac{5}{9}\left(\frac{9}{5}C\right)$

$\frac{5}{9}(F - 32) = C$

27. $y = mx + b$

$y - b = mx$

$\frac{1}{m}(y - b) = \frac{1}{m}(mx)$

$\frac{y - b}{m} = x$

29. $y - y_1 = m(x - x_1)$

$y - y_1 = mx - mx_1$

$y - y_1 + mx_1 = mx$

$\frac{1}{m}(y - y_1 + mx_1) = \frac{1}{m}(mx)$

$\frac{y - y_1 + mx_1}{m} = x$

31. $a(x + b) = b(x - c)$

$ax + ab = bx - bc$

$ax = bx - bc - ab$

$ax - bx = -bc - ab$

$x(a - b) = -bc - ab$

$\frac{1}{(a - b)}[x(a - b)] = \frac{1}{(a - b)}[-bc - ab]$

$x = \frac{-bc - ab}{a - b}$

$x = \frac{-1(-bc - ab)}{-1(a - b)}$

$x = \frac{bc + ab}{-a + b}$

$x = \frac{ab + bc}{b - a}$

33. $\frac{x - a}{b} = c$

$b\left(\frac{x - a}{b}\right) = b(c)$

$x - a = bc$

$x = bc + a$

35. $\frac{1}{3}x + a = \frac{1}{2}b$

$6\left(\frac{1}{3}x + a\right) = 6\left(\frac{1}{2}b\right)$

$2x + 6a = 3b$

$2x = 3b - 6a$

$\frac{1}{2}(2x) = \frac{1}{2}(3b-6a)$

$x = \frac{3b-6a}{2}$

37. $2x-5y=7$ for x

$2x = 5y+7$

$\frac{1}{2}(2x) = \frac{1}{2}(5y+7)$

$x = \frac{5y+7}{2}$

39. $-7x-y=4$ for y

$-y = 4+7x$

$-1(-y) = -1(4+7x)$

$y = -4-7x$

$y = -7x-4$

41. $3(x-2y)=4$

$3x-6y=4$

$3x = 6y+4$

$x = \frac{6y+4}{3}$

43. $\frac{y-a}{b} = \frac{x+b}{c}$

$bc\left(\frac{y-a}{b}\right) = bc\left(\frac{x+b}{c}\right)$

$c(y-a) = b(x+b)$

$cy-ac = bx+b^2$

$cy-ac-b^2 = bx$

$\frac{1}{b}(cy-ac-b^2) = \frac{1}{b}(bx)$

$\frac{cy-ac-b^2}{b} = x$

45. $(y+1)(a-3) = x-2$

$ay-3y+a-3 = x-2$

$ay-3y = x-2-a+3$

$(a-3)y = x-a+1$

$\frac{1}{(a-3)}[(a-3)y] = \frac{1}{(a-3)}(x-a+1)$

$y = \frac{x-a+1}{a-3}$

47. Let $x =$ width.
$4x-2 =$ length.
$P = 2l+2w$

$56 = 2(4x-2)+2x$

$56 = 8x-4+2x$

$56 = 10x-4$

$60 = 10x$

$6 = x$

width $= 6$ meters
length $= 4(6)-2 = 24-2 = 22$ meters

The width is 6 meters and the length is 22 meters.

49. Let $500 =$ principal.
$1000 =$ double the principal.

$A = P+Prt$

$1000 = 500+500(9\%)t$

$1000 = 500+500(.09)t$

$1000 = 500+45t$

$500 = 45t$

$\frac{500}{45} = t$

$11\frac{1}{9} = t$

The principal will double in $11\frac{1}{9}$ years at 9%.

51. Let $P =$ principal.
$2P =$ double the principal.

$A = P+Prt$

$2P = P+P(.9\%)t$

$2P = P+P(.09)t$

$2P = P+.09Pt$

$P = .09Pt$

$\frac{1}{.09P}(P) = \frac{1}{.09P}(.09Pt)$

$11.\overline{1} = t$

$11\frac{1}{9} = t$

The principal will double in $11\frac{1}{9}$ years at 9%.

53. Let $t =$ time traveling.

$d = rt$

$d_1 = 450t$

$d_2 = 550t$

$450 + 550t = 4000$

$1000t = 4000$

$t = 4$ hours

It will take 4 hours.

55. Let $t =$ Time traveling for Dennis.
$t - \frac{3}{2} =$ Time traveling for Cathy.

$d = rt$

Distance for Dennis = Distance for Cathy

$4t = 6\left(t - \frac{3}{2}\right)$

$4t = 6t - 9$

$-2t = -9$

$t = \frac{9}{2} = 4.5$

Time traveling for Dennis
$= 4.5$ hours
Time traveling for Cathy
$= 4.5 - 1.5 = 3.0$ hours

It takes Cathy 3 hours
to catch up with Dennis.

57. Let $t =$ time traveling at 20 mph.
$4.5 - t =$ time traveling at 12 mph.

$d = rt$

$20t + 12(4.5 - t) = 70$

$20t + 54 - 12t = 70$

$8t + 54 = 70$

$8t = 16$

$t = 2$

distance $= 20(2) = 40$ miles

Bret has traveled for 40 miles
before he slowed down.

59. Let $x =$ amount of pure acid.

$(100\%)(x) + (30\%)(150) = (40\%)(x + 150)$

$1x + 0.30(150) = 0.40(x + 150)$

$100[1x + 0.30(150)] = [0.40(x + 150)]$

$100x + 30(150) = 40(x + 150)$

$100x + 4500 = 40x + 6000$

$100x = 40x + 1500$

$60x = 1500$

$x = 25$

There must be 25 ml of pure acid.

61. Let $x =$ quarts of 30% solution.
$20 - x =$ quarts of 70% solution.

$(30\%)(x) + (70\%)(20 - x) = (40\%)(20)$

$0.30x + 0.70(20 - x) = 0.40(20)$

$100[0.30x + 0.70(20 - x)] = 100[0.40(20)]$

$30x + 70(20 - x) = 40(20)$

$30x + 1400 - 70x = 800$

$-40x + 1400 = 800$

$-40x = -600$

$x = 15$

quarts of 30% solution $= 15$
quarts of 70% solution $= 20 - 15 = 5$

There should be 15 quarts of
the 30% solution and 5 quarts
of the 70% solution.

Problem Set 2.5 Inequalities

1. $x > 1$

$(1, \infty)$

1

3. $x \geq -1$

$[-1, \infty)$

-1

5. $x < -2$

$(-\infty, -2)$

7. $x \leq 2$

$(-\infty, 2]$

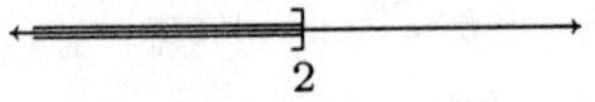

9. $(-\infty, 4)$

$x < 4$

11. $(-\infty, -7]$

$x \leq -7$

13. $(8, \infty)$

$x > 8$

15. $[-7, \infty)$

$x \geq -7$

17. $x - 3 > -2$

$x - 3 + 3 > -2 + 3$

$x > 1$

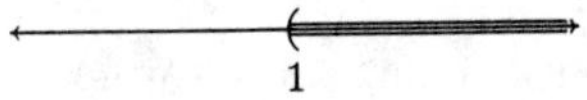

19. $-2x \geq 8$

$-\frac{1}{2}(-2x) \leq -\frac{1}{2}(8)$

$x \leq -4$

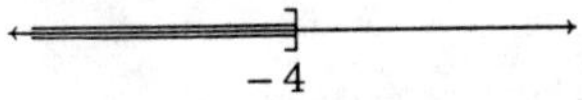

21. $5x \leq -10$

$\frac{1}{5}(5x) \leq \frac{1}{5}(-10)$

$x \leq -2$

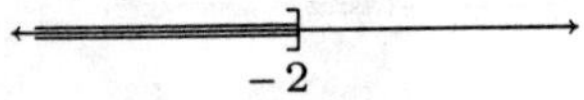

23. $2x + 1 < 5$

$2x < 4$

$\frac{1}{2}(2x) < \frac{1}{2}(4)$

$x < 2$

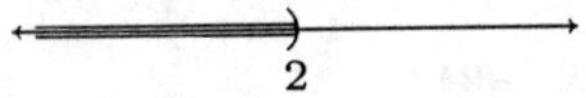

25. $3x - 2 > -5$

$3x - 2 + 2 > -5 + 2$

$3x > -3$

$\frac{1}{3}(3x) > \frac{1}{3}(-3)$

$x > -1$

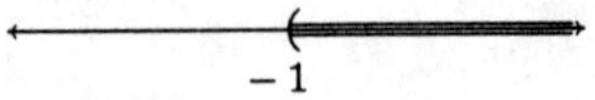

27. $-7x - 3 \leq 4$

$-7x \leq 7$

$-\frac{1}{7}(-7x) \geq -\frac{1}{7}(7)$

$x \geq -1$

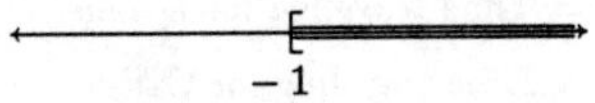

29. $2 + 6x > -10$

$6x > -12$

$\frac{1}{6}(6x) > \frac{1}{6}(-12)$

$x > -2$

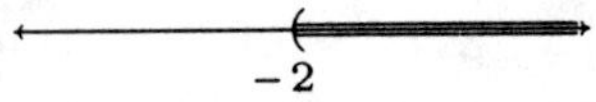

31. $5 - 3x < 11$

$-3x < 6$

$-\frac{1}{3}(-3x) > -\frac{1}{3}(6)$

$x > -2$

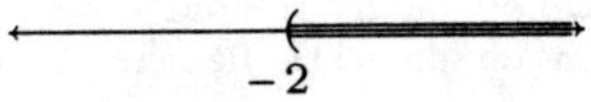

33. $15 < 1 - 7x$

$14 < -7x$

$-\frac{1}{7}(14) > -\frac{1}{7}(-7x)$

$-2 > x$

$x < -2$

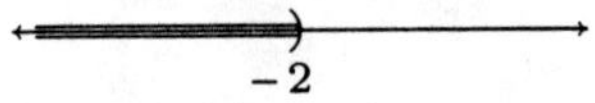

35. $-10 \leq 2 + 4x$

$-12 \leq 4x$

$\frac{1}{4}(-12) \leq \frac{1}{4}(4x)$

$-3 \leq x$

$x \geq -3$

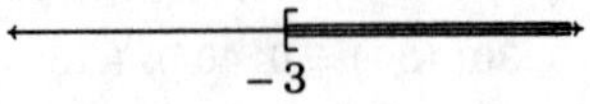

37. $3(x+2) > 6$

$3x + 6 > 6$

$3x > 0$

$\frac{1}{3}(3x) > \frac{1}{3}(0)$

$x > 0$

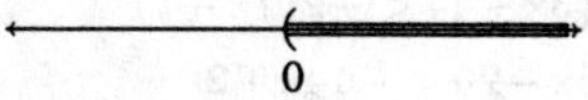

39. $5x + 2 \geq 4x + 6$

$5x \geq 4x + 4$

$x \geq 4$

41. $2x - 1 > 6$

$2x > 7$

$x > \frac{7}{2}$

$\left(\frac{7}{2}, \infty\right)$

43. $-5x - 2 < -14$

$-5x < -12$

$-\frac{1}{5}(-5x) > -\frac{1}{5}(-12)$

$x > \frac{12}{5}$

$\left(\frac{12}{5}, \infty\right)$

45. $-3(2x+1) \geq 12$

$-6x - 3 \geq 12$

$-6x \geq 15$

$-\frac{1}{6}(-6x) \leq -\frac{1}{6}(15)$

$x \leq -\frac{15}{6}$

$x \leq -\frac{5}{2}$

$\left(-\infty, -\frac{5}{2}\right]$

47. $4(3x-2) \geq -3$

$12x - 8 \geq -3$

$12x \geq 5$

$\frac{1}{12}(12x) \geq \frac{1}{12}(5)$

$x \geq \frac{5}{12}$

$\left[\frac{5}{12}, \infty\right)$

49. $6x - 2 > 4x - 14$

$6x > 4x - 12$

$2x > -12$

$\frac{1}{2}(2x) > \frac{1}{2}(-12)$

$x > -6$

$(-6, \infty)$

51. $2x - 7 < 6x + 13$

$2x < 6x + 20$

$-4x < 20$

$-\frac{1}{4}(-4x) > -\frac{1}{4}(20)$

$x > -5$

$(-5, \infty)$

53. $4(x-3) \leq -2(x+1)$

$4x - 12 \leq -2x - 2$

$4x \leq -2x + 10$

$6x \leq 10$

$\frac{1}{6}(6x) \leq \frac{1}{6}(10)$

$x \leq \frac{10}{6}$

$x \leq \frac{5}{3}$

$\left(-\infty, \frac{5}{3}\right]$

55. $5(x-4) - 6(x+2) < 4$

$5x - 20 - 6x - 12 < 4$

$-x - 32 < 4$

$-x < 36$

$-1(-x) > -1(36)$

$x > -36$

$(-36, \infty)$

57. $-3(3x+2)-2(4x+1)\geq 0$

$-9x-6-8x-2\geq 0$

$-17x-8\geq 0$

$-17x\geq 8$

$-\frac{1}{17}(-17x)\leq -\frac{1}{17}(8)$

$x\leq -\frac{8}{17}$

$\left(-\infty, -\frac{8}{17}\right]$

59. $-(x-3)+2(x-1)<3(x+4)$

$-x+3+2x-2<3x+12$

$x+1<3x+12$

$x<3x+11$

$-2x<11$

$-\frac{1}{2}(-2x)>-\frac{1}{2}(11)$

$x>-\frac{11}{2}$

$\left(-\frac{11}{2}, \infty\right)$

61. $7(x+1)-8(x-2)<0$

$7x+7-8x+16<0$

$-x+23<0$

$-x<-23$

$-1(-x)>-1(-23)$

$x>23$

$(23, \infty)$

63. $-5(x-1)+3>3x-4-4x$

$-5x+5+3>-x-4$

$-5x+8>-x-4$

$-5x>-x-12$

$-4x>-12$

$-\frac{1}{4}(-4x)<-\frac{1}{4}(-12)$

$x<3$

$(-\infty, 3)$

65. $3(x-2)-5(2x-1)\geq 0$

$3x-6-10x+5\geq 0$

$-7x-1\geq 0$

$-7x\geq 1$

$-\frac{1}{7}(-7x)\leq -\frac{1}{7}(1)$

$x\leq -\frac{1}{7}$

$\left(-\infty, -\frac{1}{7}\right]$

67. $-5(3x+4)<-2(7x-1)$

$-15x-20<-14x+2$

$-15x<-14x+22$

$-x<22$

$-1(-x)>-1(22)$

$x>-22$

$(-22, \infty)$

69. $-3(x+2)>2(x-6)$

$-3x-6>2x-12$

$-3x>2x-6$

$-5x>-6$

$-\frac{1}{5}(-5x)<-\frac{1}{5}(-6)$

$x<\frac{6}{5}$

$\left(-\infty, \frac{6}{5}\right)$

Problem Set 2.6 More on Inequalities

1. $\frac{2}{5}x+\frac{1}{3}x>\frac{44}{15}$

$15\left(\frac{2}{5}x+\frac{1}{3}x\right)>15\left(\frac{44}{15}\right)$

$15\left(\frac{2}{5}x\right)+15\left(\frac{1}{3}x\right)>44$

$6x+5x>44$

$11x>44$

$x>4$

$(4, \infty)$

3. $x-\frac{5}{6}<\frac{x}{2}+3$

$6\left(x-\frac{5}{6}\right)<6\left(\frac{x}{2}+3\right)$

$6(x)-6\left(\frac{5}{6}\right)<6\left(\frac{x}{2}\right)+6(3)$

$6x - 5 < 3x + 18$

$6x < 3x + 23$

$3x < 23$

$x < \frac{23}{3}$

$\left(-\infty, \frac{23}{3}\right)$

5. $\frac{x-2}{3} + \frac{x+1}{4} \geq \frac{5}{2}$

$12\left(\frac{x-2}{3} + \frac{x+1}{4}\right) \geq 12\left(\frac{5}{2}\right)$

$12\left(\frac{x-2}{3}\right) + 12\left(\frac{x+1}{4}\right) \geq 30$

$4(x-2) + 3(x+1) \geq 30$

$4x - 8 + 3x + 3 \geq 30$

$7x - 5 \geq 30$

$7x \geq 35$

$x \geq 5$

$[5, \infty)$

7. $\frac{3-x}{6} + \frac{x+2}{7} \leq 1$

$42\left(\frac{3-x}{6} + \frac{x+2}{7}\right) \leq 42(1)$

$42\left(\frac{3-x}{6}\right) + 42\left(\frac{x+2}{7}\right) \leq 42$

$7(3-x) + 6(x+2) \leq 42$

$21 - 7x + 6x + 12 \leq 42$

$-x + 33 \leq 42$

$-x \leq 9$

$-1(-x) \geq -1(9)$

$x \geq -9$

$[-9, \infty)$

9. $\frac{x+3}{8} - \frac{x+5}{5} \geq \frac{3}{10}$

$40\left(\frac{x+3}{8} - \frac{x+5}{5}\right) \geq 40\left(\frac{3}{10}\right)$

$40\left(\frac{x+3}{8}\right) - 40\left(\frac{x+5}{5}\right) \geq 12$

$5(x+3) - 8(x+5) \geq 12$

$5x + 15 - 8x - 40 \geq 12$

$-3x - 25 \geq 12$

$-3x \geq 37$

$-\frac{1}{3}(-3x) \leq -\frac{1}{3}(37)$

$x \leq -\frac{37}{3}$

$\left(-\infty, -\frac{37}{3}\right]$

11. $\frac{4x-3}{6} - \frac{2x-1}{12} < -2$

$12\left(\frac{4x-3}{6} - \frac{2x-1}{12}\right) < 12(-2)$

$12\left(\frac{4x-3}{6}\right) - 12\left(\frac{2x-1}{12}\right) < -24$

$2(4x-3) - (2x-1) < -24$

$8x - 6 - 2x + 1 < -24$

$6x - 5 < -24$

$6x < -19$

$x < -\frac{19}{6}$

$\left(-\infty, -\frac{19}{6}\right)$

13. $.06x + .08(250 - x) \geq 19$

$100[.06x + .08(250 - x)] \geq 100(19)$

$100(.06x) + 100[.08(250 - x)] \geq 1900$

$6x + 8(250 - x) \geq 1900$

$6x + 2000 - 8x \geq 1900$

$-2x + 2000 \geq 1900$

$-2x \geq -100$

$-\frac{1}{2}(-2x) \leq -\frac{1}{2}(-100)$

$x \leq 50$

$(-\infty, 50]$

15. $.09x + .1(x + 200) > 77$

$100[.09x + .1(x + 200)] > 100(77)$

$100(.09x) + 100[.1(x + 200)] > 7700$

$9x + 10(x + 200) > 7700$

$9x + 10x + 2000 > 7700$

$19x + 2000 > 7700$

$19x > 5700$

$x > 300$

$(300, \infty)$

17. $x \geq 3.4 + .15x$

$100(x) \geq 100(3.4 + .15x)$

$100x \geq 100(3.4) + 100(.15x)$

$100x \geq 340 + 15x$

$85x \geq 340$

$x \geq 4$

$[4, \infty)$

19. $x > -1$ and $x < 2$

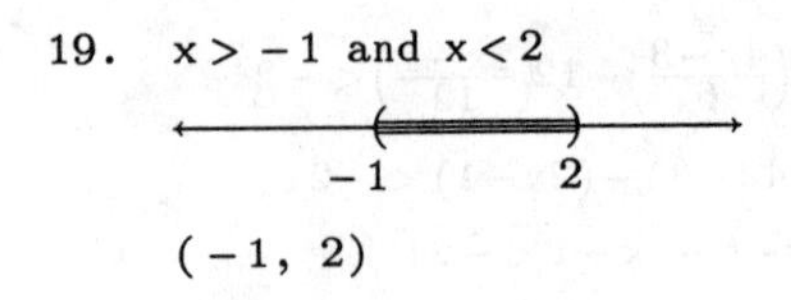

$(-1, 2)$

21. $x \leq 2$ and $x > -1$

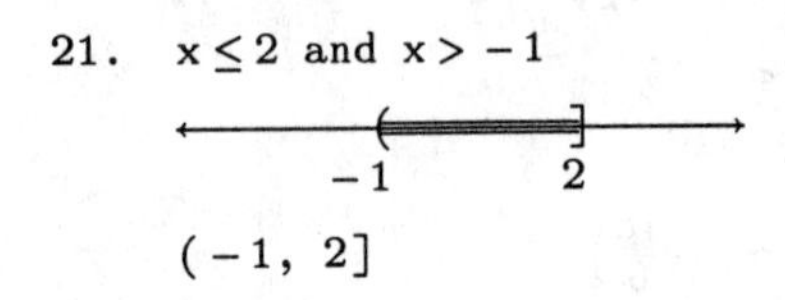

$(-1, 2]$

23. $x > 2$ or $x < -1$

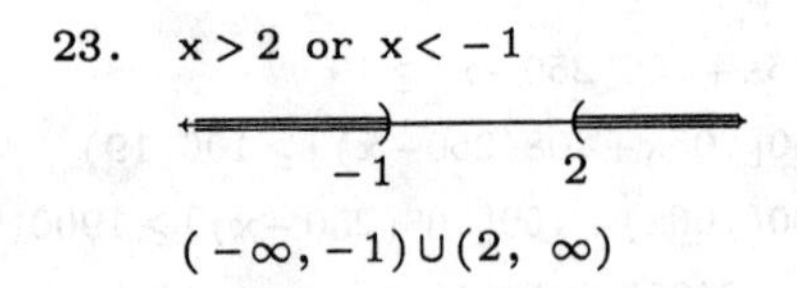

$(-\infty, -1) \cup (2, \infty)$

25. $x \leq 1$ or $x > 3$

1 3

$(-\infty, 1] \cup (3, \infty)$

27. $x > 0$ and $x > -1$

0

$(0, \infty)$

29. $x < 0$ and $x > 4$

$\emptyset$

31. $x > -2$ or $x < 3$

$(-\infty, \infty)$

33. $x > -1$ or $x > 2$

−1

$(-1, \infty)$

35. $x - 2 > -1$ and $x - 2 < 1$

$x > 1$ and $x < 3$

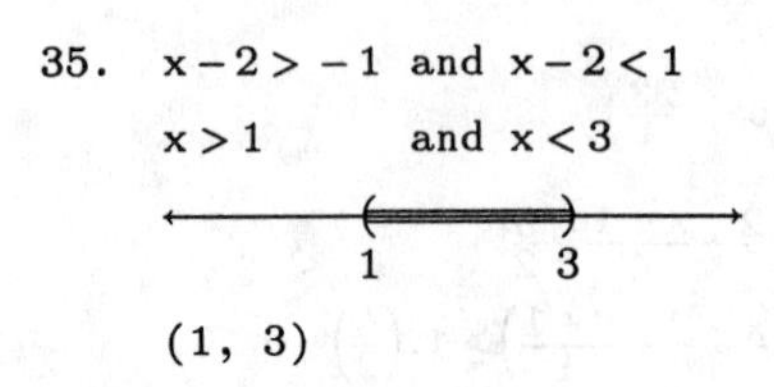

$(1, 3)$

37. $x + 2 < -3$ or $x + 2 > 3$

$x < -5$ or $x > 1$

−5 1

$(-\infty, -5) \cup (1, \infty)$

39. $2x - 1 \geq 5$ and $x > 0$

$2x \geq 6$ and $x > 0$

$x \geq 3$ and $x > 0$

3

$[3, \infty)$

41. $5x - 2 < 0$ and $3x - 1 > 0$

$5x < 2$ and $3x > 1$

$x < \frac{2}{5}$ and $x > \frac{1}{3}$

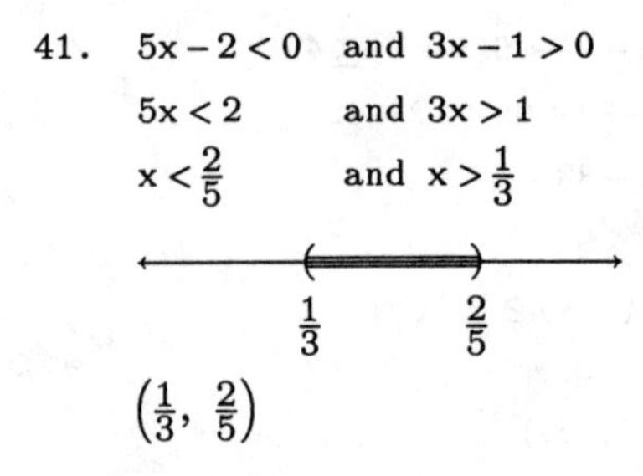

$\left(\frac{1}{3}, \frac{2}{5}\right)$

43. $3x + 2 < -1$ or $3x + 2 > 1$

$3x < -3$ or $3x > -1$

$x < -1$ or $x > -\frac{1}{3}$

-1 $-\frac{1}{3}$

$(-\infty, -1) \cup \left(-\frac{1}{3}, \infty\right)$

45. $-3 < 2x + 1 < 5$

$-4 < 2x < 4$

$-2 < x < 2$

$(-2,\ 2)$

47. $-17 \leq 3x - 2 \leq 10$

$-15 \leq 3x \leq 12$

$-5 \leq x \leq 4$

$[-5,\ 4]$

49. $1 < 4x + 3 < 9$

$-2 < 4x < 6$

$-\frac{2}{4} < x < \frac{6}{4}$

$-\frac{1}{2} < x < \frac{3}{2}$

$\left(-\frac{1}{2},\ \frac{3}{2}\right)$

51. $-6 < 4x - 5 < 6$

$-1 < 4x < 11$

$-\frac{1}{4} < x < \frac{11}{4}$

$\left(-\frac{1}{4},\ \frac{11}{4}\right)$

53. $-4 \leq \frac{x-1}{3} \leq 4$

$3(-4) \leq 3\left(\frac{x-1}{3}\right) \leq 3(4)$

$-12 \leq x - 1 \leq 12$

$-11 \leq x \leq 13$

$[-11,\ 13]$

55. $-3 < 2 - x < 3$

$-5 < -x < 1$

$-1(-5) > -1(-x) > -1(1)$

$5 > x > -1$

$(-1,\ 5)$

57. Let $r = \text{rate}$.

$(9\%)(300) + r(200) > 47$

$.09(300) + r(200) > 47$

$100[.09(300) + r(200)] > 100(47)$

$100[.09(300)] + 100r(200) > 4700$

$9(300) + 20000r > 4700$

$2700 + 20000r > 4700$

$20000r > 2000$

$r > \frac{2000}{20000}$

$r > .1$

$r > 10\%$

The rate must be more than 10%.

59. Let x = average height of the two guards.

6 ft. 8 in. $= 6\frac{8}{12}$ feet $= 6\frac{2}{3}$ feet

6 ft. 4in. $= 6\frac{4}{12}$ feet $= 6\frac{1}{3}$ feet

$\frac{3\left(6\frac{2}{3}\right) + 2(x)}{5} \geq 6\frac{1}{3}$

$\frac{20 + 2x}{5} \geq \frac{19}{3}$

$15\left(\frac{20 + 2x}{5}\right) \geq 15\left(\frac{19}{3}\right)$

$3(20 + 2x) \geq 5(19)$

$60 + 6x \geq 95$

$6x \geq 35$

$x \geq \frac{35}{6}$

$x \geq 5\frac{5}{6}$

$x \geq 5\frac{10}{12}$

The height should be 5 ft. 10 inches or better.

61. Let x = score for the third game.

$\frac{142 + 170 + x}{3} \geq 160$

$\frac{312 + x}{3} \geq 160$

$3\left(\frac{312 + x}{3}\right) \geq 3(160)$

$312 + x \geq 480$

$x \geq 168$

The score should be 168 or better.

63. Let $x = \text{score on the fifth day}$.

$\frac{82+84+78+79+x}{5} \leq 80$

$\frac{323+x}{5} \leq 80$

$5\left(\frac{323+x}{5}\right) \leq 5(80)$

$323 + x \leq 400$

$x \leq 77$

The score should be 77 or less.

65. $325 \leq \frac{9}{5}C + 32 \leq 425$

$293 \leq \frac{9}{5}C \leq 393$

$\frac{5}{9}(293) \leq \frac{5}{9}\left(\frac{9}{5}C\right) \leq \frac{5}{9}(393)$

$163 \leq C \leq 218$

The temperature would be between 163°C and 218°C inclusive.

67. $70 \leq \frac{100M}{C} \leq 125$

$70 \leq \frac{100M}{9} \leq 125$

$\frac{9}{100}(70) \leq \frac{9}{100}\left(\frac{100M}{9}\right) \leq \frac{9}{100}(125)$

$6.3 \leq M \leq 11.25$

Problem Set 2.7 Equations and Inequalities Involving Absolute Value

1. $|x| < 5$

$-5 < x < 5$

The solution set is $(-5, 5)$.

3. $|x| \leq 2$

$-2 \leq x \leq 2$

The solution set is $[-2, 2]$.

5. $|x| > 2$

$x < -2$ or $x > 2$

The solution set is $(-\infty, -2) \cup (2, \infty)$.

7. $|x - 1| < 2$

$-2 < x - 1 < 2$

$-1 < x < 3$

The solution set is $(-1, 3)$.

9. $|x + 2| \leq 4$

$-4 \leq x + 2 \leq 4$

$-6 \leq x \leq 2$

The solution set is $[-6, 2]$.

11. $|x + 2| > 1$

$x + 2 < -1$ or $x + 2 > 1$

$x < -3$ or $x > -1$

The solution set is $(-\infty, -3) \cup (-1, \infty)$.

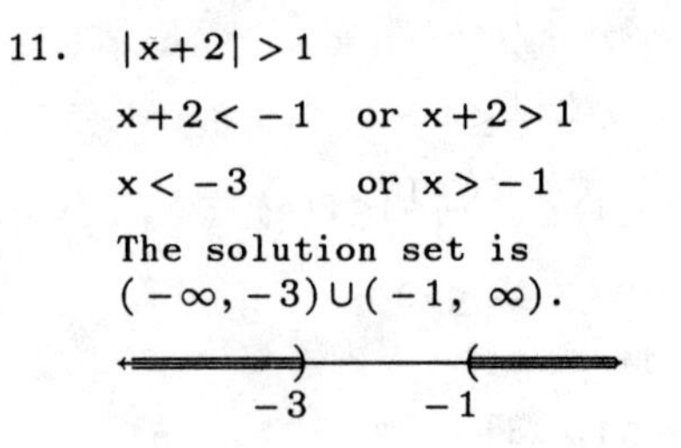

13. $|x - 3| \geq 2$

$x - 3 \leq -2$ or $x - 3 \geq 2$

$x \leq 1$ or $x \geq 5$

The solution set is $(-\infty, 1] \cup [5, \infty)$.

15. $|x - 1| = 8$

$x - 1 = 8$ or $x - 1 = -8$

$x = 9$ or $x = -7$

The solution set is $\{-7, 9\}$.

17. $|x-2|>6$

$x-2<-6$ or $x-2>6$

$x<-4$ or $x>8$

The solution set is $(-\infty,-4)\cup(8,\infty)$.

19. $|x+3|<5$

$-5<x+3<5$

$-8<x<2$

The solution set is $(-8, 2)$.

21. $|2x-4|=6$

$2x-4=6$ or $2x-4=-6$

$2x=10$ or $2x=-2$

$x=5$ or $x=-1$

The solution set is $\{-1, 5\}$.

23. $|2x-1|\leq 9$

$-9\leq 2x-1\leq 9$

$-8\leq 2x\leq 10$

$-4\leq x\leq 5$

The solution set is $[-4, 5]$.

25. $|4x+2|\geq 12$

$4x+2\leq -12$ or $4x+2\geq 12$

$4x\leq -14$ or $4x\geq 10$

$x\leq -\frac{14}{4}$ or $x\geq \frac{10}{4}$

$x\leq -\frac{7}{2}$ or $x\geq \frac{5}{2}$

The solution set is $\left(-\infty,-\frac{7}{2}\right]\cup\left[\frac{5}{2},\infty\right)$.

27. $|3x+4|=11$

$3x+4=11$ or $3x+4=-11$

$3x=7$ or $3x=-15$

$x=\frac{7}{3}$ or $x=-5$

The solution set is $\left\{-5, \frac{7}{3}\right\}$.

29. $|4-2x|=6$

$4-2x=6$ or $4-2x=-6$

$-2x=2$ or $-2x=-10$

$x=-1$ or $x=5$

The solution set is $\{-1, 5\}$.

31. $|2-x|>4$

$2-x<-4$ or $2-x>4$

$-x<-6$ or $-x>2$

$-1(-x)>-1(-6)$ or $-1(-x)<-1(2)$

$x>6$ or $x<-2$

The solution set is $(-\infty,-2)\cup(6,\infty)$.

33. $|1-2x|<2$

$-2<1-2x<2$

$-3<-2x<1$

$-\frac{1}{2}(-3)>-\frac{1}{2}(-2x)>-\frac{1}{2}(1)$

$\frac{3}{2}>x>-\frac{1}{2}$

The solution set is $\left(-\frac{1}{2}, \frac{3}{2}\right)$.

35. $|5x+9|\leq 16$

$-16\leq 5x+9\leq 16$

$-25\leq 5x\leq 7$

$-5\leq x\leq \frac{7}{5}$

The solution set is $\left[-5, \frac{7}{5}\right]$.

37. $\left|x-\frac{3}{4}\right|=\frac{2}{3}$

$x-\frac{3}{4}=\frac{2}{3}$ or $x-\frac{3}{4}=-\frac{2}{3}$

$x=\frac{2}{3}+\frac{3}{4}$ or $x=-\frac{2}{3}+\frac{3}{4}$

$x=\frac{8}{12}+\frac{9}{12}$ or $x=-\frac{8}{12}+\frac{9}{12}$

$x=\frac{17}{12}$ or $x=\frac{1}{12}$

The solution set is $\left\{\frac{1}{12}, \frac{17}{12}\right\}$.

39. $|-2x+7| \leq 13$

$-13 \leq -2x+7 \leq 13$

$-20 \leq -2x \leq 6$

$-\frac{1}{2}(-20) \geq -\frac{1}{2}(-2x) \geq -\frac{1}{2}(6)$

$10 \geq x \geq -3$

The solution set is $[-3, 10]$.

41. $\left|\frac{x-3}{4}\right| < 2$

$-2 < \frac{x-4}{4} < 2$

$4(-2) < 4\left(\frac{x-3}{4}\right) < 4(2)$

$-8 < x-3 < 8$

$-5 < x < 11$

The solution set is $(-5, 11)$.

43. $\left|\frac{2x+1}{2}\right| > 1$

$\frac{2x+1}{2} < -1$ or $\frac{2x+1}{2} > 1$

$2\left(\frac{2x+1}{2}\right) < -1(2)$ or $2\left(\frac{2x+1}{2}\right) \geq 1(2)$

$2x+1 < -2$ or $2x+1 > 2$

$2x < -3$ or $2x > 1$

$x < -\frac{3}{2}$ or $x > \frac{1}{2}$

The solution set is $\left(-\infty, -\frac{3}{2}\right) \cup \left(\frac{1}{2}, \infty\right)$.

45. $|2x-3| + 2 = 5$

$|2x-3| = 3$

$2x-3 = 3$ or $2x-3 = -3$

$2x = 6$ or $2x = 0$

$x = 3$ or $x = 0$

The solution set is $\{0, 3\}$.

47. $|x+7| - 3 \geq 4$

$|x+7| \geq 7$

$x+7 \leq -7$ or $x+7 \geq 7$

$x \leq -14$ or $x \geq 0$

The solution set is $(-\infty, -14] \cup [0, \infty)$.

49. $|2x-1| + 1 \leq 6$

$|2x-1| \leq 5$

$-5 \leq 2x-1 \leq 5$

$-4 \leq 2x \leq 6$

$-2 \leq x \leq 3$

The solution set is $[-2, 3]$.

51. $|2x+1| = -4$

The solution set is $\emptyset$.
Absolute value is never negative.

53. $|3x-1| > -2$

The solution set is $(-\infty, \infty)$.
Absolute value is always greater than a negative value.

55. $|5x-2| = 0$

$5x-2 = 0$

$5x = 2$

$x = \frac{2}{5}$

The solution set is $\left\{\frac{2}{5}\right\}$.

57. $|4x-1| < -1$

The solution set is $\emptyset$.
Absolute value is never less than a negative value.

59. $|x+4| < 0$

The solution set is $\emptyset$.
Absolute value is never less than zero.

Chapter 2 Review

1. $5(x-6) = 3(x+2)$

$5x - 30 = 3x + 6$

$5x - 30 + 30 = 3x + 6 + 30$

$5x = 3x + 36$

$5x - 3x = 3x - 3x + 36$

$2x = 36$

$\frac{1}{2}(2x) = \frac{1}{2}(36)$

$x = 18$

The solution set is $\{18\}$.

3. $-(2n-1)+3(n+2)=7$

$-2n+1+3n+6=7$

$n+7=7$

$n+7-7=7-7$

$n=0$

The solution set is $\{0\}$.

5. $\frac{3t-2}{4} = \frac{2t+1}{3}$

$12\left(\frac{3t-2}{4}\right) = 12\left(\frac{2t+1}{3}\right)$

$3(3t-2) = 4(2t+1)$

$9t-6 = 8t+4$

$9t = 8t+10$

$t = 10$

The solution set is $\{10\}$.

7. $1 - \frac{2x-1}{6} = \frac{3x}{8}$

$24\left(1 - \frac{2x-1}{6}\right) = 24\left(\frac{3x}{8}\right)$

$24(1) - 24\left(\frac{2x-1}{6}\right) = 3(3x)$

$24 - 4(2x-1) = 9x$

$24 - 8x + 4 = 9x$

$28 - 8x = 9x$

$28 = 17x$

$\frac{28}{17} = x$

The solution set is $\left\{\frac{28}{17}\right\}$.

9. $\frac{3n-1}{2} - \frac{2n+3}{7} = 1$

$14\left(\frac{3n-1}{2} - \frac{2n+3}{7}\right) = 14(1)$

$14\left(\frac{3n-1}{2}\right) - 14\left(\frac{2n+3}{7}\right) = 14$

$7(3n-1) - 2(2n+3) = 14$

$21n - 7 - 4n - 6 = 14$

$17n - 13 = 14$

$17n = 27$

$n = \frac{27}{17}$

The solution set is $\left\{\frac{27}{17}\right\}$.

11. $.06x + .08(x+100) = 15$

$100[.06x + .08(x+100)] = 100(15)$

$100(.06x) + 100[.08(x+100)] = 1500$

$6x + 8(x+100) = 1500$

$6x + 8x + 800 = 1500$

$14x + 800 = 1500$

$14x = 700$

$x = 50$

The solution set is $\{50\}$.

13. $.1(n+300) = .09n + 32$

$100[.1(n+300)] = 100[.09n+32]$

$10(n+300) = 100(.09n) + 100(32)$

$10n + 3000 = 9n + 3200$

$10n = 9n + 200$

$n = 200$

The solution set is $\{200\}$.

15. $|2n+3| = 4$

$2n+3=4$ or $2n+3=-4$

$2n=1$ or $2n=-7$

$n=\frac{1}{2}$ or $n=-\frac{7}{2}$

The solution set is $\left\{\frac{1}{2}, -\frac{7}{2}\right\}$.

17. $ax = bx + c$

$ax - bx = c$

$x(a-b) = c$

$x = \frac{c}{a-b}$

19. $5x - 7y = 11$

$5x = 11 + 7y$

$x = \frac{11+7y}{5}$

21. $A = \pi r^2 + \pi rs$

$A - \pi r^2 = \pi rs$

$\frac{A - \pi r^2}{\pi r} = s$

23. $R = \frac{R_1 R_2}{R_1 + R_2}$

$(R_1 + R_2)R = (R_1 + R_2)\left(\frac{R_1 R_2}{R_1 + R_2}\right)$

$RR_1 + RR_2 = R_1 R_2$

$RR_2 = R_1 R_2 - RR_1$

$RR_2 = R_1(R_2 - R)$

$\frac{RR_2}{R_2 - R} = R_1$

25. $5x - 2 \geq 4x - 7$

$5x \geq 4x - 5$

$x \geq -5$

$[-5, \infty)$

27. $2(3x - 1) - 3(x - 3) > 0$

$6x - 2 - 3x + 9 > 0$

$3x + 7 > 0$

$3x > -7$

$x > -\frac{7}{3}$

$\left(-\frac{7}{3}, \infty\right)$

29. $\frac{5}{6}n - \frac{1}{3}n < \frac{1}{6}$

$6\left(\frac{5}{6}n - \frac{1}{3}n\right) < 6\left(\frac{1}{6}\right)$

$6\left(\frac{5}{6}n\right) - 6\left(\frac{1}{3}n\right) < 1$

$5n - 2n < 1$

$3n < 1$

$n < \frac{1}{3}$

$\left(-\infty, \frac{1}{3}\right)$

31. $s \geq 4.5 + .25s$

$100(s) \geq 100(4.5 + .25s)$

$100s \geq 100(4.5) + 100(.25s)$

$100s \geq 450 + 25s$

$75s \geq 450$

$s \geq 6$

$[6, \infty)$

33. $|2x - 1| < 11$

$-11 < 2x - 1 < 11$

$-10 < 2x < 12$

$-5 < x < 6$

$(-5, 6)$

35. $-3(2t - 1) - (t + 2) > -6(t - 3)$

$-6t + 3 - t - 2 > -6t + 18$

$-7t + 1 > -6t + 18$

$-7t > -6t + 17$

$-t > 17$

$-1(-t) < -1(17)$

$t < -17$

$(-\infty, -17)$

37. $x > -1$ and $x < 1$

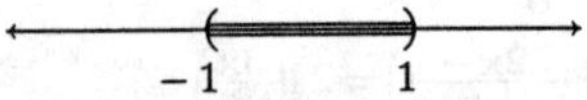

(number line: open interval from -1 to 1)

39. $x > 2$ and $x > 3$

(number line: open at 3, shaded to the right)

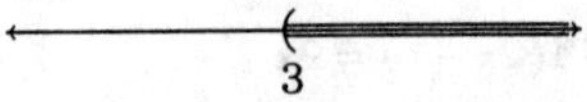

41. $2x + 1 > 3$ or $2x + 1 < -3$

$2x > 2$ or $2x < -4$

$x > 1$ or $x < -2$

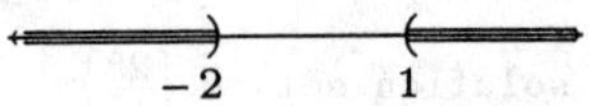

(number line: shaded left of -2 (open) and right of 1 (open))

43. $-1 < 4x - 3 \leq 9$

$2 < 4x \leq 12$

$\frac{2}{4} < x \leq 3$

$\frac{1}{2} < x \leq 3$

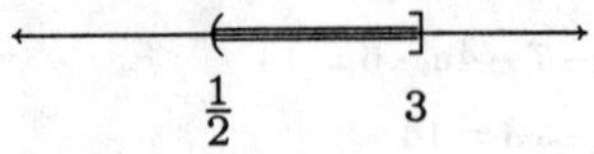

(number line: open at $\frac{1}{2}$, closed at 3)

45. Let $x = \text{length}$

$\frac{1}{3}x + 2 = \text{width}$

$P = 2l + 2w$

$44 = 2(x) + 2\left(\frac{1}{3}x + 2\right)$

$44 = 2x + \frac{2}{3}x + 4$

$40 = 2x + \frac{2}{3}x$

$3(40) = 3\left(2x + \frac{2}{3}x\right)$

$120 = 3(2x) + 3\left(\frac{2}{3}x\right)$

$120 = 6x + 2x$

$120 = 8x$

$15 = x$

length = 15 meters

width $= \frac{1}{3}(15) + 2 = 7$ meters

The length is 15 meters and the width is 7 meters.

47. Let x = score on the 4th exam.

$\frac{3(84) + x}{4} \geq 85$

$\frac{252 + x}{4} \geq 85$

$4\left(\frac{252 + x}{4}\right) \geq 4(85)$

$252 + x \geq 340$

$x \geq 88$

The score should be 88 or better.

49. Let x = normal hourly rate.

$\frac{3}{2}x$ = over 36 hours rate.

Pat worked 36 hours at the normal hourly rate and 6 hours at the overtime rate.

$36x + 6\left(\frac{3}{2}x\right) = 472.50$

$36x + 9x = 472.50$

$45x = 472.50$

$x = 10.50$

The normal hourly rate is \$10.50.

51. Let x = angle

$90 - x$ = complement of the angle.

$180 - x$ = supplement of the angle.

$90 - x = \frac{1}{10}(180 - x)$

$10(90 - x) = 10\left[\frac{1}{10}(180 - x)\right]$

$900 - 10x = 1(180 - x)$

$900 - 10x = 180 - x$

$-10x = -720 - x$

$-9x = -720$

$x = 80$

The angle is 80°.

53. Let x = score in the 5th game.

$\frac{16 + 22 + 18 + 14 + x}{5} \geq 20$

$\frac{70 + x}{5} \geq 20$

$5\left(\frac{70 + x}{5}\right) \geq 5(20)$

$70 + x \geq 100$

$x \geq 30$

She should score 30 or more.

55. Let t = time Sonya rode.

$t + \frac{5}{4}$ = time Rita rode.

Sonya's distance $= 16t$

Rita's distance $= 12\left(t + \frac{5}{4}\right)$

Rita's distance = Sonya's distance + 2

$12\left(t + \frac{5}{4}\right) = 16t + 2$

$12t + 12\left(\frac{5}{4}\right) = 16t + 2$

$12t + 15 = 16t + 2$

$12t = 16t - 13$

$-4t = -13$

$t = \frac{-13}{-4} = 3\frac{1}{4}$

Sonya's time $= 3\frac{1}{4}$ hours

Rita's time $= \frac{13}{4} + \frac{5}{4} = \frac{18}{4} = 4\frac{1}{2}$ hours.

Sonya rides for $3\frac{1}{4}$ hours and Rita rides for $4\frac{1}{2}$ hours.

Chapter 2 Test

1. $5x-2=2x-11$

   $5x=2x-9$

   $3x=-9$

   $x=-3$

   The solution set is $\{-3\}$.

3. $-3(x+4)=3(x-5)$

   $-3x-12=3x-15$

   $-3x=3x-3$

   $-6x=-3$

   $x=\frac{-3}{-6}=\frac{1}{2}$

   The solution set is $\left\{\frac{1}{2}\right\}$.

5. $\frac{3t-2}{4}=\frac{5t+1}{5}$

   $20\left(\frac{3t-2}{4}\right)=20\left(\frac{5t+1}{5}\right)$

   $5(3t-2)=4(5t+1)$

   $15t-10=20t+4$

   $15t=20t+14$

   $-5t=14$

   $t=-\frac{14}{5}$

   The solution is $\left\{-\frac{14}{5}\right\}$.

7. $|4x-3|=9$

   $4x-3=9$ or $4x-3=-9$

   $4x=12$ or $4x=-6$

   $x=3$ or $x=-\frac{6}{4}=-\frac{3}{2}$

   The solution set is $\left\{3,-\frac{3}{2}\right\}$.

9. $2-\frac{3x-1}{5}=-4$

   $5\left(2-\frac{3x-1}{5}\right)=5(-4)$

   $5(2)-5\left(\frac{3x-1}{5}\right)=-20$

   $10-(3x-1)=-20$

   $10-3x+1=-20$

   $-3x+11=-20$

   $-3x=-31$

   $x=\frac{-31}{-3}=\frac{31}{3}$

   The solution set is $\left\{\frac{31}{3}\right\}$.

11. $\frac{2}{3}x-\frac{3}{4}y=2$

    $12\left(\frac{2}{3}x-\frac{3}{4}y\right)=12(2)$

    $12\left(\frac{2}{3}x\right)-12\left(\frac{3}{4}y\right)=24$

    $8x-9y=24$

    $-9y=-8x+24$

    $y=\frac{-8x+24}{-9}$

    $y=\frac{-1(-8x+24)}{-1(-9)}$

    $y=\frac{8x-24}{9}$

13. $7x-4>5x-8$

    $7x>5x-4$

    $2x>-4$

    $x>-2$

    $(-2,\infty)$

15. $2(x-1)-3(3x+1)\geq-6(x-5)$

    $2x-2-9x-3\geq-6x+30$

    $-7x-5\geq-6x+30$

    $-7x\geq-6x+35$

    $-x\geq35$

    $-1(-x)\leq-1(35)$

    $x\leq-35$

    $(-\infty,-35]$

17. $\frac{x-2}{6}-\frac{x+3}{9}>-\frac{1}{2}$

    $18\left(\frac{x-2}{6}-\frac{x+3}{9}\right)>18\left(-\frac{1}{2}\right)$

$18\left(\frac{x-2}{6}\right) - 18\left(\frac{x+3}{9}\right) > -9$

$3(x-2) - 2(x+3) > -9$

$3x - 6 - 2x - 6 > -9$

$x - 12 > -9$

$x > 3$

$(3, \infty)$

19. $|6x-4| < 10$

$-10 < 6x - 4 < 10$

$-6 < 6x < 14$

$-1 < x < \frac{14}{6}$

$-1 < x < \frac{7}{3}$

$\left(-1, \frac{7}{3}\right)$

21. Let $x =$ original price

$(100\%)(x) - (20\%)(x) = 57.60$

$1.00x - .20x = 57.60$

$100[1.00x - .20x] = 100[57.60]$

$100x - 20x = 5760$

$80x = 5760$

$x = 72$

The original price is \$72.

23. Let $x =$ number of cups of grapefruit juice.

$100\%(x) + (8\%)(30) = (10\%)(x+30)$

$1.00x + .08(30) = .10(x+30)$

$100[1.00x + .08(30)] = 100[.10(x+30)]$

$100(1.00x) + 100[.08(30)] = 10(x+30)$

$100x + 8(30) = 10x + 300$

$100x + 240 = 10x + 300$

$100x = 10x + 60$

$90x = 60$

$x = \frac{60}{90} = \frac{2}{3}$

There should be $\frac{2}{3}$ cup of grapefruit juice added.

25. Let $x =$ angle
$90 - x =$ complement of the angle
$180 - x =$ supplement of the angle

$90 - x = \frac{2}{11}(180 - x)$

$11(90 - x) = 11\left[\frac{2}{11}(180 - x)\right]$

$11(90) - 11(x) = 2(180 - x)$

$990 - 11x = 360 - 2x$

$-11x = -630 - 2x$

$-9x = -630$

$x = 70$

The angle is 70°.

Chapter 3 Polynomials

Problem Set 3.1 Polynomials: Sums and Differences

1. $7xy+6y$

   The degree is the sum of the exponents of the 7xy term. The degree is 2.

3. $-x^2y+2xy^2-xy$

   The degree is the sum of the exponents of either the $-x^2y$ or $2x^2y$ term. The degree is 3.

5. $5x^2-7x-2$

   The degree is the exponent of the $5x^2$ term. The degree is 2.

7. $8x^6+9$

   The degree is the exponent of the $8x^6$ term. The degree is 6.

9. $-12=-12x^0$

   The degree is the exponent. The degree is 0.

11. $(3x-7)+(7x+4)$

    $3x-7+7x+4$

    $3x+7x-7+4$

    $10x-3$

13. $(-5t-4)+(-6t+9)$

    $-5t-4-6t+9$

    $-5t-6t-4+9$

    $-11t+5$

15. $(3x^2-5x-1)+(-4x^2+7x-1)$

    $3x^2-5x-1-4x^2+7x-1$

    $3x^2-4x^2-5x+7x-1-1$

    $-x^2+2x-2$

17. $(12a^2b^2-9ab)+(5a^2b^2+4ab)$

    $12a^2b^2-9ab+5a^2b^2+4ab$

    $12a^2b^2+5a^2b^2-9ab+4ab$

    $17a^2b^2-5ab$

19. $(2x-4)+(-7x+2)+(-4x+9)$

    $2x-4-7x+2-4x+9$

    $2x-7x-4x-4+2+9$

    $-9x+7$

21. $(3x+4)-(5x-2)$

    $3x+4-5x+2$

    $3x-5x+4+2$

    $-2x+6$

23. $(6a+2)-(-4a-5)$

    $6a+2+4a+5$

    $6a+4a+2+5$

    $10a+7$

25. $(7x^2+9x+8)-(3x^2-x+2)$

    $7x^2+9x+8-3x^2+x-2$

    $7x^2-3x^2+9x+x+8-2$

    $4x^2+10x+6$

27. $(-4a^2+6a+10)-(2a^2-6a-4)$

    $-4a^2+6a+10-2a^2+6a+4$

    $-4a^2-2a^2+6a+6a+10+4$

    $-6a^2+12a+14$

29. $(5x^3+2x^2+6x-13)-(2x^3+x^2-7x-2)$

    $5x^3+2x^2+6x-13-2x^3-x^2+7x+2$

    $5x^3-2x^3+2x^2-x^2+6x+7x-13+2$

    $3x^3+x^2+13x-11$

31. $$\begin{array}{r} 12x+6 \\ -(5x-2) \\ \hline 7x+8 \end{array}$$

33. $$\begin{array}{r} -7x-9 \\ -(-4x+7) \\ \hline -3x-16 \end{array}$$

35.
$$\begin{array}{r} 4x^2 - x - 2 \\ -(2x^2 + x + 6) \\ \hline 2x^2 - 2x - 8 \end{array}$$

37.
$$\begin{array}{r} -2x^3 + 6x^2 - 3x + 8 \\ -(x^3 + x^2 - x - 1) \\ \hline -3x^3 + 5x^2 - 2x + 9 \end{array}$$

39.
$$\begin{array}{r} 2x - 1 \\ -(-5x^2 + 6x - 12) \\ \hline 5x^2 - 4x + 11 \end{array}$$

41. $(x^2+9x-4)+(-5x^2-7x+10)-(2x^2-7x-1)$

$x^2 + 9x - 4 - 5x^2 - 7x + 10 - 2x^2 + 7x + 1$

$x^2 - 5x^2 - 2x^2 + 9x - 7x + 7x - 4 + 10 + 1$

$-6x^2 + 9x + 7$

43. $(4x^2 + 3) + (-7x^2 + 2x) - (-x^2 - 7x - 1)$

$4x^2 + 3 - 7x^2 + 2x + x^2 + 7x + 1$

$4x^2 - 7x^2 + x^2 + 2x + 7x + 3 + 1$

$-2x^2 + 9x + 4$

45. $(-12n^2-n+9)-[(5n^2-3n-2)+(-7n^2+n+2)]$

$-12n^2-n+9-[5n^2-3n-2-7n^2+n+2]$

$-12n^2 - n + 9 - (-2n^2 - 2n)$

$-12n^2 - n + 9 + 2n^2 + 2n$

$-12n^2 + 2n^2 - n + 2n + 9$

$-10n^2 + n + 9$

47. $(5x + 2) + (7x - 1) + (-4x - 3)$

$5x + 2 + 7x - 1 - 4x - 3$

$5x + 7x - 4x + 2 - 1 - 3$

$8x - 2$

49. $(12x - 9) - (-3x + 4) - (7x + 1)$

$12x - 9 + 3x - 4 - 7x - 1$

$12x + 3x - 7x - 9 - 4 - 1$

$8x - 14$

51. $(2x^2-7x-1)+(-4x^2-x+6)+(-7x^2-4x-1)$

$2x^2 - 7x - 1 - 4x^2 - x + 6 - 7x^2 - 4x - 1$

$2x^2 - 4x^2 - 7x^2 - 7x - x - 4x - 1 + 6 - 1$

$-9x^2 - 12x + 4$

53. $(7x^2-x-4)-(9x^2-10x+8)+(12x^2+4x-6)$

$7x^2 - x - 4 - 9x^2 + 10x - 8 + 12x^2 + 4x - 6$

$7x^2 - 9x^2 + 12x^2 - x + 10x + 4x - 4 - 8 - 6$

$10x^2 + 13x - 18$

55. $(n^2 - 7n - 9) - (-3n + 4) - (2n^2 - 9)$

$n^2 - 7n - 9 + 3n - 4 - 2n^2 + 9$

$n^2 - 2n^2 - 7n + 3n - 9 - 4 + 9$

$-n^2 - 4n - 4$

57. $3x - [5x - (x + 6)]$

$3x - [5x - x - 6]$

$3x - [4x - 6]$

$3x - 4x + 6$

$-x + 6$

59. $2x^2 - [-3x^2 - (x^2 - 4)]$

$2x^2 - [-3x^2 - x^2 + 4]$

$2x^2 - [-4x^2 + 4]$

$2x^2 + 4x^2 - 4$

$6x^2 - 4$

61. $-2n^2 - [n^2 - (-4n^2 + n + 6)]$

$-2n^2 - [n^2 + 4n^2 - n - 6]$

$-2n^2 - [5n^2 - n - 6]$

$-2n^2 - 5n^2 + n + 6$

$-7n^2 + n + 6$

63. $[4t^2 - (2t + 1) + 3] - [3t^2 + (2t - 1) - 5]$

$[4t^2 - 2t - 1 + 3] - [3t^2 + 2t - 1 - 5]$

$4t^2 - 2t + 2 - (3t^2 + 2t - 6)$

$4t^2 - 2t + 2 - 3t^2 - 2t + 6$

$4t^2 - 3t^2 - 2t - 2t + 2 + 6$

$t^2 - 4t + 8$

65. $[2n^2 - (2n^2 - n + 5)] + [3n^2 + (n^2 - 2n - 7)]$

$[2n^2 - 2n^2 + n - 5] + [3n^2 + n^2 - 2n - 7]$

$(n - 5) + (4n^2 - 2n - 7)$

$n - 5 + 4n^2 - 2n - 7$

$4n^2+n-2n-5-7$

$4n^2-n-12$

67. $[7xy-(2x-3xy+y)]-[3x-(x-10xy-y)]$

$[7xy-2x+3xy-y]-[3x-x+10xy+y]$

$(10xy-2x-y)-(2x+10xy+y)$

$10xy-2x-y-2x-10xy-y$

$10xy-10xy-2x-2x-y-y$

$-4x-2y$

69. $[4x^3-(2x^2-x-1)]-[5x^3-(x^2+2x-1)]$

$[4x^3-2x^2+x+1]-[5x^3-x^2-2x+1]$

$4x^3-2x^2+x+1-5x^3+x^2+2x-1$

$4x^3-5x^3-2x^2+x^2+x+2x+1-1$

$-x^3-x^2+3x$

71a. Rectangle

$P=2l+2w$

$P=2(3x-2)+2(x+4)$

$P=6x-4+2x+8$

$P=8x+4$

b. $P=3x+x+3+x+x+1+2x+x+2+4$

$P=9x+10$

c. Triangle

$P=s+s+s$

$P=4x+2+4x+2+4x+2$

$P=12x+6$

73. $S=2\pi(4)^2+2\pi(4)h$

$S=32\pi+8\pi h$

a. $h=5$

$S=32(3.14)+8(3.14)(5)$

$S=100.48+125.6$

$S=226.08\approx 226.1$

b. $h=7$

$S=32(3.14)+8(3.14)(7)$

$S=100.48+175.84$

$S=276.32\approx 276.3$

c. $h=14$

$S=32(3.14)+8(3.14)(14)$

$S=100.48+351.68$

$S=452.16\approx 452.2$

d. $h=18$

$S=32(3.14)+8(3.14)(18)$

$S=100.48+452.16$

$S=552.64\approx 552.6$

Problem Set 3.2 Products and Quotients of Monomials

1. $(4x^3)(9x)$

$36x^{3+1}$

$36x^4$

3. $(-2x^2)(6x^3)$

$-12x^{2+3}$

$-12x^5$

5. $(-a^2b)(-4ab^3)$

$4a^{2+1}b^{1+3}$

$4a^3b^4$

7. $(x^2yz^2)(-3xyz^4)$

$-3x^{2+1}y^{1+1}z^{2+4}$

$-3x^3y^2z^6$

9. $(5xy)(-6y^3)$

$-30xy^{1+3}$

$-30xy^4$

11. $(3a^2b)(9a^2b^4)$

$27a^{2+2}b^{1+4}$

$27a^4b^5$

13. $(m^2n)(-mn^2)$

$-m^{2+1}n^{1+2}$

$-m^3n^3$

15. $\left(\frac{2}{5}xy^2\right)\left(\frac{3}{4}x^2y^4\right)$

$\frac{6}{20}x^{1+2}y^{2+4}$

$\frac{3}{10}x^3y^6$

17. $\left(-\frac{3}{4}ab\right)\left(\frac{1}{5}a^2b^3\right)$

$-\frac{3}{20}a^{1+2}b^{1+3}$

$-\frac{3}{20}a^3b^4$

19. $\left(-\frac{1}{2}xy\right)\left(\frac{1}{3}x^2y^3\right)$

$-\frac{1}{6}x^{1+2}y^{1+3}$

$-\frac{1}{6}x^3y^4$

21. $(3x)(-2x^2)(-5x^3)$

$30x^{1+2+3}$

$30x^6$

23. $(-6x^2)(3x^3)(x^4)$

$-18x^{2+3+4}$

$-18x^9$

25. $(x^2y)(-3xy^2)(x^3y^3)$

$-3x^{2+1+3}y^{1+2+3}$

$-3x^6y^6$

27. $(-3y^2)(-2y^2)(-4y^5)$

$-24y^{2+2+5}$

$-24y^9$

29. $(4ab)(-2a^2b)(7a)$

$-56a^{1+2+1}b^{1+1}$

$-56a^4b^2$

31. $(-ab)(-3ab)(-6ab)$

$-18a^{1+1+1}b^{1+1+1}$

$-18a^3b^3$

33. $\left(\frac{2}{3}xy\right)(-3x^2y)(5x^4y^5)$

$-10x^{1+2+4}y^{1+1+5}$

$-10x^7y^7$

35. $(12y)(-5x)\left(-\frac{5}{6}x^4y\right)$

$50x^{1+4}y^{1+1}$

$50x^5y^2$

37. $(3xy^2)^3$

$3^3x^{1(3)}y^{2(3)}$

$27x^3y^6$

39. $(-2x^2y)^5$

$(-2)^5x^{2(5)}y^{1(5)}$

$-32x^{10}y^5$

41. $(-x^4y^5)^4$

$(-1)^4x^{4(4)}y^{5(4)}$

$x^{16}y^{20}$

43. $(ab^2c^3)^6$

$a^{1(6)}b^{2(6)}c^{3(6)}$

$a^6b^{12}c^{18}$

45. $(2a^2b^3)^6$

$2^{1(6)}a^{2(6)}b^{3(6)}$

$2^6a^{12}b^{18}$

$64a^{12}b^{18}$

47. $(9xy^4)^2$

$9^2x^{1(2)}y^{4(2)}$

$81x^2y^8$

49. $(-3ab^3)^4$

$(-3)^4a^{1(4)}b^{3(4)}$

$81a^4b^{12}$

51. $-(2ab)^4$

$-[2^4a^{1(4)}b^{1(4)}]$

$-(16a^4b^4)$

$-16a^4b^4$

53. $-(xy^2z^3)^6$

$-[x^6y^{2(6)}z^{3(6)}]$

$-[x^6y^{12}z^{18}]$

$-x^6y^{12}z^{18}$

55. $(-5a^2b^2c)^3$
$(-5)^3a^{2(3)}b^{2(3)}c^{1(3)}$
$-125a^6b^6c^3$

57. $(-xy^4z^2)^7$
$(-1)^7x^{1(7)}y^{4(7)}z^{2(7)}$
$-x^7y^{28}z^{14}$

59. $\frac{9x^4y^5}{3xy^2}$
$3x^{4-1}y^{5-2}$
$3x^3y^3$

61. $\frac{25x^5y^6}{-5x^2y^4}$
$-5x^{5-2}y^{6-4}$
$-5x^3y^2$

63. $\frac{-54ab^2c^3}{-6abc}$
$9a^{1-1}b^{2-1}c^{3-1}$
$9a^0bc^2$
$9(1)bc^2$
$9bc^2$

65. $\frac{-18x^2y^2z^6}{xyz^2}$
$-18x^{2-1}y^{2-1}z^{6-2}$
$-18xyz^4$

67. $\frac{a^3b^4c^7}{-abc^5}$
$-a^{3-1}b^{4-1}c^{7-5}$
$-a^2b^3c^2$

69. $\frac{-72x^2y^4}{-8x^2y^4}$
$9x^{2-2}y^{4-4}$
$9x^0y^0$
$9(1)(1)$
9

71. $\frac{14ab^3}{-14ab}$
$-1a^{1-1}b^{3-1}$
$-a^0b^2$
$-(1)(b^2)$
$-b^2$

73. $\frac{-36x^3y^5}{2y^5}$
$-18x^3y^{5-5}$
$-18x^3y^0$
$-18x^3(1)$
$-18x^3$

75. $(2x^n)(3x^{2n})$
$6x^{n+2n}$
$6x^{3n}$

77. $(a^{2n-1})(a^{3n+4})$
$a^{2n-1+3n+4}$
a^{5n+3}

79. $(x^{3n-2})(x^{n+2})$
$x^{3n-2+n+2}$
x^{4n}

81. $(a^{5n-2})(a^3)$
a^{5n-2+3}
a^{5n+1}

83. $(2x^n)(-5x^n)$
$-10x^{n+n}$
$-10x^{2n}$

85. $(-3a^2)(-4a^{n+2})$
$12a^{2+n+2}$
$12a^{n+4}$

87. $(x^n)(2x^{2n})(3x^2)$
$6x^{n+2n+2}$
$6x^{3n+2}$

89. $(3x^{n-1})(x^{n+1})(4x^{2-n})$

$12x^{n-1+n+1+2-n}$

$12x^{n+2}$

91. length $= 2x$
width $= 3x$
height $= x$

$SA = 2lw + 2lh + 2wh$

$SA = 2(2x)(3x) + 2(2x)(x) + 2(3x)(x)$

$SA = 12x^2 + 4x^2 + 6x^2$

$SA = 22x^2$

$V = lwh$

$V = (2x)(3x)(x)$

$V = 6x^3$

93. Area of larger circle $= \pi r^2$

Area of smaller circle $= \pi(6)^2 = 36\pi$

Area of shaded region $= \pi r^2 - 36\pi$

Problem Set 3.3 Multiplying Polynomials

1. $2xy(5xy^2 + 3x^2y^3)$

$2xy(5xy^2) + 2xy(3x^2y^3)$

$10x^2y^3 + 6x^3y^4$

3. $-3a^2b(4ab^2 - 5a^3)$

$-3a^2b(4ab^2) - 3a^2b(-5a^3)$

$-12a^3b^3 + 15a^5b$

5. $8a^3b^4(3ab - 2ab^2 + 4a^2b^2)$

$8a^3b^4(3ab) + 8a^3b^4(-2ab^2) + 8a^3b^4(4a^2b^2)$

$24a^4b^5 - 16a^4b^6 + 32a^5b^6$

7. $-x^2y(6xy^2 + 3x^2y^3 - x^3y)$

$-x^2y(6xy^2) + (-x^2y)(3x^2y^3) + (-x^2y)(-x^3y)$

$-6x^3y^3 - 3x^4y^4 + x^5y^2$

9. $(a + 2b)(x + y)$

$a(x + y) + 2b(x + y)$

$ax + ay + 2bx + 2by$

11. $(a - 3b)(c + 4d)$

$a(c + 4d) - 3b(c + 4d)$

$ac + 4ad - 3bc - 12bd$

13. $(x + 6)(x + 10)$

$x(x + 10) + 6(x + 10)$

$x^2 + 10x + 6x + 60$

$x^2 + 16x + 60$

15. $(y - 5)(y + 11)$

$y(y + 11) - 5(y + 11)$

$y^2 + 11y - 5y - 55$

$y^2 + 6y - 55$

17. $(n + 2)(n - 7)$

$n(n - 7) + 2(n - 7)$

$n^2 - 7n + 2n - 14$

$n^2 - 5n - 14$

19. $(x + 6)(x - 6)$

$x(x - 6) + 6(x - 6)$

$x^2 - 6x + 6x - 36$

$x^2 - 36$

21. $(x - 6)^2$

$(x)^2 + 2(x)(-6) + (-6)^2$

$x^2 - 12x + 36$

23. $(x - 6)(x - 8)$

$x(x - 8) - 6(x - 8)$

$x^2 - 8x - 6x + 48$

$x^2 - 14x + 48$

25. $(x + 1)(x - 2)(x - 3)$

$(x + 1)[x(x - 3) - 2(x - 3)]$

$(x + 1)[x^2 - 3x - 2x + 6]$

$(x + 1)(x^2 - 5x + 6)$

$x(x^2 - 5x + 6) + 1(x^2 - 5x + 6)$

$x^3 - 5x^2 + 6x + x^2 - 5x + 6$

$x^3 - 4x^2 + x + 6$

27. $(x-3)(x+3)(x-1)$
$(x-3)[x(x-1)+3(x-1)]$
$(x-3)[x^2-x+3x-3]$
$(x-3)(x^2+2x-3)$
$x(x^2+2x-3)-3(x^2+2x-3)$
$x^3+2x^2-3x-3x^2-6x+9$
x^3-x^2-9x+9

29. $(t+9)^2$
$(t)^2+2(t)(9)+(9)^2$
$t^2+18t+81$

31. $(y-7)^2$
$(y)^2+2(y)(-7)+(-7)^2$
$y^2-14y+49$

33. $(4x+5)(x+7)$
$4x(x+7)+5(x+7)$
$4x^2+28x+5x+35$
$4x^2+33x+35$

35. $(3y-1)(3y+1)$
$3y(3y+1)-1(3y+1)$
$9y^2+3y-3y-1$
$9y^2-1$

37. $(7x-2)(2x+1)$
$7x(2x+1)-2(2x+1)$
$14x^2+7x-4x-2$
$14x^2+3x-2$

39. $(1+t)(5-2t)$
$1(5-2t)+t(5-2t)$
$5-2t+5t-2t^2$
$5+3t-2t^2$

41. $(3t+7)^2$
$(3t)^2+2(3t)(7)+(7)^2$
$9t^2+42t+49$

43. $(2-5x)(2+5x)$
$2(2+5x)-5x(2+5x)$
$4+10x-10x-25x^2$
$4-25x^2$

45. $(7x-4)^2$
$(7x)^2+2(7x)(-4)+(-4)^2$
$49x^2-56x+16$

47. $(6x+7)(3x-10)$
$6x(3x-10)+7(3x-10)$
$18x^2-60x+21x-70$
$18x^2-39x-70$

49. $(2x-5y)(x+3y)$
$2x(x+3y)-5y(x+3y)$
$2x^2+6xy-5xy-15y^2$
$2x^2+xy-15y^2$

51. $(5x-2a)(5x+2a)$
$5x(5x+2a)-2a(5x+2a)$
$25x^2+10ax-10ax-4a^2$
$25x^2-4a^2$

53. $(t+3)(t^2-3t-5)$
$t(t^2-3t-5)+3(t^2-3t-5)$
$t^3-3t^2-5t+3t^2-9t-15$
$t^3-14t-15$

55. $(x-4)(x^2+5x-4)$
$x(x^2+5x-4)-4(x^2+5x-4)$
$x^3+5x^2-4x-4x^2-20x+16$
$x^3+x^2-24x+16$

57. $(2x-3)(x^2+6x+10)$
$2x(x^2+6x+10)-3(x^2+6x+10)$
$2x^3+12x^2+20x-3x^2-18x-30$
$2x^3+9x^2+2x-30$

59. $(4x-1)(3x^2-x+6)$
$4x(3x^2-x+6)-1(3x^2-x+6)$
$12x^3-4x^2+24x-3x^2+x-6$
$12x^3-7x^2+25x-6$

61. $(x^2+2x+1)(x^2+3x+4)$

$x^2(x^2+3x+4)+2x(x^2+3x+4)+1(x^2+3x+4)$

$x^4+3x^3+4x^2+2x^3+6x^2+8x+x^2+3x+4$

$x^4+5x^3+11x^2+11x+4$

63. $(2x^2+3x-4)(x^2-2x-1)$

$2x^2(x^2-2x-1)+3x(x^2-2x-1)-4(x^2-2x-1)$

$2x^4-4x^3-2x^2+3x^3-6x^2-3x-4x^2+8x+4$

$2x^4-x^3-12x^2+5x+4$

65. $(x+2)^3$

$(x)^3+3(x)^2(2)+3(x)(2)^2+(2)^3$

$x^3+6x^2+12x+8$

67. $(x-4)^3$

$(x)^3+3(x)^2(-4)+3(x)(-4)^2+(-4)^3$

$x^3-12x^2+48x-64$

69. $(2x+3)^3$

$(2x)^3+3(2x)^2(3)+3(2x)(3)^2+(3)^3$

$8x^3+3(4x^2)(3)+3(2x)(9)+27$

$8x^3+36x^2+54x+27$

71. $(4x-1)^3$

$(4x)^3+3(4x)^2(-1)+3(4x)(-1)^2+(-1)^3$

$64x^3+3(16x^2)(-1)+3(4x)(1)-1$

$64x^3-48x^2+12x-1$

73. $(5x+2)^3$

$(5x)^3+3(5x)^2(2)+3(5x)(2)^2+(2)^3$

$125x^3+3(25x^2)(2)+3(5x)(4)+8$

$125x^3+150x^2+60x+8$

75. $(x^n-4)(x^n+4)$

$x^n(x^n+4)-4(x^n+4)$

$x^{2n}+4x^n-4x^n-16$

$x^{2n}-16$

77. $(x^a+6)(x^a-2)$

$x^a(x^a-2)+6(x^a-2)$

$x^{2a}-2x^a+6x^a-12$

$x^{2a}+4x^a-12$

79. $(2x^n+5)(3x^n-7)$

$2x^n(3x^n-7)+5(3x^n-7)$

$6x^{2n}-14x^n+15x^n-35$

$6x^{2n}+x^n-35$

81. $(x^{2a}-7)(x^{2a}-3)$

$x^{2a}(x^{2a}-3)-7(x^{2a}-3)$

$x^{4a}-3x^{2a}-7x^{2a}+21$

$x^{4a}-10x^{2a}+21$

83. $(2x^n+5)^2$

$(2x^n)^2+2(2x^n)(5)+(5)^2$

$4x^{2n}+20x^n+25$

85. length $= x+2$
width $= x+6$

$A = lw$

$A = (x+2)(x+6)$

There are four rectangles with the following dimensions and areas.

x by x $\quad A = x^2$

x by 6 $\quad A = 6x$

2 by x $\quad A = 2x$

2 by 6 $\quad A = 12$

The sum of the four rectangles equals the area of the figure.

$A = x^2+6x+2x+12$

From above $A = (x+2)(x+6)$, therefore

$(x+2)(x+6) = x^2+6x+2x+12$

$(x+2)(x+6) = x^2+8x+12$

87. Area of larger rectangle

$A = lw$

$A = (2x+3)(x)$

$A = 2x^2+3x$

Area of smaller rectangle

$A = lw$

$A = 3(x-2)$

$A = 3x-6$

Area of shaded portion

$A = (2x^2+3x)-(3x-6)$

$A = 2x^2+3x-3x+6$

$A = 2x^2+6$

89. Dimensions of the box

length $= 16-2x$
width $= 16-2x$
height $= x$

$V = lwh$

$V = (16-2x)(16-2x)(x)$

$V = (16-2x)(16x-2x^2)$

$V = 16(16x-2x^2)-2x(16x-2x^2)$

$V = 256x-32x^2-32x^2+4x^3$

$V = 4x^3-64x^2+256x$

Surface Area = area of the bottom plus the area of the four sides.

Dimensions of the bottom

length $= 16-2x$
width $= 16-2x$

Area $= (16-2x)(16-2x)$

Area $= 256-64x+4x^2$

Dimensions of the sides

length $= 16-2x$
width $= x$

Area $= (16-2x)(x)$

Area $= 16x-2x^2$

Surface Area

$SA = (256-64x+4x^2)+4(16x-2x^2)$

$SA = 256-64x+4x^2+64x-8x^2$

$SA = 256-4x^2$

Problem Set 3.4 Factoring: Use of the Distributive Property

1. $63 = 9 \bullet 7$ composite

3. 59 prime

5. $51 = 3 \bullet 17$ composite

7. $91 = 7 \bullet 13$ composite

9. 71 prime

11. 28

$4 \bullet 7$

$2 \bullet 2 \bullet 7$

$2^2 \bullet 7$

13. 44

$4 \bullet 11$

$2 \bullet 2 \bullet 11$

$2^2 \bullet 11$

15. 56

$8 \bullet 7$

$2 \bullet 4 \bullet 7$

$2 \bullet 2 \bullet 2 \bullet 7$

$2^3 \bullet 7$

17. 72

$8 \bullet 9$

$2 \bullet 4 \bullet 3 \bullet 3$

$2 \bullet 2 \bullet 2 \bullet 3 \bullet 3$

$2^3 \bullet 3^2$

19. 87

$3 \bullet 29$

21. $6x+3y$

$3(2x)+3(y)$

$3(2x+y)$

23. $6x^2+14x$

$2x(3x)+2x(7)$

$2x(3x+7)$

25. $28y^2 - 4y$
$4y(7y) + 4y(-1)$
$4y(7y - 1)$

27. $20xy - 15x$
$5x(4y) + 5x(-3)$
$5x(4y - 3)$

29. $7x^3 + 10x^2$
$x^2(7x) + x^2(10)$
$x^2(7x + 10)$

31. $18a^2b + 27ab^2$
$9ab(2a) + 9ab(3b)$
$9ab(2a + 3b)$

33. $12x^3y^4 - 39x^4y^3$
$3x^3y^3(4y) + 3x^3y^3(-3x)$
$3x^3y^3(4y - 3x)$

35. $8x^4 + 12x^3 - 24x^2$
$4x^2(2x^2) + 4x^2(3x) + 4x^2(-6)$
$4x^2(2x^2 + 3x - 6)$

37. $5x + 7x^2 + 9x^4$
$x(5) + x(7x) + x(9x^3)$
$x(5 + 7x + 9x^3)$

39. $15x^2y^3 + 20xy^2 + 35x^3y^4$
$5xy^2(3xy) + 5xy^2(4) + 5xy^2(7x^2y^2)$
$5xy^2(3xy + 4 + 7x^2y^2)$

41. $x(y + 2) + 3(y + 2)$
$(y + 2)(x + 3)$

43. $3x(2a + b) - 2y(2a + b)$
$(2a + b)(3x - 2y)$

45. $x(x + 2) + 5(x + 2)$
$(x + 2)(x + 5)$

47. $ax + 4x + ay + 4y$
$x(a + 4) + y(a + 4)$
$(a + 4)(x + y)$

49. $ax - 2bx + ay - 2by$
$x(a - 2b) + y(a - 2b)$
$(a - 2b)(x + y)$

51. $3ax - 3bx - ay + by$
$3x(a - b) - y(a - b)$
$(a - b)(3x - y)$

53. $2ax + 2x + ay + y$
$2x(a + 1) + y(a + 1)$
$(a + 1)(2x + y)$

55. $ax^2 - x^2 + 2a - 2$
$x^2(a - 1) + 2(a - 1)$
$(a - 1)(x^2 + 2)$

57. $2ac + 3bd + 2bc + 3ad$
$2ac + 2bc + 3bd + 3ad$
$2c(a + b) + 3d(b + a)$
$2c(a + b) + 3d(a + b)$
$(a + b)(2c + 3d)$

59. $ax - by + bx - ay$
$ax + bx - ay - by$
$x(a + b) - y(a + b)$
$(a + b)(x - y)$

61. $x^2 + 9x + 6x + 54$
$x(x + 9) + 6(x + 9)$
$(x + 9)(x + 6)$

63. $2x^2 + 8x + x + 4$
$2x(x + 4) + 1(x + 4)$
$(x + 4)(2x + 1)$

65. $x^2 + 7x = 0$
$x(x + 7) = 0$
$x = 0$ or $x + 7 = 0$
$x = 0$ or $x = -7$
The solution set is $\{0, -7\}$.

67. $x^2 - x = 0$

$x(x-1) = 0$

$x = 0$ or $x - 1 = 0$

$x = 0$ or $x = 1$

The solution set is $\{0, 1\}$.

69. $a^2 = 5a$

$a^2 - 5a = 0$

$a(a-5) = 0$

$a = 0$ or $a - 5 = 0$

$a = 0$ or $a = 5$

The solution set is $\{0, 5\}$.

71. $-2y = 4y^2$

$0 = 4y^2 + 2y$

$0 = 2y(2y+1)$

$2y = 0$ or $2y + 1 = 0$

$y = 0$ or $2y = -1$

$y = 0$ or $y = -\frac{1}{2}$

The solution set is $\left\{-\frac{1}{2}, 0\right\}$.

73. $3x^2 + 7x = 0$

$x(3x+7) = 0$

$x = 0$ or $3x + 7 = 0$

$x = 0$ or $3x = -7$

$x = 0$ $x = -\frac{7}{3}$

The solution set is $\left\{-\frac{7}{3}, 0\right\}$.

75. $4x^2 = 5x$

$4x^2 - 5x = 0$

$x(4x-5) = 0$

$x = 0$ or $4x - 5 = 0$

$x = 0$ or $4x = 5$

$x = 0$ or $x = \frac{5}{4}$

The solution set is $\left\{0, \frac{5}{4}\right\}$.

77. $x - 4x^2 = 0$

$x(1-4x) = 0$

$x = 0$ or $1 - 4x = 0$

$x = 0$ or $1 = 4x$

$x = 0$ or $\frac{1}{4} = x$

The solution set is $\left\{0, \frac{1}{4}\right\}$.

79. $12a = -a^2$

$12a + a^2 = 0$

$a(12+a) = 0$

$a = 0$ or $12 + a = 0$

$a = 0$ or $a = -12$

The solution set is $\{-12, 0\}$.

81. $5bx^2 - 3ax = 0$

$x(5bx - 3a) = 0$

$x = 0$ or $5bx - 3a = 0$

$x = 0$ or $5bx = 3a$

$x = 0$ or $x = \frac{3a}{5b}$

The solution set is $\left\{0, \frac{3a}{5b}\right\}$.

83. $2by^2 = -3ay$

$2by^2 + 3ay = 0$

$y(2by + 3a) = 0$

$y = 0$ or $2by + 3a = 0$

$y = 0$ or $2by = -3a$

$y = 0$ or $y = -\frac{3a}{2b}$

The solution set is $\left\{-\frac{3a}{2b}, 0\right\}$.

85. $y^2 - ay + 2by - 2ab = 0$

$y(y-a) + 2b(y-a) = 0$

$(y-a)(y+2b) = 0$

$y - a = 0$ or $y + 2b = 0$

$y = a$ or $y = -2b$

The solution set is $\{-2b, a\}$.

87. Let n = number

$n^2 = 7n$

$n^2 - 7n = 0$

$n(n-7)=0$

$n=0$ or $n-7=0$

$n=0$ or $n=7$

The numbers are 0 and 7.

89. $A=\pi r^2$; $C=2\pi r$

$\pi r^2=3(2\pi r)$

$\pi r^2=6\pi r$

$\pi r^2-6\pi r=0$

$\pi r(r-6)=0$

$\pi r=0$ or $r-6=0$

$r=0$ or $r=6$

Discard the root $r=0$, therefore the radius should be 6 units.

91. Area of circle $=\pi x^2$
Perimeter of square $=4x$

$\pi x^2=4x$

$\pi x^2-4x=0$

$x(\pi x-4)=0$

$x=0$ or $\pi x-4=0$

$x=0$ or $\pi x=4$

$x=0$ or $x=\frac{4}{\pi}$

Discard the root $x=0$, therefore the side of the square should be $\frac{4}{\pi}$ units.

93. Area of square $=x^2$
Area of rectangular lot $=x(50)$

$x^2=2[x(50)]$

$x^2=100x$

$x^2-100x=0$

$x(x-100)=0$

$x=0$ or $x-100=0$

$x=0$ or $x=100$

The square is 100 feet by 100 feet and the rectangular lot is 50 feet by 100 feet.

95. Volume of sphere $=\frac{4}{3}\pi r^3$
Surface area of sphere $=4\pi r^2$

$\frac{4}{3}\pi r^3=2(4\pi r^2)$

$\frac{4}{3}\pi r^3=8\pi r^2$

$\frac{4}{3}\pi r^3-8\pi r^2=0$

$4\pi r^2\left(\frac{1}{3}r-2\right)=0$

$4\pi r^2=0$ or $\frac{1}{3}r-2=0$

$r^2=0$ or $\frac{1}{3}r=2$

$r=0$ or $r=6$

Discard the root $r=0$, therefore the radius should be 6 units.

Problem Set 3.5 Factoring: Difference of Two Squares and Sum or Difference of Two Cubes

1. x^2-1

$(x)^2-(1)^2$

$(x+1)(x-1)$

3. $16x^2-25$

$(4x)^2-(5)^2$

$(4x+5)(4x-5)$

5. $9x^2-25y^2$

$(3x)^2-(5y)^2$

$(3x+5y)(3x-5y)$

7. $25x^2y^2-36$

$(5xy)^2-(6)^2$

$(5xy+6)(5xy-6)$

9. $4x^2-y^4$

$(2x)^2-(y^2)^2$

$(2x+y^2)(2x-y^2)$

11. $1-144n^2$

$(1)^2-(12n)^2$

$(1+12n)(1-12n)$

13. $(x+2)^2-y^2$
$(x+2)^2-(y)^2$
$(x+2+y)(x+2-y)$

15. $4x^2-(y+1)^2$
$(2x)^2-(y+1)^2$
$[2x+(y+1)][2x-(y+1)]$
$(2x+y+1)(2x-y-1)$

17. $9a^2-(2b+3)^2$
$(3a)^2-(2b+3)^2$
$[3a+(2b+3)][3a-(2b+3)]$
$(3a+2b+3)(3a-2b-3)$

19. $(x+2)^2-(x+7)^2$
$[(x+2)+(x+7)][(x+2)-(x+7)]$
$(x+2+x+7)(x+2-x-7)$
$(2x+9)(-5)$
$-5(2x+9)$

21. $9x^2-36$
$9(x^2-4)$
$9(x+2)(x-2)$

23. $5x^2+5$
$5(x^2+1)$

25. $8y^2-32$
$8(y^2-4)$
$8(y+2)(y-2)$

27. a^3b-9ab
$ab(a^2-9)$
$ab(a+3)(a-3)$

29. $16x^2+25$
Not factorable

31. n^4-81
$(n^2+9)(n^2-9)$
$(n^2+9)(n+3)(n-3)$

33. $3x^3+27x$
$3x(x^2+9)$

35. $4x^3y-64xy^3$
$4xy(x^2-16y^2)$
$4xy(x+4y)(x-4y)$

37. $6x-6x^3$
$6x(1-x^2)$
$6x(1+x)(1-x)$

39. $1-x^4y^4$
$(1+x^2y^2)(1-x^2y^2)$
$(1+x^2y^2)(1+xy)(1-xy)$

41. $4x^2-64y^2$
$4(x^2-16y^2)$
$4(x+4y)(x-4y)$

43. $3x^4-48$
$3(x^4-16)$
$3(x^2+4)(x^2-4)$
$3(x^2+4)(x+2)(x-2)$

45. a^3-64
$(a)^3-(4)^3$
$(a-4)(a^2+4a+16)$

47. x^3+1
$(x)^3+(1)^3$
$(x+1)(x^2-x+1)$

49. $27x^3+64y^3$
$(3x)^3+(4y)^3$
$(3x+4y)(9x^2-12xy+16y^2)$

51. $1-27a^3$
$(1)^3-(3a)^3$
$(1-3a)(1+3a+9a^2)$

53. x^3y^3-1
$(xy)^3-(1)^3$
$(xy-1)(x^2y^2+xy+1)$

55. $x^6 - y^6$

$(x^2)^3 - (y^2)^3$

$(x^2 - y^2)(x^4 + x^2y^2 + y^4)$

$(x + y)(x - y)(x^4 + x^2y^2 + y^4)$

57. $x^2 - 25 = 0$

$(x + 5)(x - 5) = 0$

$x + 5 = 0$ or $x - 5 = 0$

$x = -5$ or $x = 5$

The solution set is $\{-5, 5\}$.

59. $9x^2 - 49 = 0$

$(3x + 7)(3x - 7) = 0$

$3x + 7 = 0$ or $3x - 7 = 0$

$3x = -7$ or $3x = 7$

$x = -\frac{7}{3}$ or $x = \frac{7}{3}$

The solution set is $\left\{-\frac{7}{3}, \frac{7}{3}\right\}$.

61. $8x^2 - 32 = 0$

$8(x^2 - 4) = 0$

$8(x + 2)(x - 2) = 0$

$x + 2 = 0$ or $x - 2 = 0$

$x = -2$ or $x = 2$

The solution set is $\{-2, 2\}$.

63. $3x^3 = 3x$

$3x^3 - 3x = 0$

$3x(x^2 - 1) = 0$

$3x(x + 1)(x - 1) = 0$

$3x = 0$ or $x + 1 = 0$ or $x - 1 = 0$

$x = 0$ or $x = -1$ or $x = 1$

The solution set is $\{-1, 0, 1\}$.

65. $20 - 5x^2 = 0$

$5(4 - x^2) = 0$

$5(2 + x)(2 - x) = 0$

$2 + x = 0$ or $2 - x = 0$

$x = -2$ or $2 = x$

The solution set is $\{-2, 2\}$.

67. $x^4 - 81 = 0$

$(x^2 + 9)(x^2 - 9) = 0$

$(x^2 + 9)(x + 3)(x - 3) = 0$

$x^2 + 9 = 0$ or $x + 3 = 0$ or $x - 3 = 0$

$x^2 = -9$ or $x = -3$ or $x = 3$

not a real number $x = -3$ or $x = 3$

The solution set is $\{-3, 3\}$.

69. $6x^3 + 24x = 0$

$6x(x^2 + 4) = 0$

$6x = 0$ or $x^2 + 4 = 0$

$x = 0$ or $x^2 = -4$

$x = 0$ not a real number

The solution set is $\{0\}$.

71. Let $x =$ number

$x^3 = 9x$

$x^3 - 9x = 0$

$x(x^2 - 9) = 0$

$x(x + 3)(x - 3) = 0$

$x = 0$ or $x + 3 = 0$ or $x - 3 = 0$

$x = 0$ or $x = -3$ or $x = 3$

The solution set is $\{-3, 0, 3\}$.

73. Let $r =$ radius of first circle
$2r =$ radius of second circle
Area of first circle $= \pi r^2$
Area of second circle $= \pi(2r)^2 = 4\pi r^2$

$\pi r^2 + 4\pi r^2 = 80\pi$

$5\pi r^2 = 80\pi$

$5\pi r^2 - 80\pi = 0$

$5\pi(r^2 - 16) = 0$

$r^2 - 16 = 0$

$(r + 4)(r - 4) = 0$

$r + 4 = 0$ or $r - 4 = 0$

$r = -4$ or $r = 4$

Discard the root $r = -4$. The radius of the first circle is 4 centimeters and the radius of the second circle is 8 centimeters.

75. Let $x = \text{width}$
$2x = \text{length}$

$A = x(2x) = 2x^2$

$2x^2 = 50$

$2x^2 - 50 = 0$

$2(x^2 - 25) = 0$

$2(x+5)(x-5) = 0$

$x+5 = 0$ or $x-5 = 0$

$x = -5$ or $x = 5$

Discard the root $x = -5$. The width is 5 meters and the length is 10 meters.

77. Let $x = \text{radius}$
$2x = \text{altitude}$

$SA = 2\pi r^2 + 2\pi rh$

$54\pi = 2\pi(x)^2 + 2\pi(x)(2x)$

$54\pi = 2\pi x^2 + 4\pi x^2$

$54\pi = 6\pi x^2$

$9 = x^2$

$0 = x^2 - 9$

$0 = (x+3)(x-3)$

$x+3 = 0$ or $x-3 = 0$

$x = -3$ or $x = 3$

Discard the root $x = -3$. The radius should be 3 inches and the length should be 6 inches.

79. Let $x = \text{radius of circle}$
$2x = \text{side of square}$

Area of square $= (2x)^2 = 4x^2$

Area of circle $= \pi x^2$

$4x^2 + \pi x^2 = 16\pi + 64$

$4x^2 + \pi x^2 - 16\pi - 64 = 0$

$x^2(4+\pi) - 16(\pi+4) = 0$

$x^2(\pi+4) - 16(\pi+4) = 0$

$(\pi+4)(x^2 - 16) = 0$

$(\pi+4)(x+4)(x-4) = 0$

$x+4 = 0$ or $x-4 = 0$

$x = -4$ or $x = 4$

Discard the root $x = -4$. The length of the side of the square should be 8 yards.

Problem Set 3.6 Factoring Trinomials

1. $x^2 + 9x + 20$

We need two integers whose product is 20 and whose sum is 9. They are 4 and 5.

$(x+5)(x+4)$

3. $x^2 - 11x + 28$

We need two integers whose product is 28 and whose sum is -11. They are -4 and -7.

$(x-4)(x-7)$

5. $a^2 - 5a - 36$

We need two integers whose product is -36 and whose sum is 5. They are 9 and -4.

$(a+9)(a-4)$

7. $y^2 + 20y + 84$

We need two integers whose product is 84 and whose sum is 20. They are 6 and 14.

$(y+6)(y+14)$

9. $x^2 - 5x - 14$

We need two integers whose product is -14 and whose sum is -5. They are -7 and 2.

$(x-7)(x+2)$

11. $x^2 + 9x + 12$

We need two integers whose product is 12 and whose sum is 9. No such integers exist.

Not factorable

13. $6 + 5x - x^2$

$(6-x)(1+x)$

15. $x^2+15xy+36y^2$

We need two integers whose product is 36 and whose sum is 15. They are 12 and 3.

$(x+12y)(x+3y)$

17. $a^2-ab-56b^2$

We need two integers whose product is -56 and whose sum is -1. They are -8 and 7.

$(a-8b)(a+7b)$

19. $15x^2+23x+6$

$15(6)=90$

We need two integers whose product is 90 and whose sum is 23. They are 5 and 18.

$15x^2+5x+18x+6$

$5x(3x+1)+6(3x+1)$

$(3x+1)(5x+6)$

21. $12x^2-x-6$

$12(-6)=-72$

We need two integers whose product is -72 and whose sum is -1. They are -9 and 8.

$12x^2-9x+8x-6$

$3x(4x-3)+2(4x-3)$

$(4x-3)(3x+2)$

23. $8a^2-6a-27$

$8(-27)=-216$

We need two integers whose product is -216 and whose sum is -6. They are -18 and 12.

$8a^2-18a+12a-27$

$2a(4a-9)+3(4a-9)$

$(4a-9)(2a+3)$

25. $12n^2-43n-20$

$12(-20)=-240$

We need two integers whose product is -240 and whose sum is -43. They are -48 and 5.

$12n^2-48n+5n-20$

$12n(n-4)+5(n-4)$

$(n-4)(12n+5)$

27. $3x^2+10x+4$

$3(4)=12$

We need two integers whose product is 12 and whose sum is 10. No such integers exist.

Not factorable

29. $20n^2-64n-21$

$20(-21)=-420$

We need two integers whose product is -420 and whose sum is -64. They are -70 and 6.

$20n^2-70n+6n-21$

$10n(2n-7)+3(2n-7)$

$(2n-7)(10n+3)$

31. $16x^2+62x-45$

$16(-45)=-720$

We need two integers whose product is -720 and whose sum is 62. They are 72 and -10.

$16x^2+72x-10x-45$

$8x(2x+9)-5(2x+9)$

$(2x+9)(8x-5)$

33. $6-29x-42x^2$

$-1(-6+29x+42x^2)$

$-1(42x^2+29x-6)$

$42(-6)=-252$

We need two integers whose product is -252 and whose sum is 29. They are 36 and -7.

$-1(42x^2+36x-7x-6)$

$-1[6x(7x+6)-1(7x+6)]$

$-1[(7x+6)(6x-1)]$

$-1(7x+6)(6x-1)$

$-1(6x-1)(7x+6)$

$(-6x+1)(7x+6)$

$(1-6x)(6+7x)$

35. $20y^2+31y-9$

$20(-9)=-180$

We need two integers whose product is -180 and whose sum is 31. They are 36 and -5.

$20y^2+36y-5y-9$

$4y(5y+9)-1(5y+9)$

$(5y+9)(4y-1)$

37. $24n^2-2n-5$

$24(-5)=-120$

We need two integers whose product is -120 and whose sum is -2. They are -12 and 10.

$24n^2-12n+10n-5$

$12n(2n-1)+5(2n-1)$

$(2n-1)(12n+5)$

39. $5n^2+33n+18$

$5(18)=90$

We need two integers whose product is 90 and whose sum is 33. They are 30 and 3.

$5n^2+30n+3n+18$

$5n(n+6)+3(n+6)$

$(n+6)(5n+3)$

41. $x^2+25x+150$

We need two integers whose product is 150 and whose sum is 25. They are 10 and 15.

$(x+10)(x+15)$

43. $n^2-36n+320$

We need two integers whose product is 320 and whose sum is -36. They are -16 and -20.

$(n-16)(n-20)$

45. $t^2+3t-180$

We need two integers whose product is -180 and whose sum is 3. They are 15 and -12.

$(t+15)(t-12)$

47. t^4-5t^2+6

We need two integers whose product is 6 and whose sum is -5. They are -2 and -3.

$(t^2-2)(t^2-3)$

49. $10x^4+3x^2-4$

$10(-4)=-40$

We need two integers whose product is -40 and whose sum is 3. They are 8 and -5.

$10x^4+8x^2-5x^2-4$

$2x^2(5x^2+4)-1(5x^2+4)$

$(5x^2+4)(2x^2-1)$

51. x^4-9x^2+8

We need two integers whose product is 8 and whose sum is -9. They are -1 and -8.

$(x^2-1)(x^2-8)$

$(x+1)(x-1)(x^2-8)$

53. $18n^4+25n^2-3$

$18(-3)=-54$

We need two integers whose product is -54 and whose sum is 25. They are 27 and -2.

$18n^4+27n^2-2n^2-3$

$9n^2(2n^2+3)-1(2n^2+3)$

$(2n^2+3)(9n^2-1)$

$(2n^2+3)(3n+1)(3n-1)$

55. x^4-17x^2+16

We need two integers whose product is 16 and whose sum is -17. They are -16 and -1.

$(x^2-16)(x^2-1)$

$(x+4)(x-4)(x+1)(x-1)$

57. $2t^2-8$

$2(t^2-4)$

$2(t+2)(t-2)$

59. $12x^2+7xy-10y^2$

$12(-10)=-120$

We need two integers whose product is -120 and whose sum is 7. They are 15 and -8.

$12x^2+15xy-8xy-10y^2$

$3x(4x+5y)-2y(4x+5y)$

$(4x+5y)(3x-2y)$

61. $18n^3+39n^2-15n$

$3n(6n^2+13n-5)$

$6(-5)=-30$

We need two integers whose product is -30 and whose sum is 13. They are 15 and -2.

$3n(6n^2+15n-2n-5)$

$3n[3n(2n+5)-1(2n+5)]$

$3n[(2n+5)(3n-1)]$

$3n(2n+5)(3n-1)$

63. $n^2-17n+60$

We need two integers whose product is 60 and whose sum is -17. They are -5 and -12.

$n^2-5n-12n+60$

$n(n-5)-12(n-5)$

$(n-5)(n-12)$

65. $36a^2-12a+1$

$36(1)=36$

We need two integers whose product is 36 and whose sum is -12. They are -6 and -6.

$36a^2-6a-6a+1$

$6a(6a-1)-1(6a-1)$

$(6a-1)(6a-1)$

$(6a-1)^2$

67. $6x^2+54$

$6(x^2+9)$

69. $3x^2+x-5$

$3(-5)=-15$

We need two integers whose product is -15 and whose sum is 1. No such integers exist. Not factorable

71. $x^2-(y-7)^2$

$[x+(y-7)][x-(y-7)]$

$(x+y-7)(x-y+7)$

73. $1-16x^4$

$(1+4x^2)(1-4x^2)$

$(1+4x^2)(1-2x)(1+2x)$

75. $12n^2+59n+72$

$12(72)=864$

We need two integers whose product is 864 and whose sum is 59. They are 27 and 32.

$12n^2+27n+32n+72$

$3n(4n+9)+8(4n+9)$

$(4n+9)(3n+8)$

77. n^3-49n

$n(n^2-49)$

$n(n+7)(n-7)$

79. x^2-7x-8

We need two integers whose product is -8 and whose sum is -7. They are -8 and 1.

$(x-8)(x+1)$

81. $3x^4-81x$

$3x(x^3-27)$

$3x(x-3)(x^2+3x+9)$

83. x^4+6x^2+9

We need two integers whose product is 9 and whose sum is 6. They are 3 and 3.

$(x^2+3)(x^2+3)$

$(x^2+3)^2$

85. x^4-5x^2-36

We need two integers whose product is -36 and whose sum is -5. They are -9 and 4.

$(x^2-9)(x^2+4)$

$(x+3)(x-3)(x^2+4)$

87. $18w^2+9w-35$

$18(-35)=-630$

We need two integers whose product is -630 and whose sum is 9. They are -21 and 30.

$18w^2-21w+30w-35$

$3w(6w-7)+5(6w-7)$

$(6w-7)(3w+5)$

89. $25n^2+64$

Not factorable

91. $2n^3+14n^2-20n$

$2n(n^2+7n-10)$

We need two integers whose product is -10 and whose sum is 7. No such integers exist.

$2n(n^2+7n-10)$

93. $2xy+6x+y+3$

$2x(y+3)+1(y+3)$

$(y+3)(2x+1)$

Problem Set 3.7 Equations and Problem Solving

1. $x^2+4x+3=0$

$(x+3)(x+1)=0$

$x+3=0$ or $x+1=0$

$x=-3$ or $x=-1$

The solution set is $\{-3,-1\}$.

3. $x^2+18x+72=0$

$(x+12)(x+6)=0$

$x+12=0$ or $x+6=0$

$x=-12$ or $x=-6$

The solution set is $\{-12,-6\}$.

5. $n^2-13n+36=0$

$(n-9)(n-4)=0$

$n-9=0$ or $n-4=0$

$n=9$ or $n=4$

The solution set is $\{4, 9\}$.

7. $x^2+4x-12=0$

$(x+6)(x-2)=0$

$x+6=0$ or $x-2=0$

$x=-6$ or $x=2$

The solution set is $\{-6, 2\}$.

9. $w^2-4w=5$

$w^2-4w-5=0$

$(w-5)(w+1)=0$

$w-5=0$ or $w+1=0$

$w=5$ or $w=-1$

The solution set is $\{-1, 5\}$.

11. $n^2+25n+156=0$

$(n+12)(n+13)=0$

$n+12=0$ or $n+13=0$

$n=-12$ or $n=-13$

The solution set is $\{-13,-12\}$.

13. $3t^2+14t-5=0$

$(3t-1)(t+5)=0$

$3t-1=0$ or $t+5=0$

$3t=1$ or $t=-5$

$t=\frac{1}{3}$ or $t=-5$

The solution set is $\left\{-5, \frac{1}{3}\right\}$.

15. $6x^2+25x+14=0$

$(2x+7)(3x+2)=0$

$2x+7=0$ or $3x+2=0$

$2x=-7$ or $3x=-2$

$x=-\frac{7}{2}$ or $x=-\frac{2}{3}$

The solution set is $\left\{-\frac{7}{2}, -\frac{2}{3}\right\}$.

17. $3t(t-4)=0$

$3t=0$ or $t-4=0$

$t=0$ or $t=4$

The solution set is $\{0, 4\}$.

19. $-6n^2+13n-2=0$

$-1(-6n^2+13n-2)=-1(0)$

$6n^2-13n+2=0$

$(6n-1)(n-2)=0$

$6n-1=0$ or $n-2=0$

$6n=1$ or $n=2$

$n=\frac{1}{6}$ or $n=2$

The solution set is $\left\{\frac{1}{6}, 2\right\}$.

21. $2n^3=72n$

$2n^3-72n=0$

$2n(n^2-36)=0$

$2n(n+6)(n-6)=0$

$2n=0$ or $n+6=0$ or $n-6=0$

$n=0$ or $n=-6$ or $n=6$

The solution set is $\{-6, 0, 6\}$.

23. $(x-5)(x+3)=9$

$x^2-2x-15=9$

$x^2-2x-24=0$

$(x-6)(x+4)=0$

$x-6=0$ or $x+4=0$

$x=6$ or $x=-4$

The solution set is $\{-4, 6\}$.

25. $16-x^4=0$

$(4-x^2)(4+x^2)=0$

$(2-x)(2+x)(4+x^2)=0$

$2-x=0$ or $2+x=0$ or $4+x^2=0$

$2=x$ or $x=-2$ or $x^2=-4$

$x=2$ or $x=-2$ or not a real number

The solution set is $\{-2, 2\}$.

27. $n^2+7n-44=0$

$(n+11)(n-4)=0$

$n+11=0$ or $n-4=0$

$n=-11$ or $n=4$

The solution set is $\{-11, 4\}$.

29. $3x^2=75$

$3x^2-75=0$

$3(x^2-25)=0$

$3(x+5)(x-5)=0$

$x+5=0$ or $x-5=0$

$x=-5$ or $x=5$

The solution set is $\{-5, 5\}$.

31. $15x^2+34x+15=0$

$(5x+3)(3x+5)=0$

$5x+3=0$ or $3x+5=0$

$5x=-3$ or $3x=-5$

$x=-\frac{3}{5}$ or $x=-\frac{5}{3}$

The solution set is $\left\{-\frac{3}{5}, -\frac{5}{3}\right\}$.

33. $8n^2-47n-6=0$

$(8n+1)(n-6)=0$

$8n+1=0$ or $n-6=0$

$8n=-1$ or $n=6$

$n=-\frac{1}{8}$ or $n=6$

The solution set is $\left\{-\frac{1}{8}, 6\right\}$.

35. $28n^2 - 47n + 15 = 0$

$(7n - 3)(4n - 5) = 0$

$7n - 3 = 0$ or $4n - 5 = 0$

$7n = 3$ or $4n = 5$

$n = \frac{3}{7}$ or $n = \frac{5}{4}$

The solution set is $\left\{\frac{3}{7}, \frac{5}{4}\right\}$.

37. $35n^2 - 18n - 8 = 0$

$(7n + 2)(5n - 4) = 0$

$7n + 2 = 0$ or $5n - 4 = 0$

$7n = -2$ or $5n = 4$

$n = -\frac{2}{7}$ or $n = \frac{4}{5}$

The solution set is $\left\{-\frac{2}{7}, \frac{4}{5}\right\}$.

39. $-3x^2 - 19x + 14 = 0$

$-1(-3x^2 - 19x + 14) = -1(0)$

$3x^2 + 19x - 14 = 0$

$(3x - 2)(x + 7) = 0$

$3x - 2 = 0$ or $x + 7 = 0$

$3x = 2$ or $x = -7$

$x = \frac{2}{3}$ or $x = -7$

The solution set is $\left\{-7, \frac{2}{3}\right\}$.

41. $n(n + 2) = 360$

$n^2 + 2n = 360$

$n^2 + 2n - 360 = 0$

$(n + 20)(n - 18) = 0$

$n + 20 = 0$ or $n - 18 = 0$

$n = -20$ or $n = 18$

The solution set is $\{-20, 18\}$.

43. $9x^4 - 37x^2 + 4 = 0$

$(9x^2 - 1)(x^2 - 4) = 0$

$(3x + 1)(3x - 1)(x + 2)(x - 2) = 0$

$3x+1=0$ or $3x - 1=0$ or $x+2=0$ or $x - 2=0$

$3x = -1$ or $3x = 1$ or $x = -2$ or $x = 2$

$x = -\frac{1}{3}$ or $x = \frac{1}{3}$ or $x = -2$ or $x = 2$

The solution set is

$\left\{-2, -\frac{1}{3}, \frac{1}{3}, 2\right\}$.

45. $3x^4 - 46x^2 - 32 = 0$

$(3x^2 + 2)(x^2 - 16) = 0$

$(3x^2 + 2)(x + 4)(x - 4) = 0$

$3x^2 + 2 = 0$ or $x + 4 = 0$ or $x - 4 = 0$

$3x^2 = -2$ or $x = -4$ or $x = 4$

not a real number $x = -4$ or $x = 4$

The solution set is $\{-4, 4\}$.

47. $2x^4 + x^2 - 3 = 0$

$(2x^2 + 3)(x^2 - 1) = 0$

$(2x^2 + 3)(x + 1)(x - 1) = 0$

$2x^2 + 3 = 0$ or $x + 1 = 0$ or $x - 1 = 0$

$2x^2 = -3$ or $x = -1$ or $x = 1$

not a real number $x = -1$ or $x = 1$

The solution set is $\{-1, 1\}$.

49. $12x^3 + 46x^2 + 40x = 0$

$2x(6x^2 + 23x + 20) = 0$

$2x(2x + 5)(3x + 4) = 0$

$2x = 0$ or $2x + 5 = 0$ or $3x + 4 = 0$

$x = 0$ or $2x = -5$ or $3x = -4$

$x = -\frac{5}{2}$ or $x = -\frac{4}{3}$

The solution set is $\left\{-\frac{5}{2}, -\frac{4}{3}, 0\right\}$.

51. $(3x - 1)^2 - 16 = 0$

$[(3x - 1) - 4][(3x - 1) + 4] = 0$

$(3x - 5)(3x + 3) = 0$

$3x - 5 = 0$ or $3x + 3 = 0$

$3x = 5$ or $3x = -3$

$x = \frac{5}{3}$ or $x = -1$

The solution set is $\left\{-1, \frac{5}{3}\right\}$.

53. $4a(a+1)=3$

$4a^2+4a=3$

$4a^2+4a-3=0$

$(2a+3)(2a-1)=0$

$2a+3=0$ or $2a-1=0$

$2a=-3$ or $2a=1$

$a=-\frac{3}{2}$ or $a=\frac{1}{2}$

The solution set is $\left\{-\frac{3}{2}, \frac{1}{2}\right\}$.

55. Let $x = 1^{st}$ consecutive integer
$x+1=2^{nd}$ consecutive integer

$x(x+1)=72$

$x^2+x=72$

$x^2+x-72=0$

$(x+9)(x-8)=0$

$x+9=0$ or $x-8=0$

$x=-9$ or $x=8$

The integers are -9 and -8 or 8 and 9.

57. Let $x=$ first integer
$2x+1=$ second integer

$x(2x+1)=105$

$2x^2+x=105$

$2x^2+x-105=0$

$(2x+15)(x-7)=0$

$2x+15=0$ or $x-7=0$

$2x=-15$ or $x=7$

$x=-\frac{15}{2}$ or $x=7$

Not an integer or $x=7$

first integer $=7$
second integer $=2(7)+1=15$
The integers are 7 and 15.

59. Let $x=$ width

$P=2l+2w$

$32=2l+2x$

$32-2x=2l$

$\frac{1}{2}(32-2x)=\frac{1}{2}(2l)$

$16-x=l$

So $x=$ width and $16-x=$ length.

$A=lw$

$60=(16-x)(x)$

$60=16x-x^2$

$x^2-16x+60=0$

$(x-10)(x-6)=0$

$x-10=0$ or $x-6=0$

$x=10$ or $x=6$

width $=10$ or width $=6$

length $=16-10$ or length $=16-6$

length $=6$ or length $=10$

So the rectangle is 6 inches by 10 inches.

61. Let $x=1^{st}$ integer
$x+1=2^{nd}$ integer

$x^2+(x+1)^2=85$

$x^2+x^2+2x+1=85$

$2x^2+2x-84=0$

$2(x^2+x-42)=0$

$2(x+7)(x-6)=0$

$x+7=0$ or $x-6=0$

$x=-7$ or $x=6$

The integers are -7 and -6 or 6 and 7.

63. Let $x=$ side of square
$x+2=$ width of rectangle
$x+4=$ length of rectangle
Area of square $=x^2$
Area of rectangle $=(x+2)(x+4)$
$=x^2+6x+8$

$x^2+x^2+6x+8=64$

$2x^2+6x+8=64$

$2x^2+6x-56=0$

$2(x^2+3x-28)=0$

$2(x+7)(x-4)=0$

$x+7=0$ or $x-4=0$

$x = -7 \quad \text{or} \quad x = 4$

Discard the root $x = -7$.
The length of the side of the square is 4 centimeters. The dimensions of the rectangle are 6 centimeters by 8 centimeters.

65. Let $x = 1^{st}$ leg
$x + 1 = 2^{nd}$ leg
$x + 2 =$ hypotenuse

$x^2 + (x+1)^2 = (x+2)^2$

$x^2 + x^2 + 2x + 1 = x^2 + 4x + 4$

$x^2 - 2x - 3 = 0$

$(x-3)(x+1) = 0$

$x - 3 = 0 \quad \text{or} \quad x + 1 = 0$

$x = 3 \quad \text{or} \quad x = -1$

Discard the root $x = -1$.
The sides of the triangle are 3 units, 4 units, and 5 units.

67. Let $x = 1^{st}$ leg
$x + 3 = 2^{nd}$ leg
$15 =$ hypotenuse

$x^2 + (x+3)^2 = 15^2$

$x^2 + x^2 + 6x + 9 = 225$

$2x^2 + 6x - 216 = 0$

$2(x^2 + 3x - 108) = 0$

$2(x+12)(x-9) = 0$

$x + 12 = 0 \quad \text{or} \quad x - 9 = 0$

$x = -12 \quad \text{or} \quad x = 9$

Discard the root $x = -12$.
The sides of the triangle are 9 inches, 12 inches, and 15 inches.

69. Let $x =$ altitude
$3x + 2 =$ side (base)

$\frac{1}{2}x(3x+2) = 28$

$2\left[\frac{1}{2}x(3x+2)\right] = 2(28)$

$x(3x+2) = 56$

$3x^2 + 2x = 56$

$3x^2 + 2x - 56 = 0$

$(3x+14)(x-4) = 0$

$3x + 14 = 0 \quad \text{or} \quad x - 4 = 0$

$3x = -14 \quad \text{or} \quad x = 4$

$x = -\frac{14}{3} \quad \text{or} \quad x = 4$

Discard the root $x = -\frac{14}{3}$.
The altitude is 4 inches and the side is 14 inches.

Chapter 3 Review

1. $(3x-2) + (4x-6) + (-2x+5)$

$3x - 2 + 4x - 6 - 2x + 5$

$3x + 4x - 2x - 2 - 6 + 5$

$5x - 3$

3. $(6x^2-2x-1) + (4x^2+2x+5) - (-2x^2+x-1)$

$6x^2 - 2x - 1 + 4x^2 + 2x + 5 + 2x^2 - x + 1$

$6x^2 + 4x^2 + 2x^2 - 2x + 2x - x - 1 + 5 + 1$

$12x^2 - x + 5$

5. $(-2a^2)(3ab^2)(a^2b^3)$

$-6a^{2+1+2}b^{2+3}$

$-6a^5b^5$

7. $(4x-3y)(6x+5y)$

$24x^2 + 20xy - 18xy - 15y^2$

$24x^2 + 2xy - 15y^2$

9. $(4x^2y^3)^4$

$4^4x^{2(4)}y^{3(4)}$

$256x^8y^{12}$

11. $(-2x^2y^3z)^3$

$(-2)^3x^{2(3)}y^{3(3)}z^3$

$-8x^6y^9z^3$

13. $[3x - (2x - 3y + 1)] - [2y - (x - 1)]$

$[3x - 2x + 3y - 1] - [2y - x + 1]$

$[x + 3y - 1] - 2y + x - 1$

$x + 3y - 1 - 2y + x - 1$

$2x + y - 2$

15. $(7-3x)(3+5x)$

$21+35x-9x-15x^2$

$21+26x-15x^2$

17. $\left(\frac{1}{2}ab\right)(8a^3b^2)(-2a^3)$

$-8a^{1+3+3}b^{1+2}$

$-8a^7b^3$

19. $(3x+2)(2x^2-5x+1)$

$3x(2x^2-5x+1)+2(2x^2-5x+1)$

$6x^3-15x^2+3x+4x^2-10x+2$

$6x^3-11x^2-7x+2$

21. $(2x+5y)^2$

$(2x)^2+2(2x)(5y)+(5y)^2$

$4x^2+20xy+25y^2$

23. $(2x+5)^3$

$(2x+5)^2(2x+5)$

$[(2x)^2+2(2x)(5)+(5)^2](2x+5)$

$(4x^2+20x+25)(2x+5)$

$(2x+5)(4x^2+20x+25)$

$2x(4x^2+20x+25)+5(4x^2+20x+25)$

$8x^3+40x^2+50x+20x^2+100x+125$

$8x^3+60x^2+150x+125$

25. $2t^2-18$

$2(t^2-9)$

$2(t+3)(t-3)$

27. $12n^2-7n+1$

$12(1)=12$

We need two integers whose product is 12 and whose sum is -7. They are -3 and -4.

$12n^2-3n-4n+1$

$3n(4n-1)-1(4n-1)$

$(4n-1)(3n-1)$

29. x^3-6x^2-72x

$x(x^2-6x-72)$

$x(x-12)(x+6)$

31. $x-(y-1)^2$

$[x+(y-1)][x-(y-1)]$

$(x+y-1)(x-y+1)$

33. $12x^2+x-35$

$12(-35) = -420$

We need two integers whose product is -420 and whose sum is 1. They are 21 and -20.

$12x^2+21x-20x-35$

$3x(4x+7)-5(4x+7)$

$(4x+7)(3x-5)$

35. $4n^2-8n$

$4n(n-2)$

37. $20x^2+3xy-2y^2$

$20(-2) = -40$

We need two integers whose product is -40 and whose sum is 3. They are 8 and -5.

$20x^2+8xy-5xy-2y^2$

$4x(5x+2y)-y(5x+2y)$

$(5x+2y)(4x-y)$

39. $3x^3-15x^2-18x$

$3x(x^2-5x-6)$

$3x(x-6)(x+1)$

41. t^4-22t^2-75

$(t^2-25)(t^2+3)$

$(t+5)(t-5)(t^2+3)$

43. $15-14x+3x^2$

$15(3)=45$

We need two integers whose product is 45 and whose sum is -14. They are -9 and -5.

$15-9x-5x+3x^2$

$3(5-3x)-x(5-3x)$

$(5-3x)(3-x)$

45. $16x^3+250$

$2(8x^3+125)$

$2(2x+5)(4x^2-10x+25)$

47. $x^2+5x-6=0$

$(x+6)(x-1)=0$

$x+6=0$ or $x-1=0$

$x=-6$ or $x=1$

The solution set is $\{-6, 1\}$.

49. $(3x-1)(5x+2)=0$

$3x-1=0$ or $5x+2=0$

$3x=1$ or $5x=-2$

$x=\frac{1}{3}$ or $x=-\frac{2}{5}$

The solution set is $\left\{-\frac{2}{5}, \frac{1}{3}\right\}$

51. $6a^3=54a$

$6a^3-54a=0$

$6a(a^2-9)=0$

$6a(a+3)(a-3)=0$

$6a=0$ or $a+3=0$ or $a-3=0$

$a=0$ or $a=-3$ or $a=3$

The solution set is $\{-3, 0, 3\}$.

53. $-n^2+2n+63=0$

$-1(-n^2+2n+63)=-1(0)$

$n^2-2n-63=0$

$(n-9)(n+7)=0$

$n-9=0$ or $n+7=0$

$n=9$ or $n=-7$

The solution set is $\{-7, 9\}$.

55. $30w^2-w-20=0$

$30(-20)=-600$

We need two integers whose product is -600 and whose sum is -1. They are -25 and 24.

$30w^2-25w+24w-20=0$

$5w(6w-5)+4(6w-5)=0$

$(6w-5)(5w+4)=0$

$6w-5=0$ or $5w+4=0$

$6w=5$ or $5w=-4$

$w=\frac{5}{6}$ or $w=-\frac{4}{5}$

The solution set is $\left\{-\frac{4}{5}, \frac{5}{6}\right\}$.

57. $9n^2-30n+25=0$

$9(25)=225$

We need two integers whose product is 225 and whose sum is -30. They are -15 and -15.

$9n^2-15n-15n+25=0$

$3n(3n-5)-5(3n-5)=0$

$(3n-5)(3n-5)=0$

$(3n-5)^2=0$

$3n-5=0$

$3n=5$

$n=\frac{5}{3}$

The solution set is $\left\{\frac{5}{3}\right\}$.

59. $7x^2+33x-10=0$

$7(-10)=-70$

We need two integers whose product is -70 and whose sum is 33. They are 35 and -2.

$7x^2+35x-2x-10=0$

$7x(x+5)-2(x+5)=0$

$(x+5)(7x-2)=0$

$x+5=0$ or $7x-2=0$

$x=-5$ or $7x=2$

$x=-5$ or $x=\frac{2}{7}$

The solution set is $\left\{-5, \frac{2}{7}\right\}$.

61. $x^2+12x-x-12=0$

$x(x+12)-1(x+12)=0$

$(x+12)(x-1)=0$

$x+12=0$ or $x-1=0$

$x=-12$ or $x=1$

The solution is $\{-12, 1\}$.

63. $30-19x-5x^2=0$

$30(-5)=-150$

We need two integers whose product is -150 and whose sum is -19. They are -25 and 6.

$30-25x+6x-5x^2=0$

$5(6-5x)+x(6-5x)=0$

$(6-5x)(5+x)=0$

$6-5x=0$ or $5+x=0$

$6=5x$ or $x=-5$

$\frac{6}{5}=x$ or $x=-5$

The solution set is $\left\{-5, \frac{6}{5}\right\}$.

65. $-4n^2-39n+10=0$

$-1(-4n^2-39n+10)=-1(0)$

$4n^2+39n-10=0$

$4(-10)=-40$

We need two integers whose product is -40 and whose sum is 39. They are 40 and -1.

$4n^2+40n-n-10=0$

$4n(n+10)-1(n+10)=0$

$(n+10)(4n-1)=0$

$n+10=0$ or $4n-1=0$

$n=-10$ or $4n=1$

$n=-10$ or $n=\frac{1}{4}$

The solution set is $\left\{-10, \frac{1}{4}\right\}$.

67. Let $x=1^{st}$ integer
$2-x=2^{nd}$ integer

$x(2-x)=-48$

$2x-x^2=-48$

$0=x^2-2x-48$

$0=(x-8)(x+6)$

$x-8=0$ or $x+6=0$

$x=8$ or $x=-6$

The integers are 8 and -6.

69. Let $x=$ distance of the northbound car
$x+4=$ distance of the eastbound car

$x^2+(x+4)^2=20^2$

$x^2+x^2+8x+16=400$

$2x^2+8x-384=0$

$2(x^2+4x-192)=0$

$2(x+16)(x-12)=0$

$x+16=0$ or $x-12=0$

$x=-16$ or $x=12$

The northbound car traveled 12 miles and the eastbound car traveled 16 miles.

71. Let $x=$ number of rows
$2x-2=$ number of chairs per row

$x(2x-2)=144$

$2x^2-2x=144$

$2x^2-2x-144=0$

$2(x^2-x-72)=0$

$2(x-9)(x+8)=0$

$x-9=0$ or $x+8=0$

$x=9$ or $x=-8$

Discard the root $x=-8$.
Number of rows $=9$
Number of chairs per row $=$
$2(9)-2=16$
There are 9 rows of 16 chairs per row.

73. Let $x=$ width of sidewalk
Area of pool $=20(30)=600$
Area of sidewalk $=336$

For the area of the sidewalk and pool
width $=20+2x$
length $=30+2x$

$(20+2x)(30+2x)-600=336$

$600+100x+4x^2-600=336$

$4x^2+100x-336=0$

$4(x^2+25x-84)=0$

$4(x-3)(x+28)=0$

$x-3=0$ and $x+28=0$

$x=3$ and $x=-28$

Discard the root $x=-28$. The width of the walk is 3 feet.

75. Let $x=$ radius
$3x=$ height

$S.A. = 2\pi r^2+2\pi rh$

$32\pi = 2\pi x^2+2\pi x(3x)$

$32\pi = 2\pi x^2+6\pi x^2$

$\frac{1}{\pi}(32\pi)=\frac{1}{\pi}(2\pi x^2)+\frac{1}{\pi}(6\pi x^2)$

$32=2x^2+6x^2$

$32=8x^2$

$0=8x^2-32$

$0=8(x^2-4)$

$0=8(x+2)(x-2)$

$x+2=0$ or $x-2=0$

$x=-2$ or $x=2$

Discard the root $x=-2$. The length of the radius is 2 inches and the altitude is 6 inches.

Chapter 3 Test

1. $(-3x-1)+(9x-2)-(4x+8)$

$-3x-1+9x-2-4x-8$

$2x-11$

3. $(-3x^2y^4)^3$

$(-3)^3x^{2(3)}y^{4(3)}$

$-27x^6y^{12}$

5. $(3n-2)(2n-3)$

$6n^2-9n-4n+6$

$6n^2-13n+6$

7. $(x+6)(2x^2-x-5)$

$x(2x^2-x-5)+6(2x^2-x-5)$

$2x^3-x^2-5x+12x^2-6x-30$

$2x^3+11x^2-11x-30$

9. $6x^2+19x-20$

$6x^2+24x-5x-20$

$6x(x+4)-5(x+4)$

$(x+4)(6x-5)$

11. $64+t^3$

$(4+t)(16-4t+t^2)$

13. $x^2-xy+4x-4y$

$x(x-y)+4(x-y)$

$(x-y)(x+4)$

15. $x^2+8x-48=0$

$(x+12)(x-4)=0$

$x+12=0$ or $x-4=0$

$x=-12$ or $x=4$

The solution set is $\{-12, 4\}$.

17. $4x^2-12x+9=0$

$(2x-3)^2=0$

$2x-3=0$

$x=\frac{3}{2}$

The solution set is $\left\{\frac{3}{2}\right\}$.

19. $3x^3+21x^2-54x=0$

$3x(x^2+7x-18)=0$

$3x(x+9)(x-2)=0$

$3x=0$ or $x+9=0$ or $x-2=0$

$x=0$ or $x=-9$ or $x=2$

The solution set is $\{-9, 0, 2\}$.

21. $n(3n-5)=2$

$3n^2-5n=2$

$3n^2-5n-2=0$

$(3n+1)(n-2)=0$

$3n+1=0$ or $n-2=0$

$3n=-1$ or $n=2$

$n=-\frac{1}{3}$ or $n=2$

The solution set is $\left\{-\frac{1}{3}, 2\right\}$.

23. Let $x = \text{width}$
$15 - x = \text{length}$

$x(15 - x) = 54$

$15x - x^2 = 54$

$0 = x^2 - 15x + 54$

$0 = (x - 9)(x - 6)$

$x - 9 = 0$ or $x - 6 = 0$

$x = 9$ or $x = 6$

The width is 6 inches and the length is 9 inches.

25. Let $x = \text{side of square}$
$x + 3 = \text{width of rectangle}$
$x + 5 = \text{length of rectangle}$

$x^2 + (x+3)(x+5) = 57$

$x^2 + x^2 + 8x + 15 = 57$

$2x^2 + 8x - 42 = 0$

$2(x^2 + 4x - 21) = 0$

$2(x+7)(x-3) = 0$

$x + 7 = 0$ or $x - 3 = 0$

$x = -7$ or $x = 3$

Discard the root $x = -7$.
The length of the rectangle is $3 + 5 = 8$ feet.

Chapter 3 Cumulative Review

1. $x^2 - 2xy + y^2$

$(-2)^2 - 2(-2)(-4) + (-4)^2$

$4 - 16 + 16$

4

3. $2x^2 - 5x + 6$

$2(3)^2 - 5(3) + 6$

$2(9) - 15 + 6$

$18 - 15 + 6$

9

5. $-(2n - 1) + 5(2n - 3) - 6(3n + 4)$

$-2n + 1 + 10n - 15 - 18n - 24$

$-10n - 38$

$-10(4) - 38$

$-40 - 38$

-78

7. $(3x^2 - 4x - 7) - (4x^2 - 7x + 8)$

$3x^2 - 4x - 7 - 4x^2 + 7x - 8$

$-x^2 + 3x - 15$

$-(-4)^2 + 3(-4) - 15$

$-(16) - 12 - 15$

$-16 - 12 - 15$

-43

9. $5(-x^2-x+3)-(2x^2-x+6)-2(x^2+4x-6)$

$-5x^2 - 5x + 15 - 2x^2 + x - 6 - 2x^2 - 8x + 12$

$-9x^2 - 12x + 21$

$-9(2)^2 - 12(2) + 21$

$-9(4) - 24 + 21$

$-36 - 24 + 21$

$-60 + 21$

-39

11. $4(3x - 2) - 2(4x - 1) - (2x + 5)$

$12x - 8 - 8x + 2 - 2x - 5$

$2x - 11$

13. $(5x - 7)(6x + 1)$

$30x^2 + 5x - 42x - 7$

$30x^2 - 37x - 7$

15. $(-4a^2b^3)^3$

$(-4)^3 a^{2(3)} b^{3(3)}$

$-64a^6b^9$

17. $(x - 3)(x^2 - x - 4)$

$x(x^2 - x - 4) - 3(x^2 - x - 4)$

$x^3 - x^2 - 4x - 3x^2 + 3x + 12$

$x^3 - 4x^2 - x + 12$

19. $7x^2-7$

$7(x^2-1)$

$7(x+1)(x-1)$

21. $3x^2-17x-56$

$3x^2-24x+7x-56$

$3x(x-8)+7(x-8)$

$(x-8)(3x+7)$

23. $xy-5x+2y-10$

$x(y-5)+2(y-5)$

$(y-5)(x+2)$

25. $4n^4-n^2-3$

$(4n^2+3)(n^2-1)$

$(4n^2+3)(n+1)(n-1)$

27. $4x^2+36$

$4(x^2+9)$

29. $5x-2y=6$

$5x=2y+6$

$\frac{1}{5}(5x)=\frac{1}{5}(2y+6)$

$x=\frac{2y+6}{5}$

31. $V=2\pi rh+2\pi r^2$

$V-2\pi r^2=2\pi rh$

$\frac{1}{2\pi r}(V-2\pi r^2)=\frac{1}{2\pi r}(2\pi rh)$

$\frac{V-2\pi r^2}{2\pi r}=h$

33. $(x-2)(x+5)=8$

$x^2+5x-2x-10=8$

$x^2+3x-10=8$

$x^2+3x-18=0$

$(x+6)(x-3)=0$

$x+6=0$ or $x-3=0$

$x=-6$ or $x=3$

The solution set is $\{-6, 3\}$.

35. $-2(n-1)+3(2n+1)=-11$

$-2n+2+6n+3=-11$

$4n+5=-11$

$4n=-16$

$n=-4$

The solution set is $\{-4\}$.

37. $8x^2-8=0$

$8(x^2-1)=0$

$8(x+1)(x-1)=0$

$x+1=0$ or $x-1=0$

$x=-1$ or $x=1$

The solution set is $\{-1, 1\}$.

39. $.1(x-.1)-.4(x+2)=-5.31$

$100[.1(x-.1)-.4(x+2)]=100(-5.31)$

$100[.1(x-.1)]-100[.4(x+2)]=-531$

$10(x-.1)-40(x+2)=-531$

$10x-1-40x-80=-531$

$-30x-81=-531$

$-30x=-450$

$x=15$

The solution set is $\{15\}$.

41. $|3n-2|=7$

$3n-2=7$ or $3n-2=-7$

$3n=9$ or $3n=-5$

$n=3$ or $n=-\frac{5}{3}$

The solution set is $\left\{-\frac{5}{3}, 3\right\}$.

43. $.08(x+200)=.07x+20$

$100[.08(x+200)]=100[.07x+20]$

$8(x+200)=100(.07x)+100(20)$

$8x+1600=7x+2000$

$x+1600=2000$

$x=400$

The solution set is $\{400\}$.

45. $x^3 = 16x$

$x^3 - 16x = 0$

$x(x^2 - 16) = 0$

$x(x+4)(x-4) = 0$

$x = 0$ or $x + 4 = 0$ or $x - 4 = 0$

$x = 0$ or $x = -4$ or $x = 4$

The solution set is $\{-4, 0, 4\}$.

47. $-12n^2 - 29n + 8 = 0$

$-1(-12n^2 - 29n + 8) = -1(0)$

$12n^2 + 29n - 8 = 0$

$(3n+8)(4n-1) = 0$

$3n + 8 = 0$ or $4n - 1 = 0$

$3n = -8$ or $4n = 1$

$n = -\frac{8}{3}$ or $n = \frac{1}{4}$

The solution set is $\left\{-\frac{8}{3}, \frac{1}{4}\right\}$.

49. $2x^3 + 6x^2 - 20x = 0$

$2x(x^2 + 3x - 10) = 0$

$2x(x+5)(x-2) = 0$

$2x = 0$ or $x + 5 = 0$ or $x - 2 = 0$

$x = 0$ or $x = -5$ or $x = 2$

The solution set is $\{-5, 0, 2\}$.

51. $-5(3n+4) < -2(7n-1)$

$-15n - 20 < -14n + 2$

$-15n < -14n + 22$

$-n < 22$

$-1(-n) > -1(22)$

$n > -22$

The solution set is $(-22, \infty)$.

53. $|2x - 1| > 7$

$2x - 1 < -7$ or $2x - 1 > 7$

$2x < -6$ or $2x > 8$

$x < -3$ or $x > 4$

The solution set is $(-\infty, -3) \cup (4, \infty)$.

55. $.09x + .1(x+200) > 77$

$100[.09x + .1(x+200)] > 100(77)$

$100[.09x] + 100[.1(x+200)] > 7700$

$9x + 10(x+200) > 7700$

$9x + 10x + 2000 > 7700$

$19x + 2000 > 7700$

$19x > 5700$

$x > 300$

The solution set is $(300, \infty)$.

57. $-(x-1) + 2(3x-1) \geq 2(x+4) - (x-1)$

$-x + 1 + 6x - 2 \geq 2x + 8 - x + 1$

$5x - 1 \geq x + 9$

$5x \geq x + 10$

$4x \geq 10$

$x \geq \frac{10}{4}$

$x \geq \frac{5}{2}$

The solution set is $\left[\frac{5}{2}, \infty\right)$.

59. Let $x = 1^{st}$ odd integer
$x + 2 = 2^{nd}$ odd integer
$x + 4 = 3^{rd}$ odd integer

$3x - (x+2) = x + 4 + 1$

$3x - x - 2 = x + 5$

$2x - 2 = x + 5$

$2x = x + 7$

$x = 7$

The integers are 7, 9, and 11.

61. Present Ages
Joey $= x$
Mother $= 46 - x$

In four years
Joey $= x + 4$
Mother $= 46 - x + 4 = 50 - x$

$x + 4 = \frac{1}{2}(50 - x) - 3$

$2(x+4) = 2\left[\frac{1}{2}(50 - x) - 3\right]$

$2x + 8 = 2\left[\frac{1}{2}(50 - x)\right] - 2(3)$

$2x+8=50-x-6$

$2x+8=44-x$

$2x=36-x$

$3x=36$

$x=12$

Joey's present age is 12 and his mother's present age is 34.

63. Let $x=$ money invested at 8%
$x+200=$ money invested at 9%

$(8\%)(x)+(9\%)(x+200)=86$

$.08x+.09(x+200)=86$

$100[.08x+.09(x+200)]=100(86)$

$100(.08x)+100[.09(x+200)]=8600$

$8x+9(x+200)=8600$

$8x+9x+1800=8600$

$17x+1800=8600$

$17x=6800$

$x=400$

There should be \$400 invested at 8% and \$600 invested at 9%.

65. Let $x=$ time in hours Billie travels
$x+\frac{5}{6}=$ time in hours Sandy travels

$12x=$ Billie's distance

$8\left(x+\frac{5}{6}\right)=$ Sandy's distance

$12x=8\left(x+\frac{5}{6}\right)$

$12x=8x+8\left(\frac{5}{6}\right)$

$12x=8x+\frac{20}{3}$

$4x=\frac{20}{3}$

$\frac{1}{4}(4x)=\frac{1}{4}\left(\frac{20}{3}\right)$

$x=\frac{5}{3}=1\frac{2}{3}$ hours

It takes Billie 1 hour and 40 minutes to overtake Sandy.

67. Let $x=$ score of 3rd game

$\frac{152+174+x}{3}\geq 160$

$\frac{326+x}{3}\geq 160$

$3\left(\frac{326+x}{3}\right)\geq 3(160)$

$326+x\geq 480$

$x\geq 154$

The score should be 154 or better.

69. Let $x=$ side of square
$x=$ altitude
$16=$ side (base)

Area of square $=x^2$

Area of triangle $=\frac{1}{2}x(16)=8x$

$x^2=\frac{1}{2}(8x)$

$x^2=4x$

$x^2-4x=0$

$x(x-4)=0$

$x=0$ or $x-4=0$

$x=0$ or $x=4$

Discard the root $x=0$. The length of the side of the square is 4 inches.

71. Let $x=$ number of rows
$x+4=$ number of chairs per row

$x(x+4)=96$

$x^2+4x=96$

$x^2+4x-96=0$

$(x+12)(x-8)=0$

$x+12=0$ or $x-8=0$

$x=-12$ or $x=8$

Discard the root $x=-12$. The number of rows is 8 and the number of chairs per row is 12.

Chapter 4 Rational Expressions

Problem Set 4.1 Simplifying Rational Expressions

1. $\frac{27}{36} = \frac{9 \bullet 3}{9 \bullet 4} = \frac{3}{4}$

3. $\frac{45}{54} = \frac{9 \bullet 5}{9 \bullet 6} = \frac{5}{6}$

5. $\frac{24}{-60} = \frac{12 \bullet 2}{12 \bullet -5} = -\frac{2}{5}$

7. $\frac{-16}{-56} = \frac{-8 \bullet 2}{-8 \bullet 7} = \frac{2}{7}$

9. $\frac{12xy}{42y} = \frac{6y}{6y} \bullet \frac{2x}{7} = \frac{2x}{7}$

11. $\frac{18a^2}{45ab} = \frac{9a}{9a} \bullet \frac{2a}{5b} = \frac{2a}{5b}$

13. $\frac{-14y^3}{56xy^2} = \frac{14y^2}{14y^2} \bullet \frac{-y}{4x} = -\frac{y}{4x}$

15. $\frac{54c^2d}{-78cd^2} = \frac{6cd}{6cd} \bullet \frac{9c}{-13d} = -\frac{9c}{13d}$

17. $\frac{-40x^3y}{-24xy^4} = \frac{-8xy}{-8xy} \bullet \frac{5x^2}{3y^3} = \frac{5x^2}{3y^3}$

19. $\frac{x^2-4}{x^2+2x} = \frac{(x+2)(x-2)}{x(x+2)} = \frac{x-2}{x}$

21. $\frac{18x+12}{12x-6} = \frac{6(3x+2)}{6(2x-1)} = \frac{3x+2}{2x-1}$

23. $\frac{a^2+7a+10}{a^2-7a-18}$

$\frac{(a+5)(a+2)}{(a-9)(a+2)}$

$\frac{a+5}{a-9}$

25. $\frac{2n^2+n-21}{10n^2+33n-7}$

$\frac{(2n+7)(n-3)}{(2n+7)(5n-1)}$

$\frac{n-3}{5n-1}$

27. $\frac{5x^2+7}{10x}$

29. $\frac{6x^2+x-15}{8x^2-10x-3}$

$\frac{(3x+5)(2x-3)}{(4x+1)(2x-3)}$

$\frac{3x+5}{4x+1}$

31. $\frac{3x^2-12x}{x^3-64}$

$\frac{3x(x-4)}{(x-4)(x^2+4x+16)}$

$\frac{3x}{x^2+4x+16}$

33. $\frac{3x^2+17x-6}{9x^2-6x+1}$

$\frac{(3x-1)(x+6)}{(3x-1)(3x-1)}$

$\frac{x+6}{3x-1}$

35. $\frac{2x^3+3x^2-14x}{x^2y+7xy-18y}$

$\frac{x(2x^2+3x-14)}{y(x^2+7x-18)}$

$\frac{x(2x+7)(x-2)}{y(x+9)(x-2)}$

$\frac{x(2x+7)}{y(x+9)}$

37. $\frac{5y^2+22y+8}{25y^2-4}$

$\frac{(5y+2)(y+4)}{(5y+2)(5y-2)}$

$\frac{y+4}{5y-2}$

39. $\frac{15x^3-15x^2}{5x^3+5x}$

$\frac{15x^2(x-1)}{5x(x^2+1)}$

$\frac{3x(x-1)}{x^2+1}$

41. $\frac{4x^2y + 8xy^2 - 12y^3}{18x^3y - 12x^2y^2 - 6xy^3}$

$\frac{4y(x^2 + 2xy - 3y^2)}{6xy(3x^2 - 2xy - y^2)}$

$\frac{4y(x + 3y)(x - y)}{6xy(3x + y)(x - y)}$

$\frac{2(x + 3y)}{3x(3x + y)}$

43. $\frac{3n^2 + 14n - 24}{7n^2 + 44n + 12}$

$\frac{(3n - 4)(n + 6)}{(7n + 2)(n + 6)}$

$\frac{3n - 4}{7n + 2}$

45. $\frac{8 + 18x - 5x^2}{10 + 31x + 15x^2}$

$\frac{(4 - x)(2 + 5x)}{(5 + 3x)(2 + 5x)}$

$\frac{4 - x}{5 + 3x}$

47. $\frac{27x^4 - x}{6x^3 + 10x^2 - 4x}$

$\frac{x(27x^3 - 1)}{2x(3x^2 + 5x - 2)}$

$\frac{x(3x - 1)(9x^2 + 3x + 1)}{2x(3x - 1)(x + 2)}$

$\frac{9x^2 + 3x + 1}{2(x + 2)}$

49. $\frac{-40x^3 + 24x^2 + 16x}{20x^3 + 28x^2 + 8x}$

$\frac{-8x(5x^2 - 3x - 2)}{4x(5x^2 + 7x + 2)}$

$\frac{-8x(5x + 2)(x - 1)}{4x(5x + 2)(x + 1)}$

$\frac{-2(x - 1)}{x + 1}$

51. $\frac{xy + ay + bx + ab}{xy + ay + cx + ac}$

$\frac{y(x + a) + b(x + a)}{y(x + a) + c(x + a)}$

$\frac{(x + a)(y + b)}{(x + a)(y + c)}$

$\frac{y + b}{y + c}$

53. $\frac{ax - 3x + 2ay - 6y}{2ax - 6x + ay - 3y}$

$\frac{x(a - 3) + 2y(a - 3)}{2x(a - 3) + y(a - 3)}$

$\frac{(a - 3)(x + 2y)}{(a - 3)(2x + y)}$

$\frac{x + 2y}{2x + y}$

55. $\frac{5x^2 + 5x + 3x + 3}{5x^2 + 3x - 30x - 18}$

$\frac{5x(x + 1) + 3(x + 1)}{x(5x + 3) - 6(5x + 3)}$

$\frac{(x + 1)(5x + 3)}{(5x + 3)(x - 6)}$

$\frac{x + 1}{x - 6}$

57. $\frac{2st - 30 - 12s + 5t}{3st - 6 - 18s + t}$

$\frac{2st - 12s + 5t - 30}{3st - 18s + t - 6}$

$\frac{2s(t - 6) + 5(t - 6)}{3s(t - 6) + 1(t - 6)}$

$\frac{(t - 6)(2s + 5)}{(t - 6)(3s + 1)}$

$\frac{2s + 5}{3s + 1}$

59. $\frac{5x - 7}{7 - 5x}$

$\frac{-1(-5x + 7)}{(7 - 5x)}$

$\frac{-1(7 - 5x)}{(7 - 5x)}$

-1

61. $\frac{n^2 - 49}{7 - n}$

$\frac{(n + 7)(n - 7)}{-1(-7 + n)}$

$$\frac{(n+7)(n-7)}{-1(n-7)}$$

$$\frac{n+7}{-1}$$

$$-n-7$$

63. $$\frac{2y-2xy}{x^2y-y}$$

$$\frac{2y(1-x)}{y(x^2-1)}$$

$$\frac{2y(1-x)}{y(x+1)(x-1)}$$

$$\frac{2y(-1)(-1+x)}{y(x+1)(x-1)}$$

$$\frac{-2y(x-1)}{y(x+1)(x-1)}$$

$$\frac{-2y}{y(x+1)}$$

$$\frac{-2}{x+1}$$

65. $$\frac{2x^3-8x}{4x-x^3}$$

$$\frac{2x(x^2-4)}{-x(-4+x^2)}$$

$$\frac{2x(x^2-4)}{-x(x^2-4)}$$

$$-2$$

67. $$\frac{n^2-5n-24}{40+3n-n^2}$$

$$\frac{(n-8)(n+3)}{(8-n)(5+n)}$$

$$\frac{(n-8)(n+3)}{-1(-8+n)(5+n)}$$

$$\frac{(n-8)(n+3)}{-1(n-8)(n+5)}$$

$$\frac{-(n+3)}{n+5}$$

Problem Set 4.2 Multiplying and Dividing Rational Expressions

1. $$\frac{7}{12}\bullet\frac{6}{35}$$

$$\frac{7}{2\bullet 6}\bullet\frac{6}{5\bullet 7}$$

$$\frac{1}{10}$$

3. $$\frac{-4}{9}\bullet\frac{18}{30}$$

$$\frac{-4}{9}\bullet\frac{9\bullet 2}{2\bullet 15}$$

$$-\frac{4}{15}$$

5. $$\frac{3}{-8}\bullet\frac{-6}{12}$$

$$\frac{3}{-8}\bullet\frac{-6}{2\bullet 6}$$

$$\frac{3}{16}$$

7. $$\left(-\frac{5}{7}\right)\div\frac{6}{7}$$

$$-\frac{5}{7}\bullet\frac{7}{6}$$

$$-\frac{5}{6}$$

9. $$-\frac{9}{5}\div\frac{27}{10}-\frac{2}{3}$$

$$-\frac{9}{5}\bullet\frac{10}{27}-\frac{2}{3}$$

$$-\frac{9}{5}\bullet\frac{5\bullet 2}{3\bullet 9}-\frac{2}{3}$$

$$-\frac{2}{3}-\frac{2}{3}$$

$$-\frac{4}{3}$$

11. $$\frac{4}{9}\bullet\frac{6}{11}\div\frac{4}{15}$$

$$\frac{4}{9}\bullet\frac{6}{11}\bullet\frac{15}{4}$$

$$\frac{4}{3\bullet 3}\bullet\frac{2\bullet 3}{11}\bullet\frac{3\bullet 5}{4}$$

$$\frac{10}{11}$$

13. $$\frac{6xy}{9y^4}\bullet\frac{30x^3y}{-48x}$$

$$\frac{6(30)x^4y^2}{9(-48)xy^4}$$

$$\frac{-5x^3}{12y^2}$$

15. $\frac{5a^2b^2}{11ab} \bullet \frac{22a^3}{15ab^2}$

$\frac{5(22)a^5b^2}{11(15)a^2b^3}$

$\frac{2a^3}{3b}$

17. $\frac{5xy}{8y^2} \bullet \frac{18x^2y}{15}$

$\frac{5(18)x^3y^2}{8(15)y^2}$

$\frac{3x^3}{4}$

19. $\frac{5x^4}{12x^2y^3} \div \frac{9}{5xy}$

$\frac{5x^4}{12x^2y^3} \bullet \frac{5xy}{9}$

$\frac{5(5)x^5y}{12(9)x^2y^3}$

$\frac{25x^3}{108y^2}$

21. $\frac{9a^2c}{12bc^2} \div \frac{21ab}{14c^3}$

$\frac{9a^2c}{12bc^2} \bullet \frac{14c^3}{21ab}$

$\frac{9(14)a^2c^4}{12(21)ab^2c^2}$

$\frac{3(2)ac^2}{4(3)b^2}$

$\frac{ac^2}{2b^2}$

23. $\frac{9x^2y^3}{14x} \bullet \frac{21y}{15xy^2} \bullet \frac{10x}{12y^3}$

$\frac{9(21)(10)x^3y^4}{14(15)(12)x^2y^5}$

$\frac{21}{14} \bullet \frac{9}{12} \bullet \frac{10}{15} \bullet \frac{x}{y}$

$\frac{3}{2} \bullet \frac{3}{4} \bullet \frac{2}{3} \bullet \frac{x}{y}$

$\frac{3x}{4y}$

25. $\frac{3x+6}{5y} \bullet \frac{x^2+4}{x^2+10x+16}$

$\frac{3(x+2)}{5y} \bullet \frac{x^2+4}{(x+8)(x+2)}$

$\frac{3(x^2+4)}{5y(x+8)}$

27. $\frac{5a^2+20a}{a^3-2a^2} \bullet \frac{a^2-a-12}{a^2-16}$

$\frac{5a(a+4)}{a^2(a-2)} \bullet \frac{(a-4)(a+3)}{(a-4)(a+4)}$

$\frac{5(a+3)}{a(a-2)}$

29. $\frac{3n^2+15n-18}{3n^2+10n-48} \bullet \frac{12n^2-17n-40}{8n^2+2n-10}$

$\frac{3(n^2+5n-6)}{3n^2+10n-48} \bullet \frac{12n^2-17n-40}{2(4n^2+n-5)}$

$\frac{3(n+6)(n-1)}{(3n-8)(n+6)} \bullet \frac{(4n+5)(3n-8)}{2(4n+5)(n-1)}$

$\frac{3}{2}$

31. $\frac{9y^2}{x^2+12x+36} \div \frac{12y}{x^2+6x}$

$\frac{9y^2}{x^2+12x+36} \bullet \frac{x^2+6x}{12y}$

$\frac{9y^2}{(x+6)(x+6)} \bullet \frac{x(x+6)}{12y}$

$\frac{3xy}{4(x+6)}$

33. $\frac{x^2-4xy+4y^2}{7xy^2} \div \frac{4x^2-3xy-10y^2}{20x^2y+25xy^2}$

$\frac{x^2-4xy+4y^2}{7xy^2} \bullet \frac{20x^2y+25xy^2}{4x^2-3xy-10y^2}$

$\frac{(x-2y)(x-2y)}{7xy^2} \bullet \frac{5xy(4x+5y)}{(4x+5y)(x-2y)}$

$\frac{5(x-2y)}{7y}$

35. $\frac{5-14n-3n^2}{1-2n-3n^2} \bullet \frac{9+7n-2n^2}{27-15n+2n^2}$

$\frac{(5+n)(1-3n)}{(1+n)(1-3n)} \bullet \frac{(9-2n)(1+n)}{(9-2n)(3-n)}$

$\frac{5+n}{3-n}$

37. $\frac{3x^4+2x^2-1}{3x^4+14x^2-5} \bullet \frac{x^4-2x^2-35}{x^4-17x^2+70}$

$\frac{(3x^2-1)(x^2+1)}{(3x^2-1)(x^2+5)} \bullet \frac{(x^2-7)(x^2+5)}{(x^2-7)(x^2-10)}$

$\frac{x^2+1}{x^2-10}$

39. $\frac{6x^2-35x+25}{4x^2-11x-45} \div \frac{18x^2+9x-20}{24x^2+74x+45}$

$\frac{6x^2-35x+25}{4x^2-11x-45} \bullet \frac{24x^2+74x+45}{18x^2+9x-20}$

$\frac{(6x-5)(x-5)}{(4x+9)(x-5)} \bullet \frac{(4x+9)(6x+5)}{(3x+4)(6x-5)}$

$\frac{6x+5}{3x+4}$

41. $\frac{10t^3+25t}{20t+10} \bullet \frac{2t^2-t-1}{t^5-t}$

$\frac{5t(2t^2+5)}{10(2t+1)} \bullet \frac{(2t+1)(t-1)}{t(t^4-1)}$

$\frac{5t(2t^2+5)}{10(2t+1)} \bullet \frac{(2t+1)(t-1)}{t(t^2+1)(t^2-1)}$

$\frac{5t(2t^2+5)}{10(2t+1)} \bullet \frac{(2t+1)(t-1)}{t(t^2+1)(t+1)(t-1)}$

$\frac{2t^2+5}{2(t^2+1)(t+1)}$

43. $\frac{4t^2+t-5}{t^3-t^2} \bullet \frac{t^4+6t^3}{16t^2+40t+25}$

$\frac{(4t+5)(t-1)}{t^2(t-1)} \bullet \frac{t^3(t+6)}{(4t+5)(4t+5)}$

$\frac{t(t+6)}{4t+5}$

45. $\frac{nr+3n+2r+6}{nr+3n-3r-9} \bullet \frac{n^2-9}{n^3-4n}$

$\frac{n(r+3)+2(r+3)}{n(r+3)-3(r+3)} \bullet \frac{(n+3)(n-3)}{n(n^2-4)}$

$\frac{(r+3)(n+2)}{(r+3)(n-3)} \bullet \frac{(n+3)(n-3)}{n(n+2)(n-2)}$

$\frac{n+3}{n(n-2)}$

47. $\frac{x^2-x}{4y} \bullet \frac{10xy^2}{2x-2} \div \frac{3x^2+3x}{15x^2y^2}$

$\frac{x^2-x}{4y} \bullet \frac{10xy^2}{2x-2} \bullet \frac{15x^2y^2}{3x^2+3x}$

$\frac{x(x-1)}{4y} \bullet \frac{10xy^2}{2(x-1)} \bullet \frac{15x^2y^2}{3x(x+1)}$

$\frac{25x^3y^3}{4(x+1)}$

49. $\frac{a^2-4ab+4b^2}{6a^2-4ab} \bullet \frac{3a^2+5ab-2b^2}{6a^2+ab-b^2} \div \frac{a^2-4b^2}{8a+4b}$

$\frac{a^2-4ab+4b^2}{6a^2-4ab} \bullet \frac{3a^2+5ab-2b^2}{6a^2+ab-b^2} \bullet \frac{8a+4b}{a^2-4b^2}$

$\frac{(a-2b)(a-2b)}{2a(3a-2b)} \bullet \frac{(3a-b)(a+2b)}{(3a-b)(2a+b)} \bullet \frac{4(2a+b)}{(a+2b)(a-2b)}$

$\frac{2(a-2b)}{a(3a-2b)}$

Problem Set 4.3 Adding and Subtracting Rational Expressions

1. $\frac{1}{4}+\frac{5}{6}$; LCD = 12

$\frac{1}{4}\left(\frac{3}{3}\right)+\frac{5}{6}\left(\frac{2}{2}\right)$

$\frac{3}{12}+\frac{10}{12}$

$\frac{13}{12}$

3. $\frac{7}{8}-\frac{3}{5}$; LCD = 40

$\frac{7}{8}\left(\frac{5}{5}\right)-\frac{3}{5}\left(\frac{8}{8}\right)$

$\frac{35}{40}-\frac{24}{40}$

$\frac{11}{40}$

5. $\frac{6}{5}+\frac{1}{-4}$; LCD = 20

$\frac{6}{5}\left(\frac{4}{4}\right)+\left(-\frac{1}{4}\right)\left(\frac{5}{5}\right)$

$\frac{24}{20}+\left(-\frac{5}{20}\right)$

$\frac{19}{20}$

7. $\frac{8}{15}+\frac{3}{25}$; LCD = 75

$\frac{8}{15}\left(\frac{5}{5}\right)+\frac{3}{25}\left(\frac{3}{3}\right)$

$\frac{40}{75}+\frac{9}{75}$

$\frac{49}{75}$

9. $\frac{1}{5}+\frac{5}{6}-\frac{7}{15}$; LCD = 30

$\frac{1}{5}\left(\frac{6}{6}\right)+\frac{5}{6}\left(\frac{5}{5}\right)-\frac{7}{15}\left(\frac{2}{2}\right)$

$\frac{6}{30}+\frac{25}{30}-\frac{14}{30}$

$\frac{31}{30}-\frac{14}{30}$

$\frac{17}{30}$

11. $\frac{1}{3}-\frac{1}{4}-\frac{3}{14}$; LCD = 84

$\frac{1}{3}\left(\frac{28}{28}\right)-\frac{1}{4}\left(\frac{21}{21}\right)-\frac{3}{14}\left(\frac{6}{6}\right)$

$\frac{28}{84}-\frac{21}{84}-\frac{18}{84}$

$\frac{7}{84}-\frac{18}{84}$

$-\frac{11}{84}$

13. $\frac{2x}{x-1}+\frac{4}{x-1}$

$\frac{2x+4}{x-1}$

15. $\frac{4a}{a+2}+\frac{8}{a+2}$

$\frac{4a+8}{a+2}$

$\frac{4(a+2)}{a+2}$

4

17. $\frac{3(y-2)}{7y}+\frac{4(y-1)}{7y}$

$\frac{3(y-2)+4(y-1)}{7y}$

$\frac{3y-6+4y-4}{7y}$

$\frac{7y-10}{7y}$

19. $\frac{x-1}{2}+\frac{x+3}{3}$; LCD = 6

$\frac{(x-1)}{2}\left(\frac{3}{3}\right)+\frac{(x+3)}{3}\left(\frac{2}{2}\right)$

$\frac{3(x-1)}{6}+\frac{2(x+3)}{6}$

$\frac{3(x-1)+2(x+3)}{6}$

$\frac{3x-3+2x+6}{6}$

$\frac{5x+3}{6}$

21. $\frac{2a-1}{4}+\frac{3a+2}{6}$; LCD = 12

$\frac{(2a-1)}{4}\left(\frac{3}{3}\right)+\frac{(3a+2)}{6}\left(\frac{2}{2}\right)$

$\frac{3(2a-1)}{12}+\frac{2(3a+2)}{12}$

$\frac{3(2a-1)+2(3a+2)}{12}$

$\frac{6a-3+6a+4}{12}$

$\frac{12a+1}{12}$

23. $\frac{n+2}{6}-\frac{n-4}{9}$; LCD = 18

$\frac{(n+2)}{6}\left(\frac{3}{3}\right)-\frac{(n-4)}{9}\left(\frac{2}{2}\right)$

$\frac{3(n+2)}{18}-\frac{2(n-4)}{18}$

$\frac{3(n+2)-2(n-4)}{18}$

$\frac{3n+6-2n+8}{18}$

$\frac{n+14}{18}$

25. $\frac{3x-1}{3}-\frac{5x+2}{5}$; LCD = 15

$\frac{(3x-1)}{3}\left(\frac{5}{5}\right)-\frac{(5x+2)}{5}\left(\frac{3}{3}\right)$

$\frac{5(3x-1)}{15}-\frac{3(5x+2)}{15}$

$\frac{5(3x-1)-3(5x+2)}{15}$

$\frac{15x-5-15x-6}{15}$

$-\frac{11}{15}$

27. $\frac{x-2}{5}-\frac{x+3}{6}+\frac{x+1}{15}$; LCD = 30

$\frac{(x-2)}{5}\left(\frac{6}{6}\right)-\frac{(x+3)}{6}\left(\frac{5}{5}\right)+\frac{(x+1)}{15}\left(\frac{2}{2}\right)$

$\frac{6(x-2)}{30}-\frac{5(x+3)}{30}+\frac{2(x+1)}{30}$

$\frac{6(x-2)-5(x+3)+2(x+1)}{30}$

$\frac{6x-12-5x-15+2x+2}{30}$

$\frac{3x-25}{30}$

29. $\frac{3}{8x}+\frac{7}{10x}$; LCD = 40x

$\frac{3}{8x}\left(\frac{5}{5}\right)+\frac{7}{10x}\left(\frac{4}{4}\right)$

$\frac{15}{40x}+\frac{28}{40x}$

$\frac{43}{40x}$

31. $\frac{5}{7x}-\frac{11}{4y}$; LCD = 28xy

$\frac{5}{7x}\left(\frac{4y}{4y}\right)-\frac{11}{4y}\left(\frac{7x}{7x}\right)$

$\frac{20y}{28xy}-\frac{77x}{28xy}$

$\frac{20y-77x}{28xy}$

33. $\frac{4}{3x}+\frac{5}{4y}-1$; LCD = 12xy

$\frac{4}{3x}\left(\frac{4y}{4y}\right)+\frac{5}{4y}\left(\frac{3x}{3x}\right)-\left(\frac{12xy}{12xy}\right)$

$\frac{16y}{12xy}+\frac{15x}{12xy}-\frac{12xy}{12xy}$

$\frac{16y+15x-12xy}{12xy}$

35. $\frac{7}{10x^2}+\frac{11}{15x}$; LCD = $30x^2$

$\frac{7}{10x^2}\left(\frac{3}{3}\right)+\frac{11}{15x}\left(\frac{2x}{2x}\right)$

$\frac{21}{30x^2}+\frac{22x}{30x^2}$

$\frac{21+22x}{30x^2}$

37. $\frac{10}{7n}-\frac{12}{4n^2}$; LCD = $28n^2$

$\frac{10}{7n}\left(\frac{4n}{4n}\right)-\frac{12}{4n^2}\left(\frac{7}{7}\right)$

$\frac{40n}{28n^2}-\frac{84}{28n^2}$

$\frac{40n-84}{28n^2}$

$\frac{4(10n-21)}{28n^2}$

$\frac{10n-21}{7n^2}$

39. $\frac{3}{n^2}-\frac{2}{5n}+\frac{4}{3}$; LCD = $15n^2$

$\frac{3}{n^2}\left(\frac{15}{15}\right)-\frac{2}{5n}\left(\frac{3n}{3n}\right)+\frac{4}{3}\left(\frac{5n^2}{5n^2}\right)$

$\frac{45}{15n^2}-\frac{6n}{15n^2}+\frac{20n^2}{15n^2}$

$\frac{45-6n+20n^2}{15n^2}$

41. $\frac{3}{x}-\frac{5}{3x^2}-\frac{7}{6x}$; LCD = $6x^2$

$\frac{3}{x}\left(\frac{6x}{6x}\right)-\frac{5}{3x^2}\left(\frac{2}{2}\right)-\frac{7}{6x}\left(\frac{x}{x}\right)$

$\frac{18x}{6x^2}-\frac{10}{6x^2}-\frac{7x}{6x^2}$

$\frac{18x-10-7x}{6x^2}$

$\frac{11x-10}{6x^2}$

43. $\frac{6}{5t^2}-\frac{4}{7t^3}+\frac{9}{5t^3}$; LCD = $35t^3$

$\frac{6}{5t^2}\left(\frac{7t}{7t}\right)-\frac{4}{7t^3}\left(\frac{5}{5}\right)+\frac{9}{5t^3}\left(\frac{7}{7}\right)$

$\frac{42t}{35t^3}-\frac{20}{35t^3}+\frac{63}{35t^3}$

$\frac{42t-20+63}{35t^3}$

$\frac{42t+43}{35t^3}$

45. $\frac{5b}{24a^2}-\frac{11a}{32b}$; LCD $=96a^2b$

$\frac{5b}{24a^2}\left(\frac{4b}{4b}\right)-\frac{11a}{32b}\left(\frac{3a^2}{3a^2}\right)$

$\frac{20b^2}{96a^2b}-\frac{33a^3}{96a^2b}$

$\frac{20b^2-33a^3}{96a^2b}$

47. $\frac{7}{9xy^3}-\frac{4}{3x}+\frac{5}{2y^2}$; LCD $=18xy^3$

$\frac{7}{9xy^3}\left(\frac{2}{2}\right)-\frac{4}{3x}\left(\frac{6y^3}{6y^3}\right)+\frac{5}{2y^2}\left(\frac{9xy}{9xy}\right)$

$\frac{14}{18xy^3}-\frac{24y^3}{18xy^3}+\frac{45xy}{18xy^3}$

$\frac{14-24y^3+45xy}{18xy^3}$

49. $\frac{2x}{x-1}+\frac{3}{x}$; LCD $=x(x-1)$

$\frac{2x}{(x-1)}\left(\frac{x}{x}\right)+\frac{3}{x}\left(\frac{x-1}{x-1}\right)$

$\frac{2x(x)}{x(x-1)}+\frac{3(x-1)}{x(x-1)}$

$\frac{2x(x)+3(x-1)}{x(x-1)}$

$\frac{2x^2+3x-3}{x(x-1)}$

51. $\frac{a-2}{a}-\frac{3}{a+4}$; LCD $=a(a+4)$

$\frac{(a-2)}{a}\left(\frac{a+4}{a+4}\right)-\frac{3}{(a+4)}\left(\frac{a}{a}\right)$

$\frac{(a-2)(a+4)}{a(a+4)}-\frac{3a}{a(a+4)}$

$\frac{(a-2)(a+4)-3a}{a(a+4)}$

$\frac{a^2+2a-8-3a}{a(a+4)}$

$\frac{a^2-a-8}{a(a+4)}$

53. $\frac{-3}{4n+5}-\frac{8}{3n+5}$; LCD $=(4n+5)(3n+5)$

$\frac{-3}{(4n+5)}\left(\frac{3n+5}{3n+5}\right)-\frac{8}{(3n+5)}\left(\frac{4n+5}{4n+5}\right)$

$\frac{-3(3n+5)}{(4n+5)(3n+5)}-\frac{8(4n+5)}{(4n+5)(3n+5)}$

$\frac{-3(3n+5)-8(4n+5)}{(4n+5)(3n+5)}$

$\frac{-9n-15-32n-40}{(4n+5)(3n+5)}$

$\frac{-41n-55}{(4n+5)(3n+5)}$

55. $\frac{-1}{x+4}+\frac{4}{7x-1}$; LCD $=(x+4)(7x-1)$

$\frac{-1}{(x+4)}\left(\frac{7x-1}{7x-1}\right)+\frac{4}{(7x-1)}\left(\frac{x+4}{x+4}\right)$

$\frac{-1(7x-1)}{(x+4)(7x-1)}+\frac{4(x+4)}{(7x-1)(x+4)}$

$\frac{-1(7x-1)+4(x+4)}{(7x-1)(x+4)}$

$\frac{-7x+1+4x+16}{(7x-1)(x+4)}$

$\frac{-3x+17}{(7x-1)(x+4)}$

57. $\frac{7}{3x-5}-\frac{5}{2x+7}$; LCD $=(3x-5)(2x+7)$

$\frac{7}{(3x-5)}\left(\frac{2x+7}{2x+7}\right)-\frac{5}{(2x+7)}\left(\frac{3x-5}{3x-5}\right)$

$\frac{7(2x+7)}{(3x-5)(2x+7)}-\frac{5(3x-5)}{(2x+7)(3x-5)}$

$\frac{7(2x+7)-5(3x-5)}{(3x-5)(2x+7)}$

$\frac{14x+49-15x+25}{(3x-5)(2x+7)}$

$\frac{-x+74}{(3x-5)(2x+7)}$

59. $\frac{5}{3x-2}+\frac{6}{4x+5}$; LCD $=(3x-2)(4x+5)$

$\frac{5}{(3x-2)}\left(\frac{4x+5}{4x+5}\right)+\frac{6}{(4x+5)}\left(\frac{3x-2}{3x-2}\right)$

$$\frac{5(4x+5)}{(3x-2)(4x+5)}+\frac{6(3x-2)}{(4x+5)(3x-2)}$$

$$\frac{5(4x+5)+6(3x-2)}{(3x-2)(4x+5)}$$

$$\frac{20x+25+18x-12}{(3x-2)(4x+5)}$$

$$\frac{38x+13}{(3x-2)(4x+5)}$$

61. $\frac{3x}{2x+5}+1$; LCD $= 2x+5$

$$\frac{3x}{2x+5}+1\left(\frac{2x+5}{2x+5}\right)$$

$$\frac{3x}{2x+5}+\frac{2x+5}{2x+5}$$

$$\frac{3x+2x+5}{2x+5}$$

$$\frac{5x+5}{2x+5}$$

63. $\frac{4x}{x-5}-3$; LCD $= x-5$

$$\frac{4x}{x-5}-3\left(\frac{x-5}{x-5}\right)$$

$$\frac{4x}{x-5}-\frac{3(x-5)}{x-5}$$

$$\frac{4x-3(x-5)}{x-5}$$

$$\frac{4x-3x+15}{x-5}$$

$$\frac{x+15}{x-5}$$

65. $-1-\frac{3}{2x+1}$; LCD $= 2x+1$

$$-\left(\frac{2x+1}{2x+1}\right)-\frac{3}{2x+1}$$

$$\frac{-1(2x+1)}{2x+1}-\frac{3}{2x+1}$$

$$\frac{-1(2x+1)-3}{2x+1}$$

$$\frac{-2x-1-3}{2x+1}$$

$$\frac{-2x-4}{2x+1}$$

67a. $\frac{1}{x-1}-\frac{x}{x-1}$

$$\frac{1-x}{x-1}$$

$$-1$$

b. $\frac{3}{2x-3}-\frac{2x}{2x-3}$

$$\frac{3-2x}{2x-3}$$

$$-1$$

c. $\frac{4}{x-4}-\frac{x}{x-4}+1$

$$\frac{4-x}{x-4}+1$$

$$-1+1$$

$$0$$

d. $-1+\frac{2}{x-2}-\frac{x}{x-2}$

$$-1+\frac{2-x}{x-2}$$

$$-1+(-1)$$

$$-2$$

Problem Set 4.4 More on Rational Expressions and Complex Fractions

1. $\frac{2x}{x^2+4x}+\frac{5}{x}$

$$\frac{2x}{x(x+4)}+\frac{5}{x};\ \text{LCD} = x(x+4)$$

$$\frac{2x}{x(x+4)}+\frac{5}{x}\left(\frac{x+4}{x+4}\right)$$

$$\frac{2x+5(x+4)}{x(x+4)}$$

$$\frac{2x+5x+20}{x(x+4)}$$

$$\frac{7x+20}{x(x+4)}$$

3. $\frac{4}{x^2+7x}-\frac{1}{x}$

$$\frac{4}{x(x+7)}-\frac{1}{x};\ \text{LCD} = x(x+7)$$

$$\frac{4}{x(x+7)}-\frac{1}{x}\left(\frac{x+7}{x+7}\right)$$

$\frac{4-1(x+7)}{x(x+7)}$

$\frac{4-x-7}{x(x+7)}$

$\frac{-x-3}{x(x+7)}$

5. $\frac{x}{x^2-1}+\frac{5}{x+1}$

$\frac{x}{(x+1)(x-1)}+\frac{5}{x+1}$

$LCD=(x+1)(x-1)$

$\frac{x}{(x+1)(x-1)}+\frac{5}{(x+1)}\left(\frac{x-1}{x-1}\right)$

$\frac{x+5(x-1)}{(x+1)(x-1)}$

$\frac{x+5x-5}{(x+1)(x-1)}$

$\frac{6x-5}{(x+1)(x-1)}$

7. $\frac{6a+4}{a^2-1}-\frac{5}{a-1}$

$\frac{6a+4}{(a+1)(a-1)}-\frac{5}{a-1}$

$LCD=(a+1)(a-1)$

$\frac{6a+4}{(a+1)(a-1)}-\frac{5}{(a-1)}\left(\frac{a+1}{a+1}\right)$

$\frac{6a+4-5(a+1)}{(a+1)(a-1)}$

$\frac{6a+4-5a-5}{(a+1)(a-1)}$

$\frac{a-1}{(a+1)(a-1)}$

$\frac{1}{a+1}$

9. $\frac{2n}{n^2-25}-\frac{3}{4n+20}$

$\frac{2n}{(n+5)(n-5)}-\frac{3}{4(n+5)}$

$LCD=4(n+5)(n-5)$

$\frac{2n}{(n+5)(n-5)}\left(\frac{4}{4}\right)-\frac{3}{4(n+5)}\left(\frac{n-5}{n-5}\right)$

$\frac{2n(4)-3(n-5)}{4(n+5)(n-5)}$

$\frac{8n-3n+15}{4(n+5)(n-5)}$

$\frac{5n+15}{4(n+5)(n-5)}$

11. $\frac{5}{x}-\frac{5x-30}{x^2+6x}+\frac{x}{x+6}$

$\frac{5}{x}-\frac{5x-30}{x(x+6)}+\frac{x}{x+6}$

$LCD=x(x+6)$

$\frac{5}{x}\left(\frac{x+6}{x+6}\right)-\frac{5x-30}{x(x+6)}+\frac{x}{x+6}\left(\frac{x}{x}\right)$

$\frac{5(x+6)-(5x-30)+x^2}{x(x+6)}$

$\frac{5x+30-5x+30+x^2}{x(x+6)}$

$\frac{x^2+60}{x(x+6)}$

13. $\frac{3}{x^2+9x+14}+\frac{5}{2x^2+15x+7}$

$\frac{3}{(x+2)(x+7)}+\frac{5}{(2x+1)(x+7)}$

$LCD=(x+2)(x+7)(2x+1)$

$\frac{3}{(x+2)(x+7)}\left(\frac{2x+1}{2x+1}\right)+\frac{5}{(2x+1)(x+7)}\left(\frac{x+2}{x+2}\right)$

$\frac{3(2x+1)+5(x+2)}{(x+2)(2x+1)(x+7)}$

$\frac{6x+3+5x+10}{(x+2)(2x+1)(x+7)}$

$\frac{11x+13}{(x+2)(2x+1)(x+7)}$

15. $\frac{1}{a^2-3a-10}-\frac{4}{a^2+4a-45}$

$\frac{1}{(a-5)(a+2)}-\frac{4}{(a+9)(a-5)}$

$LCD=(a-5)(a+2)(a+9)$

$\frac{1}{(a-5)(a+2)}\left(\frac{a+9}{a+9}\right)-\frac{4}{(a+9)(a-5)}\left(\frac{a+2}{a+2}\right)$

$$\frac{1(a+9)-4(a+2)}{(a-5)(a+2)(a+9)}$$

$$\frac{a+9-4a-8}{(a-5)(a+2)(a+9)}$$

$$\frac{-3a+1}{(a-5)(a+2)(a+9)}$$

17. $$\frac{3a}{20a^2-11a-3}+\frac{1}{12a^2+7a-12}$$

$$\frac{3a}{(5a+1)(4a-3)}+\frac{1}{(4a-3)(3a+4)}$$

$\text{LCD} = (5a+1)(4a-3)(3a+4)$

$$\frac{3a}{(5a+1)(4a-3)}\left(\frac{3a+4}{3a+4}\right)+\frac{1}{(4a-3)(3a+4)}\left(\frac{5a+1}{5a+1}\right)$$

$$\frac{3a(3a+4)+1(5a+1)}{(5a+1)(4a-3)(3a+4)}$$

$$\frac{9a^2+12a+5a+1}{(5a+1)(4a-3)(3a+4)}$$

$$\frac{9a^2+17a+1}{(5a+1)(4a-3)(3a+4)}$$

19. $$\frac{5}{x^2+3}-\frac{2}{x^2+4x-21}$$

$$\frac{5}{x^2+3}-\frac{2}{(x+7)(x-3)}$$

$\text{LCD} = (x^2+3)(x+7)(x-3)$

$$\frac{5}{(x^2+3)}\left(\frac{x+7}{x+7}\right)\left(\frac{x-3}{x-3}\right)-\frac{2}{(x+7)(x-3)}\left(\frac{x^2+3}{x^2+3}\right)$$

$$\frac{5(x+7)(x-3)-2(x^2+3)}{(x^2+3)(x+7)(x-3)}$$

$$\frac{5(x^2+4x-21)-2x^2-6}{(x^2+3)(x+7)(x-3)}$$

$$\frac{5x^2+20x-105-2x^2-6}{(x^2+3)(x+7)(x-3)}$$

$$\frac{3x^2+20x-111}{(x^2+3)(x+7)(x-3)}$$

21. $$\frac{2}{y^2+6y-16}-\frac{4}{y+8}-\frac{3}{y-2}$$

$$\frac{2}{(y+8)(y-2)}-\frac{4}{y+8}-\frac{3}{y-2}$$

$\text{LCD} = (y+8)(y-2)$

$$\frac{2}{(y+8)(y-2)}-\frac{4}{(y+8)}\left(\frac{y-2}{y-2}\right)-\frac{3}{(y-2)}\left(\frac{y+8}{y+8}\right)$$

$$\frac{2-4(y-2)-3(y+8)}{(y+8)(y-2)}$$

$$\frac{2-4y+8-3y-24}{(y+8)(y-2)}$$

$$\frac{-7y-14}{(y+8)(y-2)}$$

23. $$x-\frac{x^2}{x-2}+\frac{3}{x^2-4}$$

$$x-\frac{x^2}{x-2}+\frac{3}{(x+2)(x-2)}$$

$\text{LCD} = (x+2)(x-2)$

$$x\left(\frac{x+2}{x+2}\right)\left(\frac{x-2}{x-2}\right)-\frac{x^2}{(x-2)}\left(\frac{x+2}{x+2}\right)+\frac{3}{(x+2)(x-2)}$$

$$\frac{x(x+2)(x-2)-x^2(x+2)+3}{(x+2)(x-2)}$$

$$\frac{x(x^2-4)-x^3-2x^2+3}{(x+2)(x-2)}$$

$$\frac{x^3-4x-x^3-2x^2+3}{(x+2)(x-2)}$$

$$\frac{-2x^2-4x+3}{(x+2)(x-2)}$$

25. $$\frac{x+3}{x+10}+\frac{4x-3}{x^2+8x-20}+\frac{x-1}{x-2}$$

$$\frac{x+3}{x+10}+\frac{4x-3}{(x+10)(x-2)}+\frac{x-1}{x-2}$$

$\text{LCD} = (x+10)(x-2)$

$$\frac{(x+3)}{(x+10)}\left(\frac{x-2}{x-2}\right)+\frac{4x-3}{(x+10)(x-2)}+\frac{(x-1)}{(x-2)}\left(\frac{x+10}{x+10}\right)$$

$$\frac{(x+3)(x-2)+4x-3+(x-1)(x+10)}{(x+10)(x-2)}$$

$$\frac{x^2+x-6+4x-3+x^2+9x-10}{(x+10)(x-2)}$$

$$\frac{2x^2+14x-19}{(x+10)(x-2)}$$

27. $$\frac{n}{n-6}+\frac{n+3}{n+8}+\frac{12n+26}{n^2+2n-48}$$

$$\frac{n}{n-6}+\frac{n+3}{n+8}+\frac{12n+26}{(n+8)(n-6)}$$

$\text{LCD} = (n-6)(n+8)$

$$\frac{n}{(n-6)}\left(\frac{n+8}{n+8}\right)+\frac{(n+3)}{(n+8)}\left(\frac{n-6}{n-6}\right)+\frac{12n+26}{(n+8)(n-6)}$$

$$\frac{n(n+8)+(n+3)(n-6)+12n+26}{(n-6)(n+8)}$$

$$\frac{n^2+8n+n^2-3n-18+12n+26}{(n-6)(n+8)}$$

$$\frac{2n^2+17n+8}{(n-6)(n+8)}$$

$$\frac{(2n+1)(n+8)}{(n-6)(n+8)}$$

$$\frac{2n+1}{n-6}$$

29. $$\frac{4x-3}{2x^2+x-1}-\frac{2x+7}{3x^2+x-2}-\frac{3}{3x-2}$$

$$\frac{4x-3}{(2x-1)(x+1)}-\frac{2x+7}{(3x-2)(x+1)}-\frac{3}{3x-2}$$

$\text{LCD}=(2x-1)(x+1)(3x-2)$

$$\frac{(4x-3)}{(2x-1)(x+1)}\left(\frac{3x-2}{3x-2}\right)-\frac{(2x+7)}{(3x-2)(x+1)}\left(\frac{2x-1}{2x-1}\right)-\frac{3}{(3x-2)}\left(\frac{2x-1}{2x-1}\right)\left(\frac{x+1}{x+1}\right)$$

$$\frac{(4x-3)(3x-2)-(2x+7)(2x-1)-3(2x-1)(x+1)}{(2x-1)(3x-2)(x+1)}$$

$$\frac{12x^2-17x+6-(4x^2+12x-7)-3(2x^2+x-1)}{(2x-1)(3x-2)(x+1)}$$

$$\frac{12x^2-17x+6-4x^2-12x+7-6x^2-3x+3}{(2x-1)(3x-2)(x+1)}$$

$$\frac{2x^2-32x+16}{(2x-1)(3x-2)(x+1)}$$

31. $$\frac{n}{n^2+1}+\frac{n^2+3n}{n^4-1}-\frac{1}{n-1}$$

$$\frac{n}{n^2+1}+\frac{n^2+3n}{(n^2+1)(n+1)(n-1)}-\frac{1}{n-1}$$

$\text{LCD}=(n^2+1)(n+1)(n-1)$

$$\frac{n}{(n^2+1)}\left(\frac{n+1}{n+1}\right)\left(\frac{n-1}{n-1}\right)+\frac{n^2+3n}{(n^2+1)(n+1)(n-1)}-\frac{1}{(n-1)}\left(\frac{n^2+1}{n^2+1}\right)\left(\frac{n+1}{n+1}\right)$$

$$\frac{n(n+1)(n-1)+n^2+3n-(n^2+1)(n+1)}{(n^2+1)(n+1)(n-1)}$$

$$\frac{n(n^2-1)+n^2+3n-(n^3+n^2+n+1)}{(n^2+1)(n+1)(n-1)}$$

$$\frac{n^3-n+n^2+3n-n^3-n^2-n-1}{(n^2+1)(n+1)(n-1)}$$

$$\frac{n-1}{(n^2+1)(n+1)(n-1)}$$

$$\frac{1}{(n^2+1)(n+1)}$$

33. $$\frac{15x^2-10}{5x^2-7x+2}-\frac{3x+4}{x-1}-\frac{2}{5x-2}$$

$$\frac{15x^2-10}{(5x-2)(x-1)}-\frac{3x+4}{x-1}-\frac{2}{5x-2}$$

$$\text{LCD}=(5x-2)(x-1)$$

$$\frac{15x^2-10}{(5x-2)(x-1)}-\frac{(3x+4)}{(x-1)}\left(\frac{5x-2}{5x-2}\right)-\frac{2}{(5x-2)}\left(\frac{x-1}{x-1}\right)$$

$$\frac{15x^2-10-(3x+4)(5x-2)-2(x-1)}{(5x-2)(x-1)}$$

$$\frac{15x^2-10-(15x^2+14x-8)-2x+2}{(5x-2)(x-1)}$$

$$\frac{15x^2-10-15x^2-14x+8-2x+2}{(5x-2)(x-1)}$$

$$\frac{-16x}{(5x-2)(x-1)}$$

35. $$\frac{t+3}{3t-1}+\frac{8t^2+8t+2}{3t^2-7t+2}-\frac{2t+3}{t-2}$$

$$\frac{t+3}{3t-1}+\frac{8t^2+8t+2}{(3t-1)(t-2)}-\frac{2t+3}{t-2}$$

$$\text{LCD}=(3t-1)(t-2)$$

$$\frac{(t+3)}{(3t-1)}\left(\frac{t-2}{t-2}\right)+\frac{8t^2+8t+2}{(3t-1)(t-2)}-\frac{(2t+3)}{(t-2)}\left(\frac{3t-1}{3t-1}\right)$$

$$\frac{(t+3)(t-2)+8t^2+8t+2-(2t+3)(3t-1)}{(3t-1)(t-2)}$$

$$\frac{t^2+t-6+8t^2+8t+2-(6t^2+7t-3)}{(3t-1)(t-2)}$$

$$\frac{t^2+t-6+8t^2+8t+2-6t^2-7t+3}{(3t-1)(t-2)}$$

$$\frac{3t^2+2t-1}{(3t-1)(t-2)}$$

$$\frac{(3t-1)(t+1)}{(3t-1)(t-2)}$$

$$\frac{t+1}{t-2}$$

37. $$\frac{\frac{1}{2}-\frac{1}{4}}{\frac{5}{8}+\frac{3}{4}}=\frac{8}{8}\bullet\frac{\left(\frac{1}{2}-\frac{1}{4}\right)}{\left(\frac{5}{8}+\frac{3}{4}\right)}=\frac{8\left(\frac{1}{2}\right)-8\left(\frac{1}{4}\right)}{8\left(\frac{5}{8}\right)+8\left(\frac{3}{4}\right)}=\frac{4-2}{5+6}=\frac{2}{11}$$

39. $\dfrac{\frac{3}{28}-\frac{5}{14}}{\frac{5}{7}+\frac{1}{4}}$

$\dfrac{28}{28}\bullet\dfrac{\left(\frac{3}{28}-\frac{5}{14}\right)}{\left(\frac{5}{7}+\frac{1}{4}\right)}$

$\dfrac{28\left(\frac{3}{28}\right)-28\left(\frac{5}{14}\right)}{28\left(\frac{5}{7}\right)+28\left(\frac{1}{4}\right)}$

$\dfrac{3-10}{20+7}$

$\dfrac{-7}{27}$

41. $\dfrac{\frac{5}{6y}}{\frac{10}{3xy}}$

$\dfrac{5}{6y}\bullet\dfrac{3xy}{10}$

$\dfrac{15xy}{60y}$

$\dfrac{x}{4}$

43. $\dfrac{\frac{3}{x}-\frac{2}{y}}{\frac{4}{y}-\frac{7}{xy}}$

$\dfrac{xy}{xy}\bullet\dfrac{\left(\frac{3}{x}-\frac{2}{y}\right)}{\left(\frac{4}{y}-\frac{7}{xy}\right)}$

$\dfrac{xy\left(\frac{3}{x}\right)-xy\left(\frac{2}{y}\right)}{xy\left(\frac{4}{y}\right)-xy\left(\frac{7}{xy}\right)}$

$\dfrac{3y-2x}{4x-7}$

45. $\dfrac{\frac{6}{a}-\frac{5}{b^2}}{\frac{12}{a^2}+\frac{2}{b}}$

$\dfrac{a^2b^2}{a^2b^2}\bullet\dfrac{\left(\frac{6}{a}-\frac{5}{b^2}\right)}{\left(\frac{12}{a^2}+\frac{2}{b}\right)}$

$\dfrac{a^2b^2\left(\frac{6}{a}\right)-a^2b^2\left(\frac{5}{b^2}\right)}{a^2b^2\left(\frac{12}{a^2}\right)+a^2b^2\left(\frac{2}{b}\right)}$

$\dfrac{6ab^2-5a^2}{12b^2+2a^2b}$

47. $\dfrac{\frac{2}{x}-3}{\frac{3}{y}+4}$

$\dfrac{xy}{xy}\bullet\dfrac{\left(\frac{2}{x}-3\right)}{\left(\frac{3}{y}+4\right)}$

$\dfrac{xy\left(\frac{2}{x}\right)-xy(3)}{xy\left(\frac{3}{y}\right)+xy(4)}$

$\dfrac{2y-3xy}{3x+4xy}$

49. $\dfrac{3+\frac{2}{n+4}}{5-\frac{1}{n+4}}$

$\dfrac{(n+4)}{(n+4)}\bullet\dfrac{\left(3+\frac{2}{n+4}\right)}{\left(5-\frac{1}{n+4}\right)}$

$\dfrac{(n+4)(3)+(n+4)\left(\frac{2}{n+4}\right)}{(n+4)(5)-(n+4)\left(\frac{1}{n+4}\right)}$

$\dfrac{3n+12+2}{5n+20-1}$

$\dfrac{3n+14}{5n+19}$

51. $\dfrac{5-\frac{2}{n-3}}{4-\frac{1}{n-3}}$

$\dfrac{(n-3)}{(n-3)}\bullet\dfrac{\left(5-\frac{2}{n-3}\right)}{\left(4-\frac{1}{n-3}\right)}$

$\dfrac{(n-3)(5)-(n-3)\left(\frac{2}{n-3}\right)}{(n-3)(4)-(n-3)\left(\frac{1}{n-3}\right)}$

$$\frac{5n-15-2}{4n-12-1}$$

$$\frac{5n-17}{4n-13}$$

53. $$\frac{\frac{-1}{y-2}+\frac{5}{x}}{\frac{3}{x}-\frac{4}{xy-2x}}$$

$$\frac{-\frac{1}{y-2}+\frac{5}{x}}{\frac{3}{x}-\frac{4}{x(y-2)}}$$

$$\frac{x(y-2)}{x(y-2)}\bullet\frac{\left(-\frac{1}{y-2}+\frac{5}{x}\right)}{\left[\frac{3}{x}-\frac{4}{x(y-2)}\right]}$$

$$\frac{x(y-2)\left(\frac{-1}{y-2}\right)+x(y-2)\left(\frac{5}{x}\right)}{x(y-2)\left(\frac{3}{x}\right)-x(y-2)\left[\frac{4}{x(y-2)}\right]}$$

$$\frac{-x+5(y-2)}{3(y-2)-4}$$

$$\frac{-x+5y-10}{3y-6-4}$$

$$\frac{-x+5y-10}{3y-10}$$

55. $$\frac{\frac{2}{x-3}-\frac{3}{x+3}}{\frac{5}{x^2-9}-\frac{2}{x-3}}$$

$$\frac{\frac{2}{x-3}-\frac{3}{x+3}}{\frac{5}{(x+3)(x-3)}-\frac{2}{x-3}}$$

$$\frac{(x+3)(x-3)}{(x+3)(x-3)}\bullet\frac{\left(\frac{2}{x-3}-\frac{3}{x+3}\right)}{\left[\frac{5}{(x+3)(x-3)}-\frac{2}{x-3}\right]}$$

$$\frac{(x+3)(x-3)\left(\frac{2}{x-3}\right)-(x+3)(x-3)\left(\frac{3}{x+3}\right)}{(x+3)(x-3)\left[\frac{5}{(x+3)(x-3)}\right]-(x+3)(x-3)\left(\frac{2}{x-3}\right)}$$

$$\frac{2(x+3)-3(x-3)}{5-2(x+3)}$$

$$\frac{2x+6-3x+9}{5-2x-6}$$

$$\frac{-x+15}{-2x-1}$$

57. $$\frac{3a}{2-\frac{1}{a}}-1$$

$$\frac{a}{a}\bullet\frac{(3a)}{\left(2-\frac{1}{a}\right)}-1$$

$$\frac{a(3a)}{a(2)-a\left(\frac{1}{a}\right)}-1$$

$$\frac{3a^2}{2a-1}-1$$

$$\frac{3a^2}{2a-1}-1\left(\frac{2a-1}{2a-1}\right)$$

$$\frac{3a^2}{2a-1}-\frac{2a-1}{2a-1}$$

$$\frac{3a^2-(2a-1)}{2a-1}$$

$$\frac{3a^2-2a+1}{2a-1}$$

59. $$2-\frac{x}{3-\frac{2}{x}}$$

$$2-\frac{x}{x}\bullet\frac{(x)}{\left(3-\frac{2}{x}\right)}$$

$$2-\frac{x(x)}{x(3)-x\left(\frac{2}{x}\right)}$$

$$2-\frac{x^2}{3x-2}$$

$$2\left(\frac{3x-2}{3x-2}\right)-\frac{x^2}{3x-2}$$

$$\frac{6x-4}{3x-2}-\frac{x^2}{3x-2}$$

$$\frac{6x-4-x^2}{3x-2}$$

$$\frac{-x^2+6x-4}{3x-2}$$

Problem Set 4.5 Dividing Polynomials

1. $\frac{9x^4+18x^3}{3x}$

$\frac{9x^4}{3x}+\frac{18x^3}{3x}$

$3x^3+6x^2$

3. $\frac{-24x^6+36x^8}{4x^2}$

$\frac{-24x^6}{4x^2}+\frac{36x^8}{4x^2}$

$-6x^4+9x^6$

5. $\frac{15a^3-25a^2-40a}{5a}$

$\frac{15a^3}{5a}-\frac{25a^2}{5a}-\frac{40a}{5a}$

$3a^2-5a-8$

7. $\frac{13x^3-17x^2+28x}{-x}$

$\frac{13x^3}{-x}+\frac{-17x^2}{-x}+\frac{28x}{-x}$

$-13x^2+17x-28$

9. $\frac{-18x^2y^2+24x^3y^2-48x^2y^3}{6xy}$

$\frac{-18x^2y^2}{6xy}+\frac{24x^3y^2}{6xy}-\frac{48x^2y^3}{6xy}$

$-3xy+4x^2y-8xy^2$

11.

$$\begin{array}{r|l} & x-13 \\ \hline x+6 & x^2-7x-78 \\ & \underline{x^2+6x} \\ & -13x-78 \\ & \underline{-13x-78} \\ & 0 \end{array}$$

$x-13$

13.

$$\begin{array}{r|l} & x+20 \\ \hline x-8 & x^2+12x-160 \\ & \underline{x^2-8x} \\ & 20x-160 \\ & \underline{20x-160} \\ & 0 \end{array}$$

$x+20$

15.

$$\begin{array}{r|l} & 2x+1 \\ \hline x-1 & 2x^2-x-4 \\ & \underline{2x^2-2x} \\ & x-4 \\ & \underline{x-1} \\ & -3 \end{array}$$

$2x+1-\frac{3}{x-1}$

17.

$$\begin{array}{r|l} & 5x-1 \\ \hline 3x+5 & 15x^2+22x-5 \\ & \underline{15x^2+25x} \\ & -3x-5 \\ & \underline{-3x-5} \\ & 0 \end{array}$$

$5x-1$

19.

$$\begin{array}{r|l} & 3x^2-2x-7 \\ \hline x+3 & 3x^3+7x^2-13x-21 \\ & \underline{3x^3+9x^2} \\ & -2x^2-13x \\ & \underline{-2x^2-6x} \\ & -7x-21 \\ & \underline{-7x-21} \\ & 0 \end{array}$$

$3x^2-2x-7$

21.

$$\begin{array}{r|l} & x^2+5x-6 \\ \hline 2x-1 & 2x^3+9x^2-17x+6 \\ & \underline{2x^3-x^2} \\ & 10x^2-17x \\ & \underline{10x^2-5x} \\ & -12x+6 \\ & \underline{-12x+6} \\ & 0 \end{array}$$

x^2+5x-6

23.

$$\begin{array}{r|l} & 4x^2+7x+12 \\ \hline x-2 & 4x^3-x^2-2x+6 \\ & \underline{4x^3-8x^2} \\ & 7x^2-2x \\ & \underline{7x^2-14x} \\ & 12x+6 \\ & \underline{12x-24} \\ & 30 \end{array}$$

$4x^2+7x+12+\frac{30}{x-2}$

25.

$$x^3-4x^2-5x+3$$
$$x-6\;\overline{)\,x^4-10x^3+19x^2+33x-18}$$
$$\underline{x^4-6x^3}$$
$$-4x^3+19x^2$$
$$\underline{-4x^3+24x^2}$$
$$-5x^2+33x$$
$$\underline{-5x^2+30x}$$
$$3x-18$$
$$\underline{3x-18}$$
$$0$$

x^3-4x^2-5x+3

27.

$$x^2+5x+25$$
$$x-5\;\overline{)\,x^3+0x^2+0x-125}$$
$$\underline{x^3-5x^2}$$
$$5x^2+0x$$
$$\underline{5x^2-25x}$$
$$25x-125$$
$$\underline{25x-125}$$
$$0$$

$x^2+5x+25$

29.

$$x^2-x+1$$
$$x+1\;\overline{)\,x^3+0x^2+0x+64}$$
$$\underline{x^3+x^2}$$
$$-x^2+0x$$
$$\underline{-x^2-x}$$
$$x+64$$
$$\underline{x+1}$$
$$63$$

$x^2-x+1+\frac{63}{x+1}$

31.

$$2x^2-4x+7$$
$$x+2\;\overline{)\,2x^3+0x^2-x-6}$$
$$\underline{2x^3+4x^2}$$
$$-4x^2-x$$
$$\underline{-4x^2-8x}$$
$$7x-6$$
$$\underline{7x+14}$$
$$-20$$

$2x^2-4x+7-\frac{20}{x+2}$

33.

$$4a-4b$$
$$a-b\;\overline{)\,4a^2-8ab+4b^2}$$
$$\underline{4a^2-4ab}$$
$$-4ab+4b^2$$
$$\underline{-4ab+4b^2}$$
$$0$$

$4a-4b$

35.

$$4x+7$$
$$x^2-3x\;\overline{)\,4x^3-5x^2+2x-6}$$
$$\underline{4x^3-12x^2}$$
$$7x^2+2x$$
$$\underline{7x^2-21x}$$
$$23x-6$$

$4x+7+\frac{23x-6}{x^2-3x}$

37.

$$8y-9$$
$$y^2+y\;\overline{)\,8y^3-y^2-y+5}$$
$$\underline{8y^3+8y^2}$$
$$-9y^2-y$$
$$\underline{-9y^2-9y}$$
$$8y+5$$

$8y-9+\frac{8y+5}{y^2+y}$

39.

$$2x-1$$
$$x^2+x-1\;\overline{)\,2x^3+x^2-3x+1}$$
$$\underline{2x^3+2x^2-2x}$$
$$-x^2-x+1$$
$$\underline{-x^2-x+1}$$
$$0$$

$2x-1$

41.

$$x-3$$
$$4x^2-x+5\;\overline{)\,4x^3-13x^2+8x-15}$$
$$\underline{4x^3-x^2+5x}$$
$$-12x^2+3x-15$$
$$\underline{-12x^2+3x-15}$$
$$0$$

$x-3$

43.

$$\begin{array}{r} 5a-8 \\ a^2+3a-4 \overline{\smash{)}\, 5a^3+7a^2-2a-9} \\ \underline{5a^3+15a^2-20a} \\ -8a^2+18a-9 \\ \underline{-8a^2-24a+32} \\ 42a-41 \end{array}$$

$$5a-8+\frac{42a-41}{a^2+3a-4}$$

45.

$$\begin{array}{r} 2n^2+3n-4 \\ n^2+1 \overline{\smash{)}\, 2n^4+3n^3-2n^2+3n-4} \\ \underline{2n^4 \quad +2n^2} \\ 3n^3-4n^2+3n \\ \underline{3n^3 \quad +3n} \\ -4n^2 \quad -4 \\ \underline{-4n^2 \quad -4} \\ 0 \end{array}$$

$$2n^2+3n-4$$

47.

$$\begin{array}{r} x^4+x^3+x^2+x+1 \\ x-1 \overline{\smash{)}\, x^5+0x^4+0x^3+0x^2+0x-1} \\ \underline{x^5-x^4} \\ x^4+0x^3 \\ \underline{x^4-x^3} \\ x^3+0x^2 \\ \underline{x^3-x^2} \\ x^2+0x \\ \underline{x^2-x} \\ x-1 \\ \underline{x-1} \\ 0 \end{array}$$

$$x^4+x^3+x^2+x+1$$

49.

$$\begin{array}{r} x^3-x^2+x-1 \\ x+1 \overline{\smash{)}\, x^4+0x^3+0x^2+0x-1} \\ \underline{x^4+x^3} \\ -x^3+0x^2 \\ \underline{-x^3-x^2} \\ x^2+0x \\ \underline{x^2+x} \\ -x-1 \\ \underline{-x-1} \\ 0 \end{array}$$

$$x^3-x^2+x-1$$

51.

$$\begin{array}{r} 3x^2+x+1 \\ x^2-1 \overline{\smash{)}\, 3x^4+x^3-2x^2-x+6} \\ \underline{3x^4 \quad -3x^2} \\ x^3+x^2-x \\ \underline{x^3 \quad -x} \\ x^2 \quad +6 \\ \underline{x^2 \quad -1} \\ 7 \end{array}$$

$$3x^2+x+1+\frac{7}{x^2-1}$$

Problem Set 4.6 Fractional Equations

1. $\frac{x+1}{4}+\frac{x-2}{6}=\frac{3}{4}$

$12\left(\frac{x+1}{4}+\frac{x-2}{6}\right)=12\left(\frac{3}{4}\right)$

$12\left(\frac{x+1}{4}\right)+12\left(\frac{x-2}{6}\right)=9$

$3(x+1)+2(x-2)=9$

$3x+3+2x-4=9$

$5x-1=9$

$5x=10$

$x=2$

The solution set is $\{2\}$.

3. $\frac{x+3}{2}-\frac{x-4}{7}=1$

$14\left(\frac{x+3}{2}-\frac{x-4}{7}\right)=14(1)$

$14\left(\frac{x+3}{2}\right)-14\left(\frac{x-4}{7}\right)=14$

$7(x+3)-2(x-4)=14$

$7x+21-2x+8=14$

$5x+29=14$

$5x=-15$

$x=-3$

The solution set is $\{-3\}$.

5. $\frac{5}{n}+\frac{1}{3}=\frac{7}{n};\ n\neq 0$

$3n\left(\frac{5}{n}+\frac{1}{3}\right)=3n\left(\frac{7}{n}\right)$

$3n\left(\frac{5}{n}\right)+3n\left(\frac{1}{3}\right)=21$

$15 + n = 21$

$n = 6$

The solution set is $\{6\}$.

7. $\frac{7}{2x} + \frac{3}{5} = \frac{2}{3x};\ x \neq 0$

$30x\left(\frac{7}{2x} + \frac{3}{5}\right) = 30x\left(\frac{2}{3x}\right)$

$30x\left(\frac{7}{2x}\right) + 30x\left(\frac{3}{5}\right) = 20$

$15(7) + 6x(3) = 20$

$105 + 18x = 20$

$18x = -85$

$x = -\frac{85}{18}$

The solution set is $\left\{-\frac{85}{18}\right\}$.

9. $\frac{3}{4x} + \frac{5}{6} = \frac{4}{3x};\ x \neq 0$

$12x\left(\frac{3}{4x} + \frac{5}{6}\right) = 12x\left(\frac{4}{3x}\right)$

$12x\left(\frac{3}{4x}\right) + 12x\left(\frac{5}{6}\right) = 16$

$3(3) + 2x(5) = 16$

$9 + 10x = 16$

$10x = 7$

$x = \frac{7}{10}$

The solution set is $\left\{\frac{7}{10}\right\}$.

11. $\frac{47 - n}{n} = 8 + \frac{2}{n};\ n \neq 0$

$n\left(\frac{47 - n}{n}\right) = n\left(8 + \frac{2}{n}\right)$

$47 - n = n(8) + n\left(\frac{2}{n}\right)$

$47 - n = 8n + 2$

$47 = 9n + 2$

$45 = 9n$

$5 = n$

The solution set is $\{5\}$.

13. $\frac{n}{65 - n} = 8 + \frac{2}{65 - n};\ n \neq 65$

$(65 - n)\left(\frac{n}{65 - n}\right) = (65 - n)\left(8 + \frac{2}{65 - n}\right)$

$n = (65 - n)(8) + (65 - n)\left(\frac{2}{65 - n}\right)$

$n = 520 - 8n + 2$

$9n = 522$

$n = 58$

The solution set is $\{58\}$.

15. $n + \frac{1}{n} = \frac{17}{4};\ n \neq 0$

$4n\left(n + \frac{1}{n}\right) = 4n\left(\frac{17}{4}\right)$

$4n(n) + 4n\left(\frac{1}{n}\right) = 17n$

$4n^2 + 4 = 17n$

$4n^2 - 17n + 4 = 0$

$(4n - 1)(n - 4) = 0$

$4n - 1 = 0$ or $n - 4 = 0$

$4n = 1$ or $n = 4$

$n = \frac{1}{4}$ or $n = 4$

The solution set is $\left\{\frac{1}{4},\ 4\right\}$.

17. $n - \frac{2}{n} = \frac{23}{5};\ n \neq 0$

$5n\left(n - \frac{2}{n}\right) = 5n\left(\frac{23}{5}\right)$

$5n(n) - 5n\left(\frac{2}{n}\right) = 23n$

$5n^2 - 10 = 23n$

$5n^2 - 23n - 10 = 0$

$(5n + 2)(n - 5) = 0$

$5n + 2 = 0$ or $n - 5 = 0$

$5n = -2$ or $n = 5$

$n = -\frac{2}{5}$ or $n = 5$

The solution set is $\left\{-\frac{2}{5},\ 5\right\}$.

19. $\frac{5}{7x - 3} = \frac{3}{4x - 5};\ x \neq \frac{3}{7},\ x \neq \frac{5}{4}$

$5(4x - 5) = 3(7x - 3)$

$20x - 25 = 21x - 9$

$20x = 21x + 16$

$-x = 16$

$x = -16$

The solution set is $\{-16\}$.

21. $\frac{-2}{x-5} = \frac{1}{x+9}$; $x \neq 5$, $x \neq -9$

$-2(x+9) = 1(x-5)$

$-2x - 18 = x - 5$

$-2x = x + 13$

$-3x = 13$

$x = -\frac{13}{3}$

The solution set is $\left\{-\frac{13}{3}\right\}$.

23. $\frac{x}{x+1} - 2 = \frac{3}{x-3}$; $x \neq -1$, $x \neq 3$

$(x+1)(x-3)\left(\frac{x}{x+1} - 2\right) = (x+1)(x-3)\left(\frac{3}{x-3}\right)$

$(x+1)(x-3)\left(\frac{x}{x+1}\right) - 2(x+1)(x-3) = 3(x+1)$

$x(x-3) - 2(x^2 - 2x - 3) = 3x + 3$

$x^2 - 3x - 2x^2 + 4x + 6 = 3x + 3$

$-x^2 + x + 6 = 3x + 3$

$-x^2 - 2x + 3 = 0$

$x^2 + 2x - 3 = 0$

$(x+3)(x-1) = 0$

$x + 3 = 0$ or $x - 1 = 0$

$x = -3$ or $x = 1$

The solution set is $\{-3, 1\}$.

25. $\frac{a}{a+5} - 2 = \frac{3a}{a+5}$; $a \neq -5$

$(a+5)\left(\frac{a}{a+5} - 2\right) = (a+5)\left(\frac{3a}{a+5}\right)$

$(a+5)\left(\frac{a}{a+5}\right) - 2(a+5) = 3a$

$a - 2a - 10 = 3a$

$-a - 10 = 3a$

$-10 = 4a$

$-\frac{10}{4} = a$

$-\frac{5}{2} = a$

The solution set is $\left\{-\frac{5}{2}\right\}$.

27. $\frac{5}{x+6}=\frac{6}{x-3};\ x\neq -6,\ x\neq 3$

$5(x-3)=6(x+6)$

$5x-15=6x+36$

$5x=6x+51$

$-x=51$

$x=-51$

The solution set is $\{-51\}$.

29. $\frac{3x-7}{10}=\frac{2}{x};\ x\neq 0$

$x(3x-7)=10(2)$

$3x^2-7x=20$

$3x^2-7x-20=0$

$(3x+5)(x-4)=0$

$3x+5=0$ or $x-4=0$

$3x=-5$ or $x=4$

$x=-\frac{5}{3}$ or $x=4$

The solution set is $\left\{-\frac{5}{3},\ 4\right\}$.

31. $\frac{x}{x-6}-3=\frac{6}{x-6};\ x\neq 6$

$(x-6)\left(\frac{x}{x-6}-3\right)=(x-6)\left(\frac{6}{x-6}\right)$

$(x-6)\left(\frac{x}{x-6}\right)-3(x-6)=6$

$x-3(x-6)=6$

$x-3x+18=6$

$-2x+18=6$

$-2x=-12$

$x=6$

The solution set is $\emptyset$.

33. $\frac{3s}{s+2}+1=\frac{35}{2(3s+1)};\ s\neq -\frac{1}{3},\ s\neq -2$

$2(s+2)(3s+1)\left[\frac{3s}{s+2}+1\right]=2(s+2)(3s+1)\left[\frac{35}{2(3s+1)}\right]$

$2(s+2)(3s+1)\left(\frac{3s}{s+2}\right)+2(s+2)(3s+1)(1)=35(s+2)$

$6s(3s+1)+2(3s^2+7s+2)=35s+70$

$18s^2+6s+6s^2+14s+4=35s+70$

$24s^2+20s+4=35s+70$

$24s^2 - 15s - 66 = 0$

$3(8s^2 - 5s - 22) = 0$

$3(8x + 11)(x - 2) = 0$

$8x + 11 = 0 \quad \text{or} \quad x - 2 = 0$

$8x = -11 \quad \text{or} \quad x = 2$

$x = -\frac{11}{8} \quad \text{or} \quad x = 2$

The solution set is $\left\{-\frac{11}{8}, 2\right\}$.

35. $2 - \frac{3x}{x-4} = \frac{14}{x+7}; \; x \neq 4, \; x \neq -7$

$(x-4)(x+7)\left(2 - \frac{3x}{x-4}\right) = (x-4)(x+7)\left(\frac{14}{x+7}\right)$

$(x-4)(x+7)(2) - (x-4)(x+7)\left(\frac{3x}{x-4}\right) = 14(x-4)$

$2(x^2 + 3x - 28) - 3x(x+7) = 14x - 56$

$2x^2 + 6x - 56 - 3x^2 - 21x = 14x - 56$

$-x^2 - 15x - 56 = 14x - 56$

$0 = x^2 + 29x$

$0 = x(x + 29)$

$x = 0 \quad \text{or} \quad x + 29 = 0$

$x = 0 \quad \text{or} \quad x = -29$

The solution set is $\{-29, 0\}$.

37. $\frac{n+6}{27} = \frac{1}{n}; \; n \neq 0$

$n(n+6) = 1(27)$

$n^2 + n = 27$

$n^2 + 6n - 27 = 0$

$(n+9)(n-3) = 0$

$n + 9 = 0 \quad \text{or} \quad n - 3 = 0$

$n = -9 \quad \text{or} \quad n = 3$

The solution set is $\{-9, 3\}$.

39. $\frac{3n}{n-1} - \frac{1}{3} = \frac{-40}{3n-18}$

$\frac{3n}{n-1} - \frac{1}{3} = \frac{-40}{3(n-6)}; \; n \neq 1, \; n \neq 6$

$3(n-1)(n-6)\left(\frac{3n}{n-1} - \frac{1}{3}\right) = 3(n-1)(n-6)\left[\frac{-40}{3(n-6)}\right]$

$3(n-1)(n-6)\left(\frac{3n}{n-1}\right) - 3(n-1)(n-6)\left(\frac{1}{3}\right) = -40(n-1)$

$9n(n-6)-(n-1)(n-6)=-40n+40$

$9n^2-54n-(n^2-7n+6)=-40n+40$

$9n^2-54n-n^2+7n-6=-40n+40$

$8n^2-47n-6=-40n+40$

$8n^2-7n-46=0$

$(8n-23)(n+2)=0$

$8n-23=0 \quad \text{or} \quad n+2=0$

$8n=23 \quad \text{or} \quad n=-2$

$n=\frac{23}{8} \quad \text{or} \quad n=-2$

The solution set is $\left\{-2, \frac{23}{8}\right\}$.

41. $\frac{-3}{4x+5}=\frac{2}{5x-7}; \; x \neq -\frac{5}{4}, \; x \neq \frac{7}{5}$

$-3(5x-7)=2(4x+5)$

$-15x+21=8x+10$

$-15x=8x-11$

$-23x=-11$

$x=\frac{-11}{-23}=\frac{11}{23}$

The solution set is $\left\{\frac{11}{23}\right\}$.

43. $\frac{2x}{x-2}+\frac{15}{x^2-7x+10}=\frac{3}{x-5}$

$\frac{2x}{x-2}+\frac{15}{(x-5)(x-2)}=\frac{3}{x-5}; \; x \neq 2, \; x \neq 5$

$(x-5)(x-2)\left[\frac{2x}{x-2}+\frac{15}{(x-5)(x-2)}\right]=(x-5)(x-2)\left(\frac{3}{x-5}\right)$

$(x-5)(x-2)\left(\frac{2x}{x-2}\right)+(x-5)(x-2)\left[\frac{15}{(x-5)(x-2)}\right]=3(x-2)$

$2x(x-5)+15=3x-6$

$2x^2-10x+15=3x-6$

$2x^2-13x+21=0$

$(2x-7)(x-3)=0$

$2x-7=0 \quad \text{or} \quad x-3=0$

$2x=7 \quad \text{or} \quad x=3$

$x=\frac{7}{2} \quad \text{or} \quad x=3$

The solution set is $\left\{3, \frac{7}{2}\right\}$.

45. Let $d=$ first amount
$1750-d=$ second amount

$\frac{d}{1750-d}=\frac{3}{4};\ d\neq 1750$

$4d=3(1750-d)$

$4d=5250-3d$

$7d=5250$

$d=750$

The amounts are \$750 and \$1000.

47. Let one angle $=x$
other angle $=180-x-60=120-x$

$\frac{x}{120-x}=\frac{2}{3};\ x\neq 120$

$3x=2(120-x)$

$3x=240-2x$

$5x=240$

$x=48$

The angles are 48° and 72°.

49. Let $n=$ number

$n+\frac{1}{n}=\frac{53}{14};\ n\neq 0$

$14n\left(n+\frac{1}{n}\right)=14n\left(\frac{53}{14}\right)$

$14n(n)+14n\left(\frac{1}{n}\right)=53n$

$14n^2+14=53n$

$14n^2-53n+14=0$

$(2n-7)(7n-2)=0$

$2n-7=0$ or $7n-2=0$

$2n=7$ or $7n=2$

$n=\frac{7}{2}$ or $n=\frac{2}{7}$

The numbers are $\frac{7}{2}$ and $\frac{2}{7}$.

51. $\frac{50,000}{900}=\frac{60,000}{x};\ x\neq 0$

$50000x=60000(900)$

$50000x=54000000$

$x=1080$

The taxes are \$1080.

53. Let $x=$ Laura's sales
$120.75-x=$ Tammy's sales

$\frac{120.75-x}{x}=\frac{4}{3};\ x\neq 0$

$4x=3(120.75-x)$

$4x=362.25-3x$

$7x=362.25$

$x=51.75$

Laura's sales are \$51.75 and Tammy's sales are \$69.00.

55. Let $x=$ smaller number
$90-x=$ larger number

$\frac{90-x}{x}=10+\frac{2}{x};\ x\neq 0$

$x\left(\frac{90-x}{x}\right)=x\left(10+\frac{2}{x}\right)$

$90-x=x(10)+x\left(\frac{2}{x}\right)$

$90-x=10x+2$

$-x=10x-88$

$-11x=-88$

$x=8$

The numbers are 8 and 82.

57. Let $x=$ first piece
$20-x=$ second piece

$\frac{x}{20-x}=\frac{7}{3};\ x\neq 20$

$3x=7(20-x)$

$3x=140-7x$

$10x=140$

$x=14$

The lengths are 14 feet and 6 feet.

59. Let $x=$ female voters
$1150-x=$ male voters

$\frac{x}{1150-x}=\frac{3}{2};\ x\neq 1150$

$2x=3(1150-x)$

$2x=3450-3x$

$5x=3450$

$x=690$

They were 690 female voters
and 460 male voters.

Problem Set 4.7 More Fractional Equations and Applications

1. $\frac{x}{4x-4}+\frac{5}{x^2-1}=\frac{1}{4}$

$\frac{x}{4(x-1)}+\frac{5}{(x+1)(x-1)}=\frac{1}{4}$; $x\neq 1$, $x\neq -1$

$4(x-1)(x+1)\left[\frac{x}{4(x-1)}+\frac{5}{(x+1)(x-1)}\right]=4(x+1)(x-1)\left(\frac{1}{4}\right)$

$4(x-1)(x+1)\left[\frac{x}{4(x-1)}\right]+4(x-1)(x+1)\left[\frac{5}{(x+1)(x-1)}\right]=(x+1)(x-1)$

$x(x+1)+4(5)=x^2-1$

$x^2+x+20=x^2-1$

$x+20=-1$

$x=-21$

The solution set is $\{-21\}$.

3. $3+\frac{6}{t-3}=\frac{6}{t^2-3t}$

$3+\frac{6}{t-3}=\frac{6}{t(t-3)}$; $t\neq 0$, $t\neq 3$

$t(t-3)\left(3+\frac{6}{t-3}\right)=t(t-3)\left[\frac{6}{t(t-3)}\right]$

$3t(t-3)+6t=6$

$3t^2-9t+6t=6$

$3t^2-3t=6$

$3t^2-3t-6=0$

$3(t^2-t-2)=0$

$3(t-2)(t+1)=0$

$t-2=0$ or $t+1=0$

$t=2$ or $t=-1$

The solution set is $\{-1, 2\}$.

5. $\frac{3}{n-5}+\frac{4}{n+7}=\frac{2n+11}{n^2+2n-35}$

$\frac{3}{n-5}+\frac{4}{n+7}=\frac{2n+11}{(n+7)(n-5)}$; $n\neq -7$, $n\neq 5$

$(n+7)(n-5)\left(\frac{3}{n-5}+\frac{4}{n+7}\right)=(n+7)(n-5)\left[\frac{2n+11}{(n+7)(n-5)}\right]$

$(n+7)(n-5)\left(\frac{3}{n-5}\right)+(n+7)(n-5)\left(\frac{4}{n+7}\right)=2n+11$

$3(n+7)+4(n-5)=2n+11$

$3n+21+4n-20=2n+11$

$7n+1=2n+11$

$7n=2n+10$

$5n=10$

$n=2$

The solution set is $\{2\}$.

7. $\frac{5x}{2x+6}-\frac{4}{x^2-9}=\frac{5}{2}$

$\frac{5x}{2(x+3)}-\frac{4}{(x+3)(x-3)}=\frac{5}{2};\ x\neq 3,\ x\neq -3$

$2(x+3)(x-3)\left[\frac{5x}{2(x+3)}-\frac{4}{(x+3)(x-3)}\right]=2(x+3)(x-3)\left(\frac{5}{2}\right)$

$2(x+3)(x-3)\left[\frac{5x}{2(x+3)}\right]-2(x+3)(x-3)\left[\frac{4}{(x+3)(x-3)}\right]=5(x+3)(x-3)$

$5x(x-3)-8=5(x^2-9)$

$5x^2-15x-8=5x^2-45$

$-15x-8=-45$

$-15x=-37$

$x=\frac{37}{15}$

The solution set is $\left\{\frac{37}{15}\right\}$.

9. $1+\frac{1}{n-1}=\frac{1}{n^2-n}$

$1+\frac{1}{n-1}=\frac{1}{n(n-1)};\ n\neq 0,\ n\neq 1$

$n(n-1)\left(1+\frac{1}{n-1}\right)=n(n-1)\left[\frac{1}{n(n-1)}\right]$

$n(n-1)+n(n-1)\left(\frac{1}{n-1}\right)=1$

$n^2-n+n=1$

$n^2=1$

$n^2-1=0$

$(n+1)(n-1)=0$

$n+1=0$ or $n-1=0$

$n=-1$ or $n=1$

The solution set is $\{-1\}$.

11. $\frac{2}{n-2}-\frac{n}{n+5}=\frac{10n+15}{n^2+3n-10}$

$\frac{2}{n-2}-\frac{n}{n+5}=\frac{10n+15}{(n+5)(n-2)}$; $x\neq -5,\ n\neq 2$

$(n+5)(n-2)\left(\frac{2}{n-2}-\frac{n}{n+5}\right)=(n+5)(n-2)\left[\frac{10n+15}{(n+5)(n-2)}\right]$

$(n+5)(n-2)\left(\frac{2}{n-2}\right)-(n+5)(n-2)\left(\frac{n}{n+5}\right)=10n+15$

$2(n+5)-n(n-2)=10n+15$

$2n+10-n^2+2n=10n+15$

$-n^2+4n+10=10n+15$

$0=n^2+6n+5$

$0=(n+5)(n+1)$

$n+5=0$ or $n+1=0$

$n=-5$ or $n=-1$

The solution set is $\{-1\}$.

13. $\frac{2}{2x-3}-\frac{2}{10x^2-13x-3}=\frac{x}{5x+1}$

$\frac{2}{2x-3}-\frac{2}{(5x+1)(2x-3)}=\frac{x}{5x+1}$; $x\neq\frac{3}{2},\ x\neq -\frac{1}{5}$

$(2x-3)(5x+1)\left[\frac{2}{2x-3}-\frac{2}{(5x+1)(2x-3)}\right]=(2x-3)(5x+1)\left(\frac{x}{5x+1}\right)$

$(2x-3)(5x+1)\left(\frac{2}{2x-3}\right)-(2x+3)(5x+1)\left[\frac{2}{(5x+1)(2x-3)}\right]=x(2x-3)$

$2(5x+1)-2=2x^2-3x$

$10x+2-2=2x^2-3x$

$0=2x^2-13x$

$0=x(2x-13)$

$x=0$ or $2x-13=0$

$x=0$ or $2x=13$

$x=0$ or $x=\frac{13}{2}$

The solution set is $\left\{0,\ \frac{13}{2}\right\}$.

15. $\frac{2x}{x+3}-\frac{3}{x-6}=\frac{29}{x^2-3x-18}$

$\frac{2x}{x+3}-\frac{3}{x-6}=\frac{29}{(x-6)(x+3)}$; $x\neq 6,\ x\neq -3$

$(x-6)(x+3)\left(\frac{2x}{x+3}-\frac{3}{x-6}\right)=(x-6)(x+3)\left[\frac{29}{(x-6)(x+3)}\right]$

$(x-6)(x+3)\left(\frac{2x}{x+3}\right)-(x-6)(x+3)\left(\frac{3}{x-6}\right)=29$

$2x(x-6)-3(x+3)=29$

$2x^2-12x-3x-9=29$

$2x^2-15x-38=0$

$(2x-19)(x+2)=0$

$2x-19=0$ or $x+2=0$

$2x=19$ or $x=-2$

$x=\frac{19}{2}$ or $x=-2$

The solution set is $\left\{-2,\ \frac{19}{2}\right\}$.

17. $\frac{a}{a-5}+\frac{2}{a-6}=\frac{2}{a^2-11a+30}$

$\frac{a}{a-5}+\frac{2}{a-6}=\frac{2}{(a-6)(a-5)}$; $a\neq 6,\ a\neq 5$

$(a-6)(a-5)\left(\frac{a}{a-5}+\frac{2}{a-6}\right)=(a-6)(a-5)\left[\frac{2}{(a-6)(a-5)}\right]$

$(a-6)(a-5)\left(\frac{a}{a-5}\right)+(a-6)(a-5)\left(\frac{2}{a-6}\right)=2$

$a(a-6)+2(a-5)=2$

$a^2-6a+2a-10=2$

$a^2-4a-12=0$

$(a-6)(a+2)=0$

$a-6=0$ or $a+2=0$

$a=6$ or $a=-2$

The solution set is $\{-2\}$.

19. $\frac{-1}{2x-5}+\frac{2x-4}{4x^2-25}=\frac{5}{6x+15}$

$\frac{-1}{2x-5}+\frac{2x-4}{(2x+5)(2x-5)}=\frac{5}{3(2x+5)}$; $x\neq\frac{5}{2},\ x\neq-\frac{5}{2}$

$3(2x+5)(2x-5)\left[\frac{-1}{2x-5}+\frac{2x-4}{(2x+5)(2x-5)}\right]=3(2x+5)(2x-5)\left[\frac{5}{3(2x+5)}\right]$

$3(2x+5)(2x-5)\left(\frac{-1}{2x-5}\right)+3(2x+5)(2x-5)\left[\frac{2x-4}{(2x+5)(2x-5)}\right]=5(2x-5)$

$-3(2x+5)+3(2x-4)=10x-25$

$-6x-15+6x-12=10x-25$

$-27=10x-25$

$-2=10x$

$-\frac{2}{10}=x$

$-\frac{1}{5}=x$

The solution set is $\left\{-\frac{1}{5}\right\}$.

21. $\frac{7y+2}{12y^2+11y-15}-\frac{1}{3y+5}=\frac{2}{4y-3}$

$\frac{7y+2}{(3y+5)(4y-3)}-\frac{1}{3y+5}=\frac{2}{4y-3};\ y\neq -\frac{5}{3},\ y\neq \frac{3}{4}$

$(3y+5)(4y-3)\left[\frac{7y+2}{(3y+5)(4y-3)}-\frac{1}{3y+5}\right]=(3y+5)(4y-3)\left(\frac{2}{4y-3}\right)$

$(3y+5)(4y-3)\left[\frac{7y+2}{(3y+5)(4y-3)}\right]-(3y+5)(4y-3)\left(\frac{1}{3y+5}\right)=2(3y+5)$

$7y+2-(4y-3)=2(3y+5)$

$7y+2-4y+3=6y+10$

$3y+5=6y+10$

$3y=6y+5$

$-3y=5$

$y=-\frac{5}{3}$

The solution set is $\emptyset$.

23. $\frac{2n}{6n^2+7n-3}-\frac{n-3}{3n^2+11n-4}=\frac{5}{2n^2+11n+12}$

$\frac{2n}{(3n-1)(2n+3)}-\frac{n-3}{(3n-1)(n+4)}=\frac{5}{(2n+3)(n+4)};\ n\neq \frac{1}{3},\ n\neq -\frac{3}{2},\ n\neq -4$

$(3n-1)(2n+3)(n+4)\left[\frac{2n}{(3n-1)(2n+3)}-\frac{n-3}{(3n-1)(n+4)}\right]=(3n-1)(2n+3)(n+4)\left[\frac{5}{(2n+3)(n+4)}\right]$

$(3n-1)(2n+3)(n+4)\left[\frac{2n}{(3n-1)(2n+3)}\right]-(3n-1)(2n+3)(n+4)\left[\frac{n-3}{(3n-1)(n+4)}\right]$

$=(3n-1)(2n+3)(n+4)\left[\frac{5}{(2n+3)(n+4)}\right]$

$2n(n+4)-(2n+3)(n-3)=5(3n-1)$

$2n^2+8n-(2n^2-3n-9)=15n-5$

$2n^2+8n-2n^2+3n+9=15n-5$

$11n+9=15n-5$

$11n=15n-14$

$-4n=-14$

$n=\frac{-14}{-4}=\frac{7}{2}$

The solution set is $\left\{\frac{7}{2}\right\}$.

25. $\frac{1}{2x^2-x-1}+\frac{3}{2x^2+x}=\frac{2}{x^2-1}$

$\frac{1}{(2x+1)(x-1)}+\frac{3}{x(2x+1)}=\frac{2}{(x+1)(x-1)}$; $x\neq-\frac{1}{2}$, $x\neq1$, $x\neq-1$, $x\neq0$

$x(2x+1)(x+1)(x-1)\left[\frac{1}{(2x+1)(x-1)}+\frac{3}{x(2x+1)}\right]=x(2x+1)(x-1)(x+1)\left[\frac{2}{(x+1)(x-1)}\right]$

$x(2x+1)(x+1)(x-1)\left[\frac{1}{(2x+1)(x-1)}\right]+x(2x+1)(x+1)(x-1)\left[\frac{3}{x(2x+1)}\right]=2x(2x+1)$

$x(x+1)+3(x+1)(x-1)=2x(2x+1)$

$x^2+x+3(x^2-1)=4x^2+2x$

$x^2+x+3x^2-3=4x^2+2x$

$4x^2+x-3=4x^2+2x$

$x-3=2x$

$-3=x$

The solution set is $\{-3\}$.

27. $\frac{x+1}{x^3-9x}-\frac{1}{2x^2+x-21}=\frac{1}{2x^2+13x+21}$

$\frac{x+1}{x(x+3)(x-3)}-\frac{1}{(2x+7)(x-3)}=\frac{1}{(2x+7)(x+3)}$; $x\neq0$, $x\neq3$, $x\neq-3$, $x\neq-\frac{7}{2}$

$x(2x+7)(x+3)(x-3)\left[\frac{x+1}{x(x+3)(x-3)}-\frac{1}{(2x+7)(x-3)}\right]=x(2x+7)(x+3)(x-3)\left[\frac{1}{(2x+7)(x+3)}\right]$

$x(2x+7)(x+3)(x-3)\left[\frac{x+1}{x(x+3)(x-3)}\right]-x(2x+7)(x+3)(x-3)\left[\frac{1}{(2x+7)(x-3)}\right]=x(x-3)$

$(2x+7)(x+1)-x(x+3)=x(x-3)$

$2x^2+9x+7-x^2-3x=x^2-3x$

$x^2+6x+7=x^2-3x$

$6x+7=-3x$

$7=-9x$

$-\frac{7}{9}=x$

The solution set is $\left\{-\frac{7}{9}\right\}$.

29. $\frac{4t}{4t^2-t-3}+\frac{2-3t}{3t^2-t-2}=\frac{1}{12t^2+17t+6}$

$\frac{4t}{(4t+3)(t-1)}+\frac{2-3t}{(3t+2)(t-1)}=\frac{1}{(4t+3)(3t+2)}$; $t\neq-\frac{3}{4}$, $t\neq1$, $t\neq-\frac{2}{3}$

$(4t+3)(t-1)(3t+2)\left[\frac{4t}{(4t+3)(t-1)}+\frac{2-3t}{(3t+2)(t-1)}\right]=(4t+3)(t-1)(3t+2)\left[\frac{1}{(4t+3)(3t+2)}\right]$

$(4t+3)(t-1)(3t+2)\left[\frac{4t}{(4t+3)(t-1)}\right]+(4t+3)(t-1)(3t+2)\left[\frac{2-3t}{(3t+2)(t-1)}\right]$

$=(4t+3)(t-1)(3t+2)\left[\frac{1}{(4t+3)(3t+2)}\right]$

$4t(3t+2)+(4t+3)(2-3t)=t-1$

$12t^2+8t-12t^2-t+6=t-1$

$7t+6=t-1$

$7t=t-7$

$6t=-7$

$t=-\frac{7}{6}$

The solution set is $\left\{-\frac{7}{6}\right\}$.

31. $y=\frac{5}{6}x+\frac{2}{9}$

$18(y)=18\left(\frac{5}{6}x+\frac{2}{9}\right)$

$18y=18\left(\frac{5}{6}x\right)+18\left(\frac{2}{9}\right)$

$18y=15x+4$

$18y-4=15x$

$\frac{1}{15}(18y-4)=\frac{1}{15}(15x)$

$\frac{18y-4}{15}=x$

33. $\frac{-2}{x-4}=\frac{5}{y-1}$

$-2(y-1)=5(x-4)$

$-2y+2=5x-20$

$-2y=5x-22$

$-\frac{1}{2}(-2y)=-\frac{1}{2}(5x-22)$

$y=\frac{-5x+22}{2}$

35. $I=\frac{100M}{C}$

$\frac{CI}{100}=M$

$M=\frac{CI}{100}$

37. $\frac{R}{S}=\frac{T}{S+T}$

$S\left(\frac{R}{S}\right)=S\left(\frac{T}{S+T}\right)$

$R=\frac{ST}{S+T}$

39. $\frac{y-1}{x-3}=\frac{b-1}{a-3}$

$(y-1)(a-3)=(x-3)(b-1)$

$y(a-3)-1(a-3)=x(b-1)-3(b-1)$

$y(a-3)-a+3=bx-x-3b+3$

$y(a-3)=bx-x-3b+3+a-3$

$y(a-3)=bx-x-3b+a$

$y=\frac{bx-x-3b+a}{a-3}$

41. $\frac{x}{a}+\frac{y}{b}=1$

$ab\left(\frac{x}{a}+\frac{y}{b}\right)=ab(1)$

$ab\left(\frac{x}{a}\right)+ab\left(\frac{y}{b}\right)=ab$

$bx+ay=ab$

$ay=ab-bx$

$y=\frac{ab-bx}{a}$

43. $\frac{y-1}{x+6}=\frac{-2}{3}$

$3(y-1)=-2(x+6)$

$3y-3=-2x-12$

$3y=-2x-9$

$y=\frac{-2x-9}{3}$

45.

	rate	• time	= distance
Kent	$x+4$	$\frac{270}{x+4}$	270
Dave	x	$\frac{250}{x}$	250

$\frac{270}{x+4}=\frac{250}{x}$

$270x=250(x+4)$

$270x=250x+1000$

$20x=1000$

$x=50$

Dave's rate is 50 mph and Kent's rate is 54 mph.

47.

	Time in minutes	Rate
Inlet	10	$\frac{1}{10}$
Outlet	12	$\frac{1}{12}$
Together	x	$\frac{1}{x}$

$$\frac{1}{10}-\frac{1}{12}=\frac{1}{x}$$

$$60x\left(\frac{1}{10}-\frac{1}{12}\right)=60x\left(\frac{1}{x}\right)$$

$$60x\left(\frac{1}{10}\right)-60x\left(\frac{1}{12}\right)=60$$

$$6x-5x=60$$

$$x=60$$

The tank will overflow in 60 minutes.

49.

	rate	• time	= words
Connie	$x+20$	$\frac{600}{x+20}$	600
Katie	x	$\frac{600}{x}$	600

$$\frac{600}{x+20}+5=\frac{600}{x}$$

$$x(x+20)\left(\frac{600}{x+20}+5\right)=x(x+20)\left(\frac{600}{x}\right)$$

$$x(x+20)\left(\frac{600}{x+20}\right)+5x(x+20)=600(x+20)$$

$$600x+5x^2+100x=600x+12000$$

$$5x^2+700x=600x+12000$$

$$5x^2+100x-12000=0$$

$$5(x^2+20x-2400)=0$$

$$5(x+60)(x-40)=0$$

$$x+60=0 \quad \text{or} \quad x-40$$

$$x=-60 \quad \text{or} \quad x=40$$

Discard the root $x=-60$. Katie's rate is 40 words per minute and Connie's rate is 60 words per minute.

51.

	rate	• time	= distance
Plane A	x	$\frac{1400}{x}$	1400
Plane B	$x+50$	$\frac{2000}{x+50}$	2000

$$\frac{1400}{x}+1=\frac{2000}{x+50}$$

$$x(x+50)\left(\frac{1400}{x}+1\right)=x(x+50)\left(\frac{2000}{x+50}\right)$$

$$x(x+50)\left(\frac{1400}{x}\right)+x(x+50)=2000x$$

$$1400(x+50)+x^2+50x=2000x$$

$$1400x+70000+x^2+50x=2000x$$

$$x^2+1450x+70000=2000x$$

$$x^2-550x+70000=0$$

$$(x-350)(x-200)=0$$

$$x-350=0 \quad \text{or} \quad x-200=0$$

$$x=350 \quad \text{or} \quad x=200$$

Case 1: Plane A could travel at 350mph for 4 hours while Plane B travels at 400 mph for 5 hours.

Case 2: Plane A could travel at 200 mph for 7 hours while Plane B travels at 250 mph for 8 hours.

53.

	Time in minutes	Rate
Amy	2x	$\frac{1}{2x}$
Nancy	x	$\frac{1}{x}$
Together	40	$\frac{1}{40}$

$$\frac{1}{2x}+\frac{1}{x}=\frac{1}{40}$$

$$40x\left(\frac{1}{2x}+\frac{1}{x}\right)=40x\left(\frac{1}{40}\right)$$

$$40x\left(\frac{1}{2x}\right)+40x\left(\frac{1}{x}\right)=x$$

$$20+40=x$$

$$60=x$$

It would take Nancy 60 minutes and Amy would take 120 minutes.

55.

	rate •	time =	money
Anticipated	$\frac{12}{x}$	x	12
Actual	$\frac{12}{x+1}$	x+1	12

$\frac{12}{x} - 1 = \frac{12}{x+1}$

$x(x+1)\left(\frac{12}{x} - 1\right) = x(x+1)\left(\frac{12}{x+1}\right)$

$x(x+1)\left(\frac{12}{x}\right) - x(x+1) = 12x$

$12(x+1) - x^2 - x = 12x$

$12x + 12 - x^2 - x = 12x$

$-x^2 + 11x + 12 = 12x$

$0 = x^2 + x - 12$

$(x+4)(x-3) = 0$

$x+4 = 0$ or $x-3 = 0$

$x = -4$ or $x = 3$

Discard the root $x = -4$.
The anticipated time was 3 hours.

57.

	rate •	time =	distance
Out	x+4	$\frac{24}{x+4}$	24
Back	x	$\frac{12}{x}$	12

$\frac{24}{x+4} - \frac{1}{2} = \frac{12}{x}$

$2x(x+4)\left(\frac{24}{x+4} - \frac{1}{2}\right) = 2x(x+4)\left(\frac{12}{x}\right)$

$2x(x+4)\left(\frac{24}{x+4}\right) - 2x(x+4)\left(\frac{1}{2}\right) = 24(x+4)$

$2x(24) - x(x+4) = 24x + 96$

$48x - x^2 - 4x = 24x + 96$

$-x^2 + 44x = 24x + 96$

$0 = x^2 - 20x + 96$

$0 = (x-12)(x-8)$

$x - 12 = 0$ or $x - 8 = 0$

$x = 12$ or $x = 8$

Case 1: 12 mph on the way back and 16 mph on the way out.
Case 2: 8 mph on the way back and 12 mph on the way out.

Chapter 4 Review

1. $\frac{26x^2y^3}{39x^4y^2} = \frac{2 \bullet 13x^2y^3}{3 \bullet 13x^4y^2} = \frac{2y}{3x^2}$

3. $\frac{n^2 - 3n - 10}{n^2 + n - 2} = \frac{(n-5)(n+2)}{(n-1)(n+2)} = \frac{n-5}{n-1}$

5. $\frac{8x^3 - 2x^2 - 3x}{12x^2 - 9x} = \frac{x(8x^2 - 2x - 3)}{3x(4x-3)}$

$\frac{x(4x-3)(2x+1)}{3x(4x-3)} = \frac{2x+1}{3}$

7. $\frac{\frac{5}{8} - \frac{1}{2}}{\frac{1}{6} + \frac{3}{4}} = \frac{24}{24} \bullet \frac{\left(\frac{5}{8} - \frac{1}{2}\right)}{\left(\frac{1}{6} + \frac{3}{4}\right)}$

$\frac{24\left(\frac{5}{8}\right) - 24\left(\frac{1}{2}\right)}{24\left(\frac{1}{6}\right) + 24\left(\frac{3}{4}\right)} = \frac{15 - 12}{4 + 18} = \frac{3}{22}$

9. $\frac{\frac{3}{x-2} - \frac{4}{x^2-4}}{\frac{2}{x+2} + \frac{1}{x-2}}$

$\frac{\frac{3}{x-2} - \frac{4}{(x+2)(x-2)}}{\frac{2}{x+2} + \frac{1}{x-2}}$

$\frac{(x+2)(x-2)}{(x+2)(x-2)} \bullet \frac{\left[\frac{3}{x-2} - \frac{4}{(x+2)(x-2)}\right]}{\left(\frac{2}{x+2} + \frac{1}{x-2}\right)}$

$\frac{(x+2)(x-2)\left(\frac{3}{x-2}\right) - (x+2)(x-2)\left[\frac{4}{(x+2)(x-2)}\right]}{(x+2)(x-2)\left(\frac{2}{x+2}\right) + (x+2)(x-2)\left(\frac{1}{x-2}\right)}$

$\frac{3(x+2) - 4}{2(x-2) + 1(x+2)}$

$\frac{3x + 6 - 4}{2x - 4 + x + 2}$

$\frac{3x+2}{3x-2}$

11. $\frac{6xy^2}{7y^3} \div \frac{15x^2y}{5x^2}$

$\frac{6xy^2}{7y^3} \bullet \frac{5x^2}{15x^2y}$

$\frac{6(5)x^3y^2}{7(15)x^2y^4}$

$\frac{2x}{7y^2}$

13. $\frac{n^2+10n+25}{n^2-n} \bullet \frac{5n^3-3n^2}{5n^2+22n-15}$

$\frac{(n+5)(n+5)}{n(n-1)} \bullet \frac{n^2(5n-3)}{(5n-3)(n+5)}$

$\frac{n(n+5)}{n-1}$

15. $\frac{2x+1}{5} + \frac{3x-2}{4}$; LCD $= 20$

$\frac{(2x+1)}{5} \bullet \frac{4}{4} + \frac{(3x-2)}{4} \bullet \frac{5}{5}$

$\frac{4(2x+1)}{20} + \frac{5(3x-2)}{20}$

$\frac{4(2x+1)+5(3x-2)}{20}$

$\frac{8x+4+15x-10}{20}$

$\frac{23x-6}{20}$

17. $\frac{3x}{x+7} - \frac{2}{x}$; LCD $= x(x+7)$

$\frac{3x}{(x+7)} \bullet \frac{x}{x} - \frac{2}{x} \bullet \frac{(x+7)}{(x+7)}$

$\frac{3x^2}{x(x+7)} - \frac{2(x+7)}{x(x+7)}$

$\frac{3x^2-2(x+7)}{x(x+7)}$

$\frac{3x^2-2x-14}{x(x+7)}$

19. $\frac{3}{n^2-5n-36} + \frac{2}{n^2+3n-4}$

$\frac{3}{(n-9)(n+4)} + \frac{2}{(n+4)(n-1)}$

LCD $= (n-9)(n+4)(n-1)$

$\frac{3}{(n-9)(n+4)} \bullet \frac{(n-1)}{(n-1)} + \frac{2}{(n+4)(n-1)} \bullet \frac{(n-9)}{(n-9)}$

$\frac{3(n-1)}{(n-9)(n+4)(n-1)} + \frac{2(n-9)}{(n-9)(n+4)(n-1)}$

$\frac{3(n-1)+2(n-9)}{(n-9)(n+4)(n-1)}$

$\frac{3n-3+2n-18}{(n-9)(n+4)(n-1)}$

$\frac{5n-21}{(n-9)(n+4)(n-1)}$

21.
$$\begin{array}{r|l} & \quad\quad\quad 6x-1 \\ 3x+2 & 18x^2+\ 9x-2 \\ & \underline{18x^2+12x} \\ & \quad\quad\ -3x-2 \\ & \quad\quad\ \underline{-3x-2} \\ & \quad\quad\quad\quad 0 \end{array}$$

$6x-1$

23. $\frac{4x+5}{3} + \frac{2x-1}{5} = 2$

$15\left(\frac{4x+5}{3} + \frac{2x-1}{5}\right) = 15(2)$

$15\left(\frac{4x+5}{3}\right) + 15\left(\frac{2x-1}{5}\right) = 30$

$5(4x+5)+3(2x-1) = 30$

$20x+25+6x-3 = 30$

$26x+22 = 30$

$26x = 8$

$x = \frac{8}{26} = \frac{4}{13}$

The solution set is $\left\{\frac{4}{13}\right\}$.

25. $\frac{a}{a-2} - \frac{3}{2} = \frac{2}{a-2}$; $a \neq 2$

$2(a-2)\left(\frac{a}{a-2} - \frac{3}{2}\right) = 2(a-2)\left(\frac{2}{a-2}\right)$

$2(a-2)\left(\frac{a}{a-2}\right) - 2(a-2)\left(\frac{3}{2}\right) = 4$

$2a-3(a-2) = 4$

$2a-3a+6 = 4$

$-a+6 = 4$

$-a = -2$

$a = 2$

The solution set is $\emptyset$.

27. $n+\frac{1}{n}=\frac{53}{14};\ n\neq 0$

$14n\left(n+\frac{1}{n}\right)=14n\left(\frac{53}{14}\right)$

$14n(n)+14n\left(\frac{1}{n}\right)=53n$

$14n^2+14=53n$

$14n^2-53n+14=0$

$(2n-7)(7n-2)=0$

$2n-7=0$ or $7n-2=0$

$2n=7$ or $7n=2$

$n=\frac{7}{2}$ or $n=\frac{2}{7}$

The solution set is $\left\{\frac{2}{7},\ \frac{7}{2}\right\}$.

29. $\frac{x}{2x+1}-1=\frac{-4}{7(x-2)};\ x\neq -\frac{1}{2},\ x\neq 2$

$7(2x+1)(x-2)\left(\frac{x}{2x+1}-1\right)=7(2x+1)(x-2)\left[\frac{-4}{7(x-2)}\right]$

$7(2x+1)(x-2)\left(\frac{x}{2x+1}\right)-1(7)(2x+1)(x-2)=-4(2x+1)$

$7x(x-2)-7(2x^2-3x-2)=-8x-4$

$7x^2-14x-14x^2+21x+14=-8x-4$

$-7x^2+7x+14=-8x-4$

$0=7x^2-15x-18$

$0=(7x+6)(x-3)$

$7x+6=0$ or $x-3=0$

$7x=-6$ or $x=3$

$x=-\frac{6}{7}$ or $x=3$

The solution set is $\left\{-\frac{6}{7},\ 3\right\}$.

31. $\frac{2n}{2n^2+11n-21}-\frac{n}{n^2+5n-14}=\frac{3}{n^2+5n-14}$

$\frac{2n}{(2n-3)(n+7)}-\frac{n}{(n+7)(n-2)}=\frac{3}{(n+7)(n-2)};\ n\neq\frac{3}{2},\ n\neq -7,\ n\neq 2$

$(2n-3)(n+7)(n-2)\left[\frac{2n}{(2n-3)(n+7)}-\frac{n}{(n+7)(n-2)}\right]=(2n-3)(n+7)(n-2)\left[\frac{3}{(n+7)(n-2)}\right]$

$(2n-3)(n+7)(n-2)\left[\frac{2n}{(2n-3)(n+7)}\right]-(2n-3)(n+7)(n-2)\left[\frac{n}{(n+7)(n-2)}\right]=3(2n-3)$

$2n(n-2)-n(2n-3)=3(2n-3)$

$2n^2-4n-2n^2+3n=6n-9$

$-n=6n-9$

$-7n = -9$

$n = \frac{-9}{-7} = \frac{9}{7}$

The solution set is $\left\{\frac{9}{7}\right\}$.

33. $\frac{y-6}{x+1} = \frac{3}{4}$

$4(y-6) = 3(x+1)$

$4y - 24 = 3x + 3$

$4y = 3x + 27$

$y = \frac{3x+27}{4}$

35. Let $x =$ one part
$1400 - x =$ other part

$\frac{x}{1400-x} = \frac{3}{5}$

$5x = 3(1400 - x)$

$5x = 4200 - 3x$

$8x = 4200$

$x = 525$

One part is $525 and the other part is $875.

37.

	rate	• time	= distance
Car A	x	$\frac{250}{x}$	250
Car B	$x+5$	$\frac{440}{x+5}$	440

$\frac{250}{x} + 3 = \frac{440}{x+5}$

$x(x+5)\left(\frac{250}{x} + 3\right) = x(x+5)\left(\frac{440}{x+5}\right)$

$x(x+5)\left(\frac{250}{x}\right) + 3x(x+5) = 440x$

$250(x+5) + 3x^2 + 15x = 440x$

$250x + 1250 + 3x^2 + 15x = 440x$

$3x^2 - 175x + 1250 = 0$

$(3x - 25)(x - 50) = 0$

$3x - 25 = 0$ or $x - 50 = 0$

$3x = 25$ or $x = 50$

$x = \frac{25}{3}$ or $x = 50$

Case 1: Car A would travel at $\frac{25}{3} = 8\frac{1}{3}$ mph and Car B would travel at $13\frac{1}{3}$ mph.

Case 2: Car A would travel at 50 mph and Car B would travel at 55 mph.

39.

	rate	• time	= Pay
Anticipated	$\frac{640}{x}$	x	640
Actual	$\frac{640}{x+20}$	$x+20$	640

$\frac{640}{x} - 1.60 = \frac{640}{x+20}$

$10x(x+20)\left(\frac{640}{x} - 1.6\right) = 10x(x+20)\left(\frac{640}{x+20}\right)$

$10x(x+20)\left(\frac{640}{x}\right) - 10x(x+20)(1.6) = 10x(640)$

$6400(x+20) - 16x(x+20) = 6400x$

$6400x + 128000 - 16x^2 - 320x = 6400x$

$-16x^2 + 6080x + 128000 = 6400x$

$-16x^2 - 320x + 128000 = 0$

$16x^2 + 320x - 128000 = 0$

$16(x^2 + 20x - 8000) = 0$

$16(x - 80)(x + 100) = 0$

$x - 80 = 0$ or $x + 100 = 0$

$x = 80$ or $x = -100$

Discard the root $x = -100$.
The anticipated would take 80 hours.

Chapter 4 Test

1. $\frac{39x^2y^3}{72x^3y} = \frac{3(13)x^2y^3}{3(24)x^3y} = \frac{13y^2}{24x}$

3. $\frac{6n^2 - 5n - 6}{3n^2 + 14n + 8} = \frac{(2n-3)(3n+2)}{(3n+2)(n+4)} = \frac{2n-3}{n+4}$

5. $\frac{5x^2y}{8x} \bullet \frac{12y^2}{20xy} = \frac{5(12)x^2y^3}{8(20)x^2y} = \frac{3y^2}{8}$

7. $\frac{3x^2+10x-8}{5x^2+19x-4} \div \frac{3x^2-23x+14}{x^2-3x-28}$

$\frac{3x^2+10x-8}{5x^2+19x-4} \bullet \frac{x^2-3x-28}{3x^2-23x+14}$

$\frac{(3x-2)(x+4)}{(5x-1)(x+4)} \bullet \frac{(x-7)(x+4)}{(3x-2)(x-7)}$

$\frac{x+4}{5x-1}$

9. $\frac{5x-6}{3} - \frac{x-12}{6}$; LCD = 6

$\frac{(5x-6)}{3} \bullet \frac{2}{2} - \frac{x-12}{6}$

$\frac{2(5x-6)}{6} - \frac{x-12}{6}$

$\frac{2(5x-6)-(x-12)}{6}$

$\frac{10x-12-x+12}{6}$

$\frac{9x}{6} = \frac{3x}{2}$

11. $\frac{3x}{x-6} + \frac{2}{x}$; LCD = $x(x-6)$

$\frac{3x}{(x-6)} \bullet \frac{x}{x} + \frac{2}{x} \bullet \frac{(x-6)}{(x-6)}$

$\frac{3x^2}{x(x-6)} + \frac{2(x-6)}{x(x-6)}$

$\frac{3x^2+2(x-6)}{x(x-6)}$

$\frac{3x^2+2x-12}{x(x-6)}$

13. $\frac{3}{2n^2+n+10} + \frac{5}{n^2+5n-14}$

$\frac{3}{(2n+5)(n-2)} + \frac{5}{(n+7)(n-2)}$

LCD = $(2n+5)(n-2)(n+7)$

$\frac{3}{(2n+5)(n-2)} \bullet \frac{(n+7)}{(n+7)} + \frac{5}{(n+7)(n-2)} \bullet \frac{(2n+5)}{(2n+5)}$

$\frac{3(n+7)}{(2n+5)(n-2)(n+7)} + \frac{5(2n+5)}{(n+7)(n+2)(2n+5)}$

$\frac{3(n+7)+5(2n+5)}{(2n+5)(n-2)(n+7)}$

$\frac{3n+21+10n+25}{(2n+5)(n-2)(n+7)}$

$\frac{13n+46}{(2n+5)(n-2)(n+7)}$

15. $\frac{\frac{3}{2x} - \frac{1}{6}}{\frac{2}{3x} + \frac{3}{4}}$

$\frac{12x}{12x} \bullet \frac{\left(\frac{3}{2x} - \frac{1}{6}\right)}{\left(\frac{2}{3x} + \frac{3}{4}\right)}$

$\frac{12x\left(\frac{3}{2x}\right) - 12x\left(\frac{1}{6}\right)}{12x\left(\frac{2}{3x}\right) + 12x\left(\frac{3}{4}\right)}$

$\frac{18-2x}{8+9x}$

17. $\frac{x-1}{2} - \frac{x+2}{5} = -\frac{3}{5}$

$10\left(\frac{x-1}{2} - \frac{x+2}{5}\right) = 10\left(-\frac{3}{5}\right)$

$10\left(\frac{x-1}{2}\right) - 10\left(\frac{x+2}{5}\right) = -6$

$5(x-1) - 2(x+2) = -6$

$5x - 5 - 2x - 4 = -6$

$3x - 9 = -6$

$3x = 3$

$x = 1$

The solution set is $\{1\}$.

19. $\frac{-3}{4n-1} = \frac{-2}{3n+11}$; $n \neq \frac{1}{4}$, $n \neq -\frac{11}{3}$

$-3(3n+11) = -2(4n-1)$

$-9n - 33 = -8n + 2$

$-9n = -8n + 35$

$-n = 35$

$n = -35$

The solution set is $\{-35\}$.

21. $\frac{6}{x-4} - \frac{4}{x+3} = \frac{8}{x-4}$; $x \neq 4$, $x \neq -3$

$(x-4)(x+3)\left(\frac{6}{x-4} - \frac{4}{x+3}\right) = (x-4)(x+3)\left(\frac{8}{x-4}\right)$

$(x-4)(x+3)\left(\frac{6}{x-4}\right) - (x-4)(x+3)\left(\frac{4}{x+3}\right) = 8(x+3)$

$6(x+3)-4(x-4)=8x+24$

$6x+18-4x+16=8x+24$

$2x+34=8x+24$

$2x=8x-10$

$-6x=-10$

$x=\frac{-10}{-6}=\frac{5}{3}$

The solution set is $\left\{\frac{5}{3}\right\}$.

23. Let $x=$ numerator
$3x-9=$ denominator

$\frac{x}{3x-9}=\frac{3}{8};\ x\neq 3$

$8x=3(3x-9)$

$8x=9x-27$

$-x=-27$

$x=27$

The number is $\frac{27}{72}$.

25.

	rate • time = distance		
Rene	$x+3$	$\frac{60}{x+3}$	60
Sue	x	$\frac{60}{x}$	60

$\frac{60}{x+3}+1=\frac{60}{x}$

$x(x+3)\left(\frac{60}{x+3}+1\right)=x(x+3)\left(\frac{60}{x}\right)$

$x(x+3)\left(\frac{60}{x+3}\right)+1x(x+3)=60(x+3)$

$60x+x^2+3x=60x+180$

$x^2+63x=60x+180$

$x^2+3x-180=0$

$(x+15)(x-12)=0$

$x+15=0$ or $x-12=0$

$x=-15$ or $x=12$

Discard the root $x=-15$.
Rene's rate is 15 mph.

Chapter 5 Exponents and Radicals

Problem Set 5.1 Using Integers as Exponents

1. $3^{-3}=\frac{1}{3^3}=\frac{1}{27}$

3. $-10^{-2}=\frac{-1}{10^2}=\frac{-1}{100}$

5. $\frac{1}{3^{-4}}=3^4=81$

7. $-\left(\frac{1}{3}\right)^{-3}$

$-\frac{(1)^{-3}}{(3)^{-3}}$

$-\frac{3^3}{1^3}=-\frac{27}{1}=-27$

9. $\left(-\frac{1}{2}\right)^{-3}$

$\frac{(-1)^{-3}}{(2)^{-3}}$

$\frac{(2)^3}{(-1)^3}=\frac{8}{-1}=-8$

11. $\left(-\frac{3}{4}\right)^0=1$

13. $\frac{1}{\left(\frac{3}{7}\right)^{-2}}=\left(\frac{3}{7}\right)^2=\frac{3^2}{7^2}=\frac{9}{49}$

15. $2^7 \bullet 2^{-3}$

2^{7-3}

$2^4=16$

17. $10^{-5} \bullet 10^2$

$10^{-5+2}=10^{-3}$

$\frac{1}{10^3}=\frac{1}{1000}$

19. $10^{-1} \bullet 10^{-2}$

$10^{-1-2}=10^{-3}$

$\frac{1}{10^3}=\frac{1}{1000}$

21. $(3^{-1})^{-3}=3^{-1(-3)}$

$3^3=27$

23. $(5^3)^{-1}=5^{3(-1)}$

$5^{-3}=\frac{1}{5^3}=\frac{1}{125}$

25. $(2^3 \bullet 3^{-2})^{-1}$

$2^{3(-1)} \bullet 3^{-2(-1)}$

$2^{-3} \bullet 3^2$

$\frac{3^2}{2^3}=\frac{9}{8}$

27. $(4^2 \bullet 5^{-1})^2$

$4^{2(2)} \bullet 5^{-1(2)}$

$4^4 \bullet 5^{-2}$

$\frac{4^4}{5^2}=\frac{256}{25}$

29. $\left(\frac{2^{-1}}{5^{-2}}\right)^{-1}$

$\frac{2^{-1(-1)}}{5^{-2(-1)}}$

$\frac{2}{5^2}=\frac{2}{25}$

31. $\left(\frac{2^{-1}}{3^{-2}}\right)^2$

$\frac{2^{-1(2)}}{3^{-2(2)}}$

$\frac{2^{-2}}{3^{-4}}=\frac{3^4}{2^2}=\frac{81}{4}$

33. $\frac{3^3}{3^{-1}}$

$3^{3-(-1)}$

$3^4=81$

35. $\frac{10^{-2}}{10^2}$

$10^{-2-2} = 10^{-4}$

$\frac{1}{10^4} = \frac{1}{10,000}$

37. $2^{-2} + 3^{-2}$

$\frac{1}{2^2} + \frac{1}{3^2}$

$\frac{1}{4} + \frac{1}{9}$; LCD = 36

$\frac{1}{4} \bullet \frac{9}{9} + \frac{1}{9} \bullet \frac{4}{4}$

$\frac{9}{36} + \frac{4}{36}$

$\frac{13}{36}$

39. $\left(\frac{1}{3}\right)^{-1} - \left(\frac{2}{5}\right)^{-1}$

$\frac{(1)^{-1}}{(3)^{-1}} - \frac{(2)^{-1}}{(5)^{-1}}$

$\frac{3}{1} - \frac{5}{2}$; LCD = 2

$\frac{3}{1} \bullet \frac{2}{2} - \frac{5}{2}$

$\frac{6}{2} - \frac{5}{2}$

$\frac{1}{2}$

41. $(2^{-3} + 3^{-2})^{-1}$

$\left(\frac{1}{2^3} + \frac{1}{3^2}\right)^{-1}$

$\left(\frac{1}{8} + \frac{1}{9}\right)^{-1}$

$\left(\frac{9}{72} + \frac{8}{72}\right)^{-1}$

$\left(\frac{17}{72}\right)^{-1}$

$\frac{(17)^{-1}}{(72)^{-1}} = \frac{72}{17}$

43. $x^2 \bullet x^{-8}$

x^{2-8}

$x^{-6} = \frac{1}{x^6}$

45. $a^3 \bullet a^{-5} \bullet a^{-1}$

a^{3-5-1}

$a^{-3} = \frac{1}{a^3}$

47. $(a^{-4})^2$

$a^{-4(2)}$

$a^{-8} = \frac{1}{a^8}$

49. $(x^2y^{-6})^{-1}$

$x^{2(-1)}y^{-6(-1)}$

$x^{-2}y^6 = \frac{y^6}{x^2}$

51. $(ab^3c^{-2})^{-4}$

$a^{1(-4)}b^{3(-4)}c^{-2(-4)}$

$a^{-4}b^{-12}c^8$

$\frac{c^8}{a^4b^{12}}$

53. $(2x^3y^{-4})^{-3}$

$2^{-3}x^{3(-3)}y^{-4(-3)}$

$2^{-3}x^{-9}y^{12}$

$\frac{y^{12}}{2^3x^9} = \frac{y^{12}}{8x^9}$

55. $\left(\frac{x^{-1}}{y^{-4}}\right)^{-3}$

$\frac{x^{-1(-3)}}{y^{-4(-3)}}$

$\frac{x^3}{y^{12}}$

57. $\left(\frac{3a^{-2}}{2b^{-1}}\right)^{-2}$

$\frac{3^{-2}a^{-2(-2)}}{2^{-2}b^{-1(-2)}}$

$\frac{3^{-2}a^4}{2^{-2}b^2}$

$\frac{2^2a^4}{3^2b^2} = \frac{4a^4}{9b^2}$

59. $\frac{x^{-6}}{x^{-4}}$

$x^{-6-(-4)}$

$x^{-2} = \frac{1}{x^2}$

61. $\frac{a^3b^{-2}}{a^{-2}b^{-4}}$

$a^{3-(-2)}b^{-2-(-4)}$

a^5b^2

63. $(2xy^{-1})(3x^{-2}y^4)$

$6x^{1-2}y^{-1+4}$

$6x^{-1}y^3$

$\frac{6y^3}{x}$

65. $(-7a^2b^{-5})(-a^{-2}b^7)$

$7a^{2-2}b^{-5+7}$

$7a^0b^2$

$7(1)b^2$

$7b^2$

67. $\frac{28x^{-2}y^{-3}}{4x^{-3}y^{-1}}$

$7x^{-2-(-3)}y^{-3-(-1)}$

$7x^1y^{-2}$

$\frac{7x}{y^2}$

69. $\frac{-72a^2b^{-4}}{6a^3b^{-7}}$

$-12a^{2-3}b^{-4-(-7)}$

$-12a^{-1}b^3$

$\frac{-12b^3}{a}$

71. $\left(\frac{35x^{-1}y^{-2}}{7x^4y^3}\right)^{-1}$

$(5x^{-1-4}y^{-2-3})^{-1}$

$(5x^{-5}y^{-5})^{-1}$

$5^{-1}x^{-5(-1)}y^{-5(-1)}$

$5^{-1}x^5y^5$

$\frac{x^5y^5}{5}$

73. $\left(\frac{-36a^{-1}b^{-6}}{4a^{-1}b^4}\right)^{-2}$

$[-9a^{-1-(-1)}b^{-6-4}]^{-2}$

$(-9a^0b^{-10})^{-2}$

$(-9b^{-10})^{-2}$

$(-9)^{-2}b^{-10(-2)}$

$(-9)^{-2}b^{20}$

$\frac{b^{20}}{(-9)^2} = \frac{b^{20}}{81}$

75. $x^{-2} + x^{-3}$

$\frac{1}{x^2} + \frac{1}{x^3}$; LCD $= x^3$

$\frac{1}{x^2} \bullet \frac{x}{x} + \frac{1}{x^3}$

$\frac{x}{x^3} + \frac{1}{x^3}$

$\frac{x+1}{x^3}$

77. $x^{-3} - y^{-1}$

$\frac{1}{x^3} - \frac{1}{y}$; LCD $= x^3y$

$\frac{1}{x^3} \bullet \frac{y}{y} - \frac{1}{y} \bullet \frac{x^3}{x^3}$

$\frac{y}{x^3y} - \frac{x^3}{x^3y}$

$\frac{y - x^3}{x^3y}$

79. $3a^{-2} + 4b^{-1}$

$\frac{3}{a^2} + \frac{4}{b}$; LCD $= a^2b$

$\frac{3}{a^2} \bullet \frac{b}{b} + \frac{4}{b} \bullet \frac{a^2}{a^2}$

$\frac{3b}{a^2b} + \frac{4a^2}{a^2b}$

$\frac{3b + 4a^2}{a^2b}$

81. $x^{-1}y^{-2} - xy^{-1}$

$\frac{1}{xy^2} - \frac{x}{y}$; LCD $= xy^2$

$\frac{1}{xy^2} - \frac{x}{y} \bullet \frac{xy}{xy}$

$\frac{1}{xy^2} - \frac{x^2y}{xy^2}$

$\frac{1 - x^2y}{xy^2}$

83. $2x^{-1} - 3x^{-2}$

$\frac{2}{x} - \frac{3}{x^2}$; LCD $= x^2$

$\frac{2}{x} \bullet \frac{x}{x} - \frac{3}{x^2}$

$\frac{2x}{x^2} - \frac{3}{x^2}$

$\frac{2x - 3}{x^2}$

Problem Set 5.2 Roots and Radicals

1. $\sqrt{64} = 8$

3. $-\sqrt{100} = -(10) = -10$

5. $\sqrt[3]{27} = 3$

7. $\sqrt[3]{-64} = -4$

9. $\sqrt[4]{81} = 3$

11. $\sqrt{\frac{16}{25}} = \frac{4}{5}$

13. $-\sqrt{\frac{36}{49}} = -\left(\frac{6}{7}\right) = -\frac{6}{7}$

15. $\sqrt{\frac{9}{36}} = \frac{3}{6} = \frac{1}{2}$

17. $\sqrt[3]{\frac{27}{64}} = \frac{3}{4}$

19. $\sqrt[3]{8^3} = 8$

21. $\sqrt{27} = \sqrt{9}\sqrt{3} = 3\sqrt{3}$

23. $\sqrt{32} = \sqrt{16}\sqrt{2} = 4\sqrt{2}$

25. $\sqrt{80} = \sqrt{16}\sqrt{5} = 4\sqrt{5}$

27. $\sqrt{160} = \sqrt{16}\sqrt{10} = 4\sqrt{10}$

29. $4\sqrt{18} = 4\sqrt{9}\sqrt{2} = 4(3)\sqrt{2} = 12\sqrt{2}$

31. $-6\sqrt{20}$

$-6\sqrt{4}\sqrt{5}$

$-6(2)\sqrt{5}$

$-12\sqrt{5}$

33. $\frac{2}{5}\sqrt{75}$

$\frac{2}{5}\sqrt{25}\sqrt{3}$

$\frac{2}{5}(5)\sqrt{3}$

$2\sqrt{3}$

35. $\frac{3}{2}\sqrt{24}$

$\frac{3}{2}\sqrt{4}\sqrt{6}$

$\frac{3}{2}(2)\sqrt{6}$

$3\sqrt{6}$

37. $-\frac{5}{6}\sqrt{28}$

$-\frac{5}{6}\sqrt{4}\sqrt{7}$

$-\frac{5}{6}(2)\sqrt{7}$

$-\frac{5}{3}\sqrt{7}$

39. $\sqrt{\frac{19}{4}} = \frac{\sqrt{19}}{\sqrt{4}} = \frac{\sqrt{19}}{2}$

41. $\sqrt{\frac{27}{16}} = \frac{\sqrt{27}}{\sqrt{16}} = \frac{\sqrt{9}\sqrt{3}}{4} = \frac{3\sqrt{3}}{4}$

43. $\sqrt{\frac{75}{81}} = \frac{\sqrt{75}}{\sqrt{81}} = \frac{\sqrt{25}\sqrt{3}}{9} = \frac{5\sqrt{3}}{9}$

45. $\sqrt{\frac{2}{7}} = \frac{\sqrt{2}}{\sqrt{7}} \bullet \frac{\sqrt{7}}{\sqrt{7}} = \frac{\sqrt{14}}{\sqrt{49}} = \frac{\sqrt{14}}{7}$

47. $\sqrt{\frac{2}{3}}=\frac{\sqrt{2}}{\sqrt{3}}\bullet\frac{\sqrt{3}}{\sqrt{3}}=\frac{\sqrt{6}}{\sqrt{9}}=\frac{\sqrt{6}}{3}$

49. $\frac{\sqrt{5}}{\sqrt{12}}=\frac{\sqrt{5}}{\sqrt{4}\sqrt{3}}=\frac{\sqrt{5}}{2\sqrt{3}}$

$\frac{\sqrt{5}}{2\sqrt{3}}\bullet\frac{\sqrt{3}}{\sqrt{3}}=\frac{\sqrt{15}}{2\sqrt{9}}=\frac{\sqrt{15}}{6}$

51. $\frac{\sqrt{11}}{\sqrt{24}}=\frac{\sqrt{11}}{\sqrt{4}\sqrt{6}}=\frac{\sqrt{11}}{2\sqrt{6}}=\frac{\sqrt{11}}{2\sqrt{6}}\bullet\frac{\sqrt{6}}{\sqrt{6}}$

$\frac{\sqrt{66}}{2\sqrt{36}}=\frac{\sqrt{66}}{2(6)}=\frac{\sqrt{66}}{12}$

53. $\frac{\sqrt{18}}{\sqrt{27}}=\sqrt{\frac{18}{27}}=\sqrt{\frac{2}{3}}=\frac{\sqrt{2}}{\sqrt{3}}$

$\frac{\sqrt{2}}{\sqrt{3}}\bullet\frac{\sqrt{3}}{\sqrt{3}}=\frac{\sqrt{6}}{\sqrt{9}}=\frac{\sqrt{6}}{3}$

55. $\frac{\sqrt{35}}{\sqrt{7}}=\sqrt{\frac{35}{7}}=\sqrt{5}$

57. $\frac{2\sqrt{3}}{\sqrt{7}}=\frac{2\sqrt{3}}{\sqrt{7}}\bullet\frac{\sqrt{7}}{\sqrt{7}}=\frac{2\sqrt{21}}{\sqrt{49}}=\frac{2\sqrt{21}}{7}$

59. $\frac{-4\sqrt{12}}{\sqrt{5}}=\frac{-4\sqrt{4}\sqrt{3}}{\sqrt{5}}=\frac{-4(2)\sqrt{3}}{\sqrt{5}}$

$\frac{-8\sqrt{3}}{\sqrt{5}}=\frac{-8\sqrt{3}}{\sqrt{5}}\bullet\frac{\sqrt{5}}{\sqrt{5}}$

$\frac{-8\sqrt{15}}{\sqrt{25}}=\frac{-8\sqrt{15}}{5}$

61. $\frac{3\sqrt{2}}{4\sqrt{3}}=\frac{3\sqrt{2}}{4\sqrt{3}}\bullet\frac{\sqrt{3}}{\sqrt{3}}=\frac{3\sqrt{6}}{4(3)}=\frac{\sqrt{6}}{4}$

63. $\frac{-8\sqrt{18}}{10\sqrt{50}}=\frac{-4}{5}\sqrt{\frac{18}{50}}=\frac{-4}{5}\sqrt{\frac{9}{25}}$

$-\frac{4}{5}\sqrt{\frac{9}{25}}=-\frac{4}{5}\left(\frac{3}{5}\right)=-\frac{12}{25}$

65. $\sqrt[3]{16}=\sqrt[3]{8}\sqrt[3]{2}=2\sqrt[3]{2}$

67. $2\sqrt[3]{81}=2\sqrt[3]{27}\sqrt[3]{3}=2(3)\sqrt[3]{3}=6\sqrt[3]{3}$

69. $\frac{2}{\sqrt[3]{9}}=\frac{2}{\sqrt[3]{9}}\bullet\frac{\sqrt[3]{3}}{\sqrt[3]{3}}=\frac{2\sqrt[3]{3}}{\sqrt[3]{27}}=\frac{2\sqrt[3]{3}}{3}$

71. $\frac{\sqrt[3]{27}}{\sqrt[3]{4}}=\frac{3}{\sqrt[3]{4}}=\frac{3}{\sqrt[3]{4}}\bullet\frac{\sqrt[3]{2}}{\sqrt[3]{2}}$

$\frac{3\sqrt[3]{2}}{\sqrt[3]{8}}=\frac{3\sqrt[3]{2}}{2}$

73. $\frac{\sqrt[3]{6}}{\sqrt[3]{4}}=\sqrt[3]{\frac{6}{4}}=\sqrt[3]{\frac{3}{2}}=\frac{\sqrt[3]{3}}{\sqrt[3]{2}}\bullet\frac{\sqrt[3]{4}}{\sqrt[3]{4}}$

$\frac{\sqrt[3]{12}}{\sqrt[3]{8}}=\frac{\sqrt[3]{12}}{2}$

75. $S=\sqrt{30Df}$

$S=\sqrt{30(150)(.4)}=\sqrt{1800}=42$ mph

$S=\sqrt{30(200)(.4)}=\sqrt{2400}=49$ mph

$S=\sqrt{30(350)(.4)}=\sqrt{4200}=65$ mph

77. $K=\sqrt{s(s-a)(s-b)(s-c)}$

$s=\frac{a+b+c}{2}=\frac{14+16+18}{2}=\frac{48}{2}=24$

$K=\sqrt{24(24-14)(24-16)(24-18)}$

$K=\sqrt{24(10)(8)(6)}$

$K=\sqrt{11520}=107$ square centimeters

79. $K=\sqrt{s(s-a)(s-b)(s-c)}$

$s=\frac{a+b+c}{2}=\frac{18+18+18}{2}=\frac{54}{4}=27$

$K=\sqrt{27(27-18)(27-18)(27-18)}$

$K=\sqrt{27(9)(9)(9)}$

$K=\sqrt{19683}=140$ square inches.

Problem Set 5.3 Combining Radicals and Simplifying Radicals That Contain Variables

1. $5\sqrt{18}-2\sqrt{2}$

$5\sqrt{9}\sqrt{2}-2\sqrt{2}$

$5(3)\sqrt{2}-2\sqrt{2}$

$15\sqrt{2}-2\sqrt{2}$

$13\sqrt{2}$

3. $7\sqrt{12}+10\sqrt{48}$

$7\sqrt{4}\sqrt{3}+10\sqrt{16}\sqrt{3}$

$7(2)\sqrt{3}+10(4)\sqrt{3}$

$14\sqrt{3}+40\sqrt{3}$

$54\sqrt{3}$

5. $-2\sqrt{50}-5\sqrt{32}$

$-2\sqrt{25}\sqrt{2}-5\sqrt{16}\sqrt{2}$

$-2(5)\sqrt{2}-5(4)\sqrt{2}$

$-10\sqrt{2}-20\sqrt{2}$

$-30\sqrt{2}$

7. $3\sqrt{20}-\sqrt{5}-2\sqrt{45}$

$3\sqrt{4}\sqrt{5}-\sqrt{5}-2\sqrt{9}\sqrt{5}$

$3(2)\sqrt{5}-\sqrt{5}-2(3)\sqrt{5}$

$6\sqrt{5}-\sqrt{5}-6\sqrt{5}$

$-\sqrt{5}$

9. $-9\sqrt{24}+3\sqrt{54}-12\sqrt{6}$

$-9\sqrt{4}\sqrt{6}+3\sqrt{9}\sqrt{6}-12\sqrt{6}$

$-9(2)\sqrt{6}+3(3)\sqrt{6}-12\sqrt{6}$

$-18\sqrt{6}+9\sqrt{6}-12\sqrt{6}$

$-21\sqrt{6}$

11. $\frac{3}{4}\sqrt{7}-\frac{2}{3}\sqrt{28}$

$\frac{3}{4}\sqrt{7}-\frac{2}{3}\sqrt{4}\sqrt{7}$

$\frac{3}{4}\sqrt{7}-\frac{2}{3}(2)\sqrt{7}$

$\frac{3}{4}\sqrt{7}-\frac{4}{3}\sqrt{7}$

$\frac{9}{12}\sqrt{7}-\frac{16}{12}\sqrt{7}$

$-\frac{7\sqrt{7}}{12}$

13. $\frac{3}{5}\sqrt{40}+\frac{5}{6}\sqrt{90}$

$\frac{3}{5}\sqrt{4}\sqrt{10}+\frac{5}{6}\sqrt{9}\sqrt{10}$

$\frac{3}{5}(2)\sqrt{10}+\frac{5}{6}(3)\sqrt{10}$

$\frac{6}{5}\sqrt{10}+\frac{5}{2}\sqrt{10}$

$\frac{12}{10}\sqrt{10}+\frac{25}{10}\sqrt{10}$

$\frac{37\sqrt{10}}{10}$

15. $\frac{3\sqrt{18}}{5}-\frac{5\sqrt{72}}{6}+\frac{3\sqrt{98}}{4}$

$\frac{3\sqrt{9}\sqrt{2}}{5}-\frac{5\sqrt{36}\sqrt{2}}{6}+\frac{3\sqrt{49}\sqrt{2}}{4}$

$\frac{3(3)\sqrt{2}}{5}-\frac{5(6)\sqrt{2}}{6}+\frac{3(7)\sqrt{2}}{4}$

$\frac{9\sqrt{2}}{5}-5\sqrt{2}+\frac{21\sqrt{2}}{4}$

$\frac{(9\sqrt{2})}{5}\bullet\frac{4}{4}-(5\sqrt{2})\bullet\frac{20}{20}+\frac{(21\sqrt{2})}{4}\bullet\frac{5}{5}$

$\frac{36\sqrt{2}}{20}-\frac{100\sqrt{2}}{20}+\frac{105\sqrt{2}}{20}$

$\frac{41\sqrt{2}}{20}$

17. $5\sqrt[3]{3}+2\sqrt[3]{24}-6\sqrt[3]{81}$

$5\sqrt[3]{3}+2\sqrt[3]{8}\sqrt[3]{3}-6\sqrt[3]{27}\sqrt[3]{3}$

$5\sqrt[3]{3}+4\sqrt[3]{3}-18\sqrt[3]{3}$

$-9\sqrt[3]{3}$

19. $-\sqrt[3]{16}+7\sqrt[3]{54}-9\sqrt[3]{2}$

$-\sqrt[3]{8}\sqrt[3]{2}+7\sqrt[3]{27}\sqrt[3]{2}-9\sqrt[3]{2}$

$-2\sqrt[3]{2}+7(3)\sqrt[3]{2}-9\sqrt[3]{2}$

$-2\sqrt[3]{2}+21\sqrt[3]{2}-9\sqrt[3]{2}$

$10\sqrt[3]{2}$

21. $\sqrt{32x}$

$\sqrt{16}\sqrt{2x}$

$4\sqrt{2x}$

23. $\sqrt{75x^2}$

$\sqrt{25x^2}\sqrt{3}$

$5x\sqrt{3}$

25. $\sqrt{20x^2y}$

$\sqrt{4x^2}\sqrt{5y}$

$2x\sqrt{5y}$

27. $\sqrt{64x^3y^7}$

$\sqrt{64x^2y^6}\sqrt{xy}$

$8xy^3\sqrt{xy}$

29. $\sqrt{54a^4b^3}$

$\sqrt{9a^4b^2}\sqrt{6b}$

$3a^2b\sqrt{6b}$

31. $\sqrt{63x^6y^8}$

$\sqrt{9x^6y^8}\sqrt{7}$

$3x^3y^4\sqrt{7}$

33. $2\sqrt{40a^3}$

$2\sqrt{4a^2}\sqrt{10a}$

$2(2a)\sqrt{10a}$

$4a\sqrt{10a}$

35. $\frac{2}{3}\sqrt{96xy^3}$

$\frac{2}{3}\sqrt{16y^2}\sqrt{6xy}$

$\frac{2}{3}(4y)\sqrt{6xy}$

$\frac{8y}{3}\sqrt{6xy}$

37. $\sqrt{\frac{2x}{5y}} = \frac{\sqrt{2x}}{\sqrt{5y}} \bullet \frac{\sqrt{5y}}{\sqrt{5y}} = \frac{\sqrt{10xy}}{\sqrt{25y^2}} = \frac{\sqrt{10xy}}{5y}$

39. $\sqrt{\frac{5}{12x^4}} = \frac{\sqrt{5}}{\sqrt{4x^4}\sqrt{3}} = \frac{\sqrt{5}}{2x^2\sqrt{3}}$

$\frac{\sqrt{5}}{2x^2\sqrt{3}} \bullet \frac{\sqrt{3}}{\sqrt{3}} = \frac{\sqrt{15}}{2x^2\sqrt{9}}$

$\frac{\sqrt{15}}{2x^2(3)} = \frac{\sqrt{15}}{6x^2}$

41. $\frac{5}{\sqrt{18y}} = \frac{5}{\sqrt{9}\sqrt{2y}} = \frac{5}{3\sqrt{2y}}$

$\frac{5}{3\sqrt{2y}} \bullet \frac{\sqrt{2y}}{\sqrt{2y}} = \frac{5\sqrt{2y}}{3\sqrt{4y^2}}$

$\frac{5\sqrt{2y}}{3(2y)} = \frac{5\sqrt{2y}}{6y}$

43. $\frac{\sqrt{7x}}{\sqrt{8y^5}} = \frac{\sqrt{7x}}{\sqrt{4y^4}\sqrt{2y}} = \frac{\sqrt{7x}}{2y^2\sqrt{2y}}$

$\frac{\sqrt{7x}}{2y^2\sqrt{2y}} \bullet \frac{\sqrt{2y}}{\sqrt{2y}} = \frac{\sqrt{14xy}}{2y^2\sqrt{4y^2}}$

$\frac{\sqrt{14xy}}{(2y^2)(2y)} = \frac{\sqrt{14xy}}{4y^3}$

45. $\frac{\sqrt{18y^3}}{\sqrt{16x}} = \frac{\sqrt{9y^2}\sqrt{2y}}{\sqrt{16}\sqrt{x}} = \frac{3y\sqrt{2y}}{4\sqrt{x}}$

$\frac{3y\sqrt{2y}}{4\sqrt{x}} \bullet \frac{\sqrt{x}}{\sqrt{x}} = \frac{3y\sqrt{2xy}}{4\sqrt{x^2}} = \frac{3y\sqrt{2xy}}{4x}$

47. $\frac{\sqrt{24a^2b^3}}{\sqrt{7ab^6}} = \frac{\sqrt{4a^2b^2}\sqrt{6b}}{\sqrt{b^6}\sqrt{7a}} = \frac{2ab\sqrt{6b}}{b^3\sqrt{7a}}$

$\frac{2a\sqrt{6b}}{b^2\sqrt{7a}} \bullet \frac{\sqrt{7a}}{\sqrt{7a}} = \frac{2a\sqrt{42ab}}{b^2\sqrt{49a^2}} = \frac{2a\sqrt{42ab}}{b^2(7a)}$

$\frac{2a\sqrt{42ab}}{7ab^2} = \frac{2\sqrt{42ab}}{7b^2}$

49. $\sqrt[3]{24y} = \sqrt[3]{8}\sqrt[3]{3y} = 2\sqrt[3]{3y}$

51. $\sqrt[3]{16x^4} = \sqrt[3]{8x^3}\sqrt[3]{2x} = 2x\sqrt[3]{2x}$

53. $\sqrt[3]{56x^6y^8} = \sqrt[3]{8x^6y^6}\sqrt[3]{7y^2} = 2x^2y^2\sqrt[3]{7y^2}$

55. $\sqrt[3]{\frac{7}{9x^2}} = \frac{\sqrt[3]{7}}{\sqrt[3]{9x^2}} \bullet \frac{\sqrt[3]{3x}}{\sqrt[3]{3x}} = \frac{\sqrt[3]{21x}}{\sqrt[3]{27x^3}} = \frac{\sqrt[3]{21x}}{3x}$

57. $\frac{\sqrt[3]{3y}}{\sqrt[3]{16x^4}} = \frac{\sqrt[3]{3y}}{\sqrt[3]{8x^3}\sqrt[3]{2x}} = \frac{\sqrt[3]{3y}}{2x\sqrt[3]{2x}}$

$\frac{\sqrt[3]{3y}}{2x\sqrt[3]{2x}} \bullet \frac{\sqrt[3]{4x^2}}{\sqrt[3]{4x^2}} = \frac{\sqrt[3]{12x^2y}}{2x\sqrt[3]{8x^3}}$

$\frac{\sqrt[3]{12x^2y}}{2x(2x)} = \frac{\sqrt[3]{12x^2y}}{4x^2}$

59. $\frac{\sqrt[3]{12xy}}{\sqrt[3]{3x^2y^5}} = \sqrt[3]{\frac{12xy}{3x^2y^5}} = \sqrt[3]{\frac{4}{xy^4}} = \frac{\sqrt[3]{4}}{\sqrt[3]{xy^4}}$

$$\frac{\sqrt[3]{4}}{\sqrt[3]{y^3}\sqrt[3]{xy}} = \frac{\sqrt[3]{4}}{y\sqrt[3]{xy}} = \frac{\sqrt[3]{4}}{y\sqrt[3]{xy}} \bullet \frac{\sqrt[3]{x^2y^2}}{\sqrt[3]{x^2y^2}}$$

$$\frac{\sqrt[3]{4x^2y^2}}{y\sqrt[3]{x^3y^3}} = \frac{\sqrt[3]{4x^2y^2}}{y(xy)} = \frac{\sqrt[3]{4x^2y^2}}{xy^2}$$

61. $\sqrt{8x+12y}$
$\sqrt{4(2x+3y)}$
$\sqrt{4}\sqrt{2x+3y}$
$2\sqrt{2x+3y}$

63. $\sqrt{16x+48y}$
$\sqrt{16(x+3y)}$
$\sqrt{16}\sqrt{x+3y}$
$4\sqrt{x+3y}$

65. $-3\sqrt{4x}+5\sqrt{9x}+6\sqrt{16x}$
$-3\sqrt{4}\sqrt{x}+5\sqrt{9}\sqrt{x}+6\sqrt{16}\sqrt{x}$
$-3(2)\sqrt{x}+5(3)\sqrt{x}+6(4)\sqrt{x}$
$-6\sqrt{x}+15\sqrt{x}+24\sqrt{x}$
$33\sqrt{x}$

67. $2\sqrt{18x}-3\sqrt{8x}-6\sqrt{50x}$
$2\sqrt{9}\sqrt{2x}-3\sqrt{4}\sqrt{2x}-6\sqrt{25}\sqrt{2x}$
$2(3)\sqrt{2x}-3(2)\sqrt{2x}-6(5)\sqrt{2x}$
$6\sqrt{2x}-6\sqrt{2x}-30\sqrt{2x}$
$-30\sqrt{2x}$

69. $5\sqrt{27n}-\sqrt{12n}-6\sqrt{3n}$
$5\sqrt{9}\sqrt{3n}-\sqrt{4}\sqrt{3n}-6\sqrt{3n}$
$5(3)\sqrt{3n}-2\sqrt{3n}-6\sqrt{3n}$
$15\sqrt{3n}-2\sqrt{3n}-6\sqrt{3n}$
$7\sqrt{3n}$

71. $7\sqrt{4ab}-\sqrt{16ab}-10\sqrt{25ab}$
$7\sqrt{4}\sqrt{ab}-\sqrt{16}\sqrt{ab}-10\sqrt{25}\sqrt{ab}$
$7(2)\sqrt{ab}-4\sqrt{ab}-10(5)\sqrt{ab}$
$14\sqrt{ab}-4\sqrt{ab}-50\sqrt{ab}$
$-40\sqrt{ab}$

73. $-3\sqrt{2x^3}+4\sqrt{8x^3}-3\sqrt{32x^3}$
$-3\sqrt{x^2}\sqrt{2x}+4\sqrt{4x^2}\sqrt{2x}-3\sqrt{16x^2}\sqrt{2x}$
$-3(x)\sqrt{2x}+4(2x)\sqrt{2x}-3(4x)\sqrt{2x}$
$-3x\sqrt{2x}+8x\sqrt{2x}-12x\sqrt{2x}$
$-7x\sqrt{2x}$

Problem Set 5.4 Products and Quotients Involving Radicals

1. $\sqrt{6}\sqrt{12}$
$\sqrt{72}=\sqrt{36}\sqrt{2}=6\sqrt{2}$

3. $(3\sqrt{3})(2\sqrt{6})$
$6\sqrt{18}$
$6\sqrt{9}\sqrt{2}=6(3)\sqrt{2}=18\sqrt{2}$

5. $(4\sqrt{2})(-6\sqrt{5})$
$-24\sqrt{10}$

7. $(-3\sqrt{3})(-4\sqrt{8})$
$12\sqrt{24}$
$12\sqrt{4}\sqrt{6}=12(2)\sqrt{6}=24\sqrt{6}$

9. $(5\sqrt{6})(4\sqrt{6})$
$20\sqrt{36}$
$20(6)=120$

11. $(2\sqrt[3]{4})(6\sqrt[3]{2})$
$12\sqrt[3]{8}$
$12(2)=24$

13. $(4\sqrt[3]{6})(7\sqrt[3]{4})$
$28\sqrt[3]{24}$
$28\sqrt[3]{8}\sqrt[3]{3}=28(2)\sqrt[3]{3}=56\sqrt[3]{3}$

15. $\sqrt{2}(\sqrt{3}+\sqrt{5})$
$\sqrt{6}+\sqrt{10}$

17. $3\sqrt{5}(2\sqrt{2}-\sqrt{7})$
$6\sqrt{10}-3\sqrt{35}$

19. $2\sqrt{6}(3\sqrt{8}-5\sqrt{12})$
$6\sqrt{48}-10\sqrt{72}$

$6\sqrt{16}\sqrt{3}-10\sqrt{36}\sqrt{2}$

$24\sqrt{3}-60\sqrt{2}$

21. $-4\sqrt{5}(2\sqrt{5}+4\sqrt{12})$

$-8\sqrt{25}-16\sqrt{60}$

$-8(5)-16\sqrt{4}\sqrt{15}$

$-40-32\sqrt{15}$

23. $3\sqrt{x}(5\sqrt{2}+\sqrt{y})$

$15\sqrt{2x}+3\sqrt{xy}$

25. $\sqrt{xy}(5\sqrt{xy}-6\sqrt{x})$

$5\sqrt{x^2y^2}-6\sqrt{x^2y}$

$5xy-6\sqrt{x^2}\sqrt{y}$

$5xy-6x\sqrt{y}$

27. $\sqrt{5y}(\sqrt{8x}+\sqrt{12y^2})$

$\sqrt{40xy}+\sqrt{60xy^3}$

$\sqrt{4}\sqrt{10xy}+\sqrt{4y^2}\sqrt{15xy}$

$2\sqrt{10xy}+2y\sqrt{15xy}$

29. $5\sqrt{3}(2\sqrt{8}-3\sqrt{18})$

$10\sqrt{24}-15\sqrt{54}$

$10\sqrt{4}\sqrt{6}-15\sqrt{9}\sqrt{6}$

$20\sqrt{6}-45\sqrt{6}$

$-25\sqrt{6}$

31. $(\sqrt{3}+4)(\sqrt{3}-7)$

$\sqrt{9}-7\sqrt{3}+4\sqrt{3}-28$

$3-3\sqrt{3}-28$

$-25-3\sqrt{3}$

33. $(\sqrt{5}-6)(\sqrt{5}-3)$

$\sqrt{25}-3\sqrt{5}-6\sqrt{5}+18$

$5-9\sqrt{5}+18$

$23-9\sqrt{5}$

35. $(3\sqrt{5}-2\sqrt{3})(2\sqrt{7}+\sqrt{2})$

$6\sqrt{35}+3\sqrt{10}-4\sqrt{21}-2\sqrt{6}$

37. $(2\sqrt{6}+3\sqrt{5})(\sqrt{8}-3\sqrt{12})$

$2\sqrt{48}-6\sqrt{72}+3\sqrt{40}-9\sqrt{60}$

$2\sqrt{16}\sqrt{3}-6\sqrt{36}\sqrt{2}+3\sqrt{4}\sqrt{10}-9\sqrt{4}\sqrt{15}$

$8\sqrt{3}-36\sqrt{2}+6\sqrt{10}-18\sqrt{15}$

39. $(2\sqrt{6}+5\sqrt{5})(3\sqrt{6}-\sqrt{5})$

$6\sqrt{36}-2\sqrt{30}+15\sqrt{30}-5\sqrt{25}$

$36+13\sqrt{30}-25$

$11+13\sqrt{30}$

41. $(3\sqrt{2}-5\sqrt{3})(6\sqrt{2}-7\sqrt{3})$

$18\sqrt{4}-21\sqrt{6}-30\sqrt{6}+35\sqrt{9}$

$36-51\sqrt{6}+105$

$141-51\sqrt{6}$

43. $(\sqrt{6}+4)(\sqrt{6}-4)$

$\sqrt{36}-4\sqrt{6}+4\sqrt{6}-16$

$6-16$

-10

45. $(\sqrt{2}+\sqrt{10})(\sqrt{2}-\sqrt{10})$

$\sqrt{4}-\sqrt{20}+\sqrt{20}-\sqrt{100}$

$2-10$

-8

47. $(\sqrt{2x}+\sqrt{3y})(\sqrt{2x}-\sqrt{3y})$

$\sqrt{4x^2}-\sqrt{6xy}+\sqrt{6xy}-\sqrt{9y^2}$

$2x-3y$

49. $2\sqrt[3]{3}(5\sqrt[3]{4}+\sqrt[3]{6})$

$10\sqrt[3]{12}+2\sqrt[3]{18}$

51. $3\sqrt[3]{4}(2\sqrt[3]{2}-6\sqrt[3]{4})$

$6\sqrt[3]{8}-18\sqrt[3]{16}$

$6(2)-18\sqrt[3]{8}\sqrt[3]{2}$

$12-36\sqrt[3]{2}$

53. $\dfrac{2}{\sqrt{7}+1}$

$\dfrac{2}{(\sqrt{7}+1)}\bullet\dfrac{(\sqrt{7}-1)}{(\sqrt{7}-1)}$

$$\frac{2(\sqrt{7}-1)}{\sqrt{49}-1}=\frac{2(\sqrt{7}-1)}{7-1}$$

$$\frac{2(\sqrt{7}-1)}{6}=\frac{\sqrt{7}-1}{3}$$

55. $$\frac{3}{\sqrt{2}-5}=\frac{3}{(\sqrt{2}-5)}\bullet\frac{(\sqrt{2}+5)}{(\sqrt{2}+5)}$$

$$\frac{3(\sqrt{2}+5)}{\sqrt{4}-25}=\frac{3(\sqrt{2}+5)}{2-25}$$

$$\frac{3(\sqrt{2}+5)}{-23}=\frac{3\sqrt{2}+15}{-23}=\frac{-3\sqrt{2}-15}{23}$$

57. $$\frac{1}{\sqrt{2}+\sqrt{7}}=\frac{1}{(\sqrt{2}+\sqrt{7})}\bullet\frac{(\sqrt{2}-\sqrt{7})}{(\sqrt{2}-\sqrt{7})}$$

$$\frac{\sqrt{2}-\sqrt{7}}{\sqrt{4}-\sqrt{49}}=\frac{\sqrt{2}-\sqrt{7}}{2-7}=\frac{\sqrt{2}-\sqrt{7}}{-5}$$

$$\frac{-\sqrt{2}+\sqrt{7}}{5}=\frac{\sqrt{7}-\sqrt{2}}{5}$$

59. $$\frac{\sqrt{2}}{\sqrt{10}-\sqrt{3}}=\frac{\sqrt{2}}{(\sqrt{10}-\sqrt{3})}\bullet\frac{(\sqrt{10}+\sqrt{3})}{(\sqrt{10}+\sqrt{3})}$$

$$\frac{\sqrt{2}(\sqrt{10}+\sqrt{3})}{\sqrt{100}-\sqrt{9}}=\frac{\sqrt{20}+\sqrt{6}}{10-3}$$

$$\frac{\sqrt{4}\sqrt{5}+\sqrt{6}}{7}=\frac{2\sqrt{5}+\sqrt{6}}{7}$$

61. $$\frac{\sqrt{3}}{2\sqrt{5}+4}=\frac{\sqrt{3}}{(2\sqrt{5}+4)}\bullet\frac{(2\sqrt{5}-4)}{(2\sqrt{5}-4)}$$

$$\frac{\sqrt{3}(2\sqrt{5}-4)}{4\sqrt{25}-16}=\frac{2\sqrt{15}-4\sqrt{3}}{20-16}$$

$$\frac{2\sqrt{15}-4\sqrt{3}}{4}=\frac{2(\sqrt{15}-2\sqrt{3})}{4}$$

$$\frac{\sqrt{15}-2\sqrt{3}}{2}$$

63. $$\frac{6}{3\sqrt{7}-2\sqrt{6}}$$

$$\frac{6}{(3\sqrt{7}-2\sqrt{6})}\bullet\frac{(3\sqrt{7}+2\sqrt{6})}{(3\sqrt{7}+2\sqrt{6})}$$

$$\frac{6(3\sqrt{7}+2\sqrt{6})}{9\sqrt{49}-4\sqrt{36}}=\frac{18\sqrt{7}+12\sqrt{6}}{9(7)-4(6)}$$

$$\frac{18\sqrt{7}+12\sqrt{6}}{63-24}=\frac{18\sqrt{7}+12\sqrt{6}}{39}$$

$$\frac{3(6\sqrt{7}+4\sqrt{6})}{39}=\frac{6\sqrt{7}+4\sqrt{6}}{13}$$

65. $$\frac{\sqrt{6}}{3\sqrt{2}+2\sqrt{3}}$$

$$\frac{\sqrt{6}}{(3\sqrt{2}+2\sqrt{3})}\bullet\frac{(3\sqrt{2}-2\sqrt{3})}{(3\sqrt{2}-2\sqrt{3})}$$

$$\frac{\sqrt{6}(3\sqrt{2}-2\sqrt{3})}{9\sqrt{4}-4\sqrt{9}}=\frac{3\sqrt{12}-2\sqrt{18}}{9(2)-4(3)}$$

$$\frac{3\sqrt{4}\sqrt{3}-2\sqrt{9}\sqrt{2}}{18-12}=\frac{6\sqrt{3}-6\sqrt{2}}{6}$$

$$\frac{6(\sqrt{3}-\sqrt{2})}{6}=\sqrt{3}-\sqrt{2}$$

67. $$\frac{2}{\sqrt{x}+4}$$

$$\frac{2}{\sqrt{x}+4}\bullet\frac{(\sqrt{x}-4)}{(\sqrt{x}-4)}$$

$$\frac{2(\sqrt{x}-4)}{\sqrt{x^2}-16}=\frac{2\sqrt{x}-8}{x-16}$$

69. $$\frac{\sqrt{x}}{\sqrt{x}-5}$$

$$\frac{\sqrt{x}}{(\sqrt{x}-5)}\bullet\frac{(\sqrt{x}+5)}{(\sqrt{x}+5)}$$

$$\frac{\sqrt{x}(\sqrt{x}+5)}{\sqrt{x^2}-25}=\frac{\sqrt{x^2}+5\sqrt{x}}{x-25}$$

$$\frac{x+5\sqrt{x}}{x-25}$$

71. $$\frac{\sqrt{x}-2}{\sqrt{x}+6}$$

$$\frac{(\sqrt{x}-2)}{(\sqrt{x}+6)}\bullet\frac{(\sqrt{x}-6)}{(\sqrt{x}-6)}$$

$$\frac{\sqrt{x^2}-6\sqrt{x}-2\sqrt{x}+12}{\sqrt{x^2}-36}$$

$$\frac{x-8\sqrt{x}+12}{x-36}$$

73. $\dfrac{\sqrt{x}}{\sqrt{x}+2\sqrt{y}}$

$\dfrac{\sqrt{x}}{(\sqrt{x}+2\sqrt{y})} \bullet \dfrac{(\sqrt{x}-2\sqrt{y})}{(\sqrt{x}-2\sqrt{y})}$

$\dfrac{\sqrt{x^2}-2\sqrt{xy}}{\sqrt{x^2}-4\sqrt{y^2}} = \dfrac{x-2\sqrt{xy}}{x-4y}$

75. $\dfrac{3\sqrt{y}}{2\sqrt{x}-3\sqrt{y}}$

$\dfrac{3\sqrt{y}}{(2\sqrt{x}-3\sqrt{y})} \bullet \dfrac{(2\sqrt{x}+3\sqrt{y})}{(2\sqrt{x}+3\sqrt{y})}$

$\dfrac{6\sqrt{xy}+9\sqrt{y^2}}{4\sqrt{x^2}-9\sqrt{y^2}} = \dfrac{6\sqrt{xy}+9y}{4x-9y}$

Problem Set 5.5 Equations Involving Radicals

1. $\sqrt{5x} = 10$

$(\sqrt{5x})^2 = 10^2$

$5x = 100$

$x = 20$

Check

$\sqrt{5(20)} \stackrel{?}{=} 10$

$\sqrt{100} \stackrel{?}{=} 10$

$10 = 10$

The solution set is $\{20\}$.

3. $\sqrt{2x} + 4 = 0$

$\sqrt{2x} = -4$

$(\sqrt{2x})^2 = (-4)^2$

$2x = 16$

$x = 8$

Check

$\sqrt{2(8)} + 4 \stackrel{?}{=} 0$

$\sqrt{16} + 4 \stackrel{?}{=} 0$

$4 + 4 \stackrel{?}{=} 0$

$8 \neq 0$

The solution set is $\emptyset$.

5. $2\sqrt{n} = 5$

$\sqrt{n} = \frac{5}{2}$

$(\sqrt{n})^2 = \left(\frac{5}{2}\right)^2$

$n = \frac{25}{4}$

Check

$2\sqrt{\frac{25}{4}} \stackrel{?}{=} 5$

$2\left(\frac{5}{2}\right) \stackrel{?}{=} 5$

$5 = 5$

The solution set is $\left\{\frac{25}{4}\right\}$.

7. $3\sqrt{n} - 2 = 0$

$3\sqrt{n} = 2$

$\sqrt{n} = \frac{2}{3}$

$(\sqrt{n})^2 = \left(\frac{2}{3}\right)^2$

$n = \frac{4}{9}$

Check

$3\sqrt{\frac{4}{9}} - 2 \stackrel{?}{=} 0$

$3\left(\frac{2}{3}\right) - 2 \stackrel{?}{=} 0$

$2 - 2 \stackrel{?}{=} 0$

$0 = 0$

The solution set is $\left\{\frac{4}{9}\right\}$.

9. $\sqrt{3y+1} = 4$

$(\sqrt{3y+1})^2 = 4^2$

$3y + 1 = 16$

$3y = 15$

$y = 5$

Check

$\sqrt{3(5)+1} \stackrel{?}{=} 4$

$\sqrt{16} \stackrel{?}{=} 4$

$4 = 4$

The solution set is $\{5\}$.

11. $\sqrt{4y-3}-6=0$
$\sqrt{4y-3}=6$
$(\sqrt{4y-3})^2=6^2$
$4y-3=36$
$4y=39$
$y=\frac{39}{4}$
Check
$\sqrt{4\left(\frac{39}{4}\right)-3}-6\stackrel{?}{=}0$
$\sqrt{39-3}-6\stackrel{?}{=}0$
$\sqrt{36}-6\stackrel{?}{=}0$
$6-6\stackrel{?}{=}0$
$0\stackrel{?}{=}0$
The solution set is $\left\{\frac{39}{4}\right\}$.

13. $\sqrt{2x-5}=-1$
$(\sqrt{2x-5})^2=(-1)^2$
$2x-5=1$
$2x=6$
$x=3$
Check
$\sqrt{2(3)-5}\stackrel{?}{=}-1$
$\sqrt{1}\stackrel{?}{=}-1$
$1\neq-1$
The solution set is $\emptyset$.

15. $\sqrt{5x+2}=\sqrt{6x+1}$
$(\sqrt{5x+2})^2=(\sqrt{6x+1})^2$
$5x+2=6x+1$
$2=x+1$
$1=x$
Check
$\sqrt{5(1)+2}\stackrel{?}{=}\sqrt{6(1)+1}$
$\sqrt{7}=\sqrt{7}$
The solution set is $\{1\}$.

17. $\sqrt{3x+1}=\sqrt{7x-5}$
$(\sqrt{3x+1})^2=(\sqrt{7x-5})^2$
$3x+1=7x-5$
$3x=7x-6$
$-4x=-6$
$x=\frac{-6}{-4}=\frac{3}{2}$
Check
$\sqrt{3\left(\frac{3}{2}\right)+1}\stackrel{?}{=}\sqrt{7\left(\frac{3}{2}\right)-5}$
$\sqrt{\frac{9}{2}+1}\stackrel{?}{=}\sqrt{\frac{21}{2}-5}$
$\sqrt{\frac{11}{2}}=\sqrt{\frac{11}{2}}$
The solution set is $\left\{\frac{3}{2}\right\}$.

19. $\sqrt{3x-2}-\sqrt{x+4}=0$
$\sqrt{3x-2}=\sqrt{x+4}$
$(\sqrt{3x-2})^2=(\sqrt{x+4})^2$
$3x-2=x+4$
$3x=x+6$
$2x=6$
$x=3$
Check
$\sqrt{3(3)-2}-\sqrt{3+4}\stackrel{?}{=}0$
$\sqrt{7}-\sqrt{7}\stackrel{?}{=}0$
$0=0$
The solution set is $\{3\}$.

21. $5\sqrt{t-1}=6$
$\sqrt{t-1}=\frac{6}{5}$
$(\sqrt{t-1})^2=\left(\frac{6}{5}\right)^2$
$t-1=\frac{36}{25}$
$t=\frac{36}{25}+1$
$t=\frac{61}{25}$
Check
$5\sqrt{\frac{61}{25}-1}\stackrel{?}{=}6$

$5\sqrt{\frac{36}{25}} \stackrel{?}{=} 6$

$5\left(\frac{6}{5}\right) \stackrel{?}{=} 6$

$6 = 6$

The solution set is $\left\{\frac{61}{25}\right\}$.

23. $\sqrt{x^2+7} = 4$

$(\sqrt{x^2+7})^2 = 4^2$

$x^2+7 = 16$

$x^2-9 = 0$

$(x+3)(x-3) = 0$

$x+3=0$ or $x-3=0$

$x=-3$ or $x=3$

Checking $x = -3$

$\sqrt{(-3)^2+7} \stackrel{?}{=} 4$

$\sqrt{16} \stackrel{?}{=} 4$

$4 = 4$

Checking $x = 3$

$\sqrt{3^2+7} \stackrel{?}{=} 4$

$\sqrt{16} \stackrel{?}{=} 4$

$4 = 4$

The solution set is $\{-3, 3\}$.

25. $\sqrt{x^2+13x+37} = 1$

$(\sqrt{x^2+13x+37})^2 = 1^2$

$x^2+13x+37 = 1$

$x^2+13x+36 = 0$

$(x+4)(x+9) = 0$

$x+4=0$ or $x+9=0$

$x=-4$ or $x=-9$

Checking $x = -4$

$\sqrt{(-4)^2+13(-4)+37} \stackrel{?}{=} 1$

$\sqrt{16-52+37} \stackrel{?}{=} 1$

$\sqrt{1} \stackrel{?}{=} 1$

$1 = 1$

Checking $x = -9$

$\sqrt{(-9)^2+13(-9)+37} \stackrel{?}{=} 1$

$\sqrt{81-117+37} \stackrel{?}{=} 1$

$\sqrt{1} \stackrel{?}{=} 1$

$1 = 1$

The solution set is $\{-9, -4\}$.

27. $\sqrt{x^2-x+1} = x+1$

$(\sqrt{x^2-x+1})^2 = (x+1)^2$

$x^2-x+1 = x^2+2x+1$

$-x+1 = 2x+1$

$-x = 2x$

$-3x = 0$

$x = 0$

Check

$\sqrt{0^2-0+1} \stackrel{?}{=} 0+1$

$\sqrt{1} \stackrel{?}{=} 1$

$1 = 1$

The solution set is $\{0\}$.

29. $\sqrt{x^2+3x+7} = x+2$

$(\sqrt{x^2+3x+7})^2 = (x+2)^2$

$x^2+3x+7 = x^2+4x+4$

$3x+7 = 4x+4$

$3x+3 = 4x$

$3 = x$

Check

$\sqrt{3^2+3(3)+7} \stackrel{?}{=} 3+2$

$\sqrt{9+9+7} \stackrel{?}{=} 5$

$\sqrt{25} \stackrel{?}{=} 5$

$5 = 5$

The solution set is $\{3\}$.

31. $\sqrt{-4x+17} = x-3$

$(\sqrt{-4x+17})^2 = (x-3)^2$

$-4x+17 = x^2-6x+9$

$0 = x^2-2x-8$

$0 = (x-4)(x+2)$

$x-4=0$ or $x+2=0$

$x=4$ or $x=-2$

Checking $x=4$

$\sqrt{-4(4)+17} \stackrel{?}{=} 4-3$

$\sqrt{1} \stackrel{?}{=} 1$

$1=1$

Checking $x=-2$

$\sqrt{-4(-2)+17} \stackrel{?}{=} -2-3$

$\sqrt{25} \stackrel{?}{=} -5$

$5 \neq -5$

The solution set is $\{4\}$.

33. $\sqrt{n+4} = n+4$

$(\sqrt{n+4})^2 = (n+4)^2$

$n+4 = n^2+8n+16$

$0 = n^2+7n+12$

$0 = (n+4)(n+3)$

$n+4=0$ or $n+3=0$

$n=-4$ or $n=-3$

Checking $n=-4$

$\sqrt{-4+4} \stackrel{?}{=} -4+4$

$\sqrt{0} \stackrel{?}{=} 0$

$0=0$

Checking $n=-3$

$\sqrt{-3+4} \stackrel{?}{=} -3+4$

$\sqrt{1} \stackrel{?}{=} 1$

$1=1$

The solution set is $\{-4,-3\}$.

35. $\sqrt{3y} = y-6$

$(\sqrt{3y})^2 = (y-6)^2$

$3y = y^2-12y+36$

$0 = y^2-15y+36$

$0 = (y-12)(y-3)$

$y-12=0$ or $y-3=0$

$y=12$ or $y=3$

Checking $y=12$

$\sqrt{3(12)} \stackrel{?}{=} 12-6$

$\sqrt{36} \stackrel{?}{=} 6$

$6=6$

Checking $y=3$

$\sqrt{3(3)} \stackrel{?}{=} 3-6$

$\sqrt{9} \stackrel{?}{=} -3$

$3 \neq -3$

The solution set is $\{12\}$.

37. $4\sqrt{x}+5 = x$

$4\sqrt{x} = x-5$

$(4\sqrt{x})^2 = (x-5)^2$

$16x = x^2-10x+25$

$0 = x^2-26x+25$

$0 = (x-25)(x-1)$

$x-25=0$ or $x-1=0$

$x=25$ or $x=1$

Checking $x=25$

$4\sqrt{25}+5 \stackrel{?}{=} 25$

$4(5)+5 \stackrel{?}{=} 25$

$25=25$

Checking $x=1$

$4\sqrt{1}+5 \stackrel{?}{=} 1$

$4+5 \stackrel{?}{=} 1$

$9 \neq 1$

The solution set is $\{25\}$.

39. $\sqrt[3]{x-2} = 3$

$(\sqrt[3]{x-2})^3 = 3^3$

$x-2=27$

$x=29$

Check

$\sqrt[3]{29-2} \stackrel{?}{=} 3$

$\sqrt[3]{27} \stackrel{?}{=} 3$

$3=3$

The solution set is $\{29\}$.

41. $\sqrt[3]{2x+3} = -3$

$(\sqrt[3]{2x+3})^3 = (-3)^3$

$2x+3 = -27$

$2x = -30$

$x = -15$

Check

$\sqrt[3]{2(-15)+3} \stackrel{?}{=} -3$

$\sqrt[3]{-27} \stackrel{?}{=} -3$

$-3 = -3$

The solution set is $\{-15\}$.

43. $\sqrt[3]{2x+5} = \sqrt[3]{4-x}$

$(\sqrt[3]{2x+5})^3 = (\sqrt[3]{4-x})^3$

$2x+5 = 4-x$

$2x = -1-x$

$3x = -1$

$x = -\frac{1}{3}$

Check

$\sqrt[3]{2\left(-\frac{1}{3}\right)+5} \stackrel{?}{=} \sqrt[3]{4-\left(-\frac{1}{3}\right)}$

$\sqrt[3]{-\frac{2}{3}+5} \stackrel{?}{=} \sqrt[3]{4+\frac{1}{3}}$

$\sqrt[3]{4\frac{1}{3}} = \sqrt[3]{4\frac{1}{3}}$

The solution set is $\left\{-\frac{1}{3}\right\}$.

45. $\sqrt{x+19} - \sqrt{x+28} = -1$

$\sqrt{x+19} = \sqrt{x+28} - 1$

$(\sqrt{x+19})^2 = (\sqrt{x+28} - 1)^2$

$x+19 = (\sqrt{x+28})^2 + 2(-1)\sqrt{x+28} + 1$

$x+19 = x+28 - 2\sqrt{x+28} + 1$

$x+19 = x+29 - 2\sqrt{x+28}$

$-10 = -2\sqrt{x+28}$

$5 = \sqrt{x+28}$

$(5)^2 = (\sqrt{x+28})^2$

$25 = x+28$

$-3 = x$

Check

$\sqrt{-3+19} - \sqrt{-3+28} \stackrel{?}{=} -1$

$\sqrt{16} - \sqrt{25} \stackrel{?}{=} -1$

$4-5 \stackrel{?}{=} -1$

$-1 = -1$

The solution set is $\{-3\}$.

47. $\sqrt{3x+1} + \sqrt{2x+4} = 3$

$\sqrt{3x+1} = 3 - \sqrt{2x+4}$

$(\sqrt{3x+1})^2 = (3 - \sqrt{2x+4})^2$

$3x+1 = 9 + 2(3)(-\sqrt{2x+4}) + (-\sqrt{2x+4})^2$

$3x+1 = 9 - 6\sqrt{2x+4} + 2x+4$

$3x+1 = 2x+13 - 6\sqrt{2x+4}$

$x-12 = -6\sqrt{2x+4}$

$(x-12)^2 = (-6\sqrt{2x+4})^2$

$x^2 - 24x + 144 = 36(2x+4)$

$x^2 - 24x + 144 = 72x + 144$

$x^2 - 96x = 0$

$x(x-96) = 0$

$x = 0 \quad \text{or} \quad x-96 = 0$

$x = 0 \quad \text{or} \quad x = 96$

Checking $x=0$

$\sqrt{3(0)+1} + \sqrt{2(0)+4} \stackrel{?}{=} 3$

$\sqrt{1} + \sqrt{4} \stackrel{?}{=} 3$

$1+2 \stackrel{?}{=} 3$

$3 = 3$

Checking $x=96$

$\sqrt{3(96)+1} + \sqrt{2(96)+4} \stackrel{?}{=} 3$

$\sqrt{289} + \sqrt{196} \stackrel{?}{=} 3$

$17+14 \stackrel{?}{=} 3$

$31 \neq 3$

The solution set is $\{0\}$.

49. $\sqrt{n-4} + \sqrt{n+4} = 2\sqrt{n-1}$

$(\sqrt{n-4} + \sqrt{n+4})^2 = (2\sqrt{n-1})^2$

$(\sqrt{n-4})^2 + 2\sqrt{n-4}\sqrt{n+4} + (\sqrt{n+4})^2 = 4(n-1)$

$n-4 + 2\sqrt{(n-4)(n+4)} + n+4 = 4n-4$

$2n+2\sqrt{n^2-16}=4n-4$

$2\sqrt{n^2-16}=2n-4$

$2\sqrt{n^2-16}=2(n-2)$

$\sqrt{n^2-16}=n-2$

$(\sqrt{n^2-16})^2=(n-2)^2$

$n^2-16=n^2-4n+4$

$-16=-4n+4$

$-20=-4n$

$5=n$

Check

$\sqrt{5-4}+\sqrt{5+4}\stackrel{?}{=}2\sqrt{5-1}$

$\sqrt{1}+\sqrt{9}\stackrel{?}{=}2\sqrt{4}$

$1+3\stackrel{?}{=}2(2)$

$4=4$

The solution set is {5}.

51. $\sqrt{t+3}-\sqrt{t-2}=\sqrt{7-t}$

$(\sqrt{t+3}-\sqrt{t-2})^2=(\sqrt{7-t})^2$

$(\sqrt{t+3})^2-2\sqrt{t+3}\sqrt{t-2}+(-\sqrt{t-2})^2=7-t$

$t+3-2\sqrt{(t+3)(t-2)}+t-2=7-t$

$2t+1-2\sqrt{t^2+t-6}=7-t$

$-2\sqrt{t^2+t-6}=-3t+6$

$(-2\sqrt{t^2+t-6})^2=(-3t+6)^2$

$4(t^2+t-6)=9t^2-36t+36$

$4t^2+4t-24=9t^2-36t+36$

$0=5t^2-40t+60$

$0=5(t^2-8t+12)$

$0=5(t-6)(t-2)$

$t-6=0$ or $t-2=0$

$t=6$ or $t=2$

Checking $t=6$

$\sqrt{6+3}-\sqrt{6-2}\stackrel{?}{=}\sqrt{7-6}$

$\sqrt{9}-\sqrt{4}\stackrel{?}{=}\sqrt{1}$

$3-2\stackrel{?}{=}1$

$1=1$

Checking $t=2$

$\sqrt{2+3}-\sqrt{2-2}\stackrel{?}{=}\sqrt{7-2}$

$\sqrt{5}-\sqrt{0}\stackrel{?}{=}\sqrt{5}$

$\sqrt{5}=\sqrt{5}$

The solution set is {2, 6}.

53. $\sqrt{30Df}=S$

$\sqrt{30D(.95)}=S$

$\sqrt{28.5D}=S$

$(\sqrt{28.5D})^2=S^2$

$28.5D=S^2$

$D=\dfrac{S^2}{28.5}$

$D=\dfrac{40^2}{28.5}=56$ feet

$D=\dfrac{55^2}{28.5}=106$ feet

$D=\dfrac{65^2}{28.5}=148$ feet

55. $L=\dfrac{32T^2}{4\pi^2}$

$L=\dfrac{32(2)^2}{4\pi^2}=3.2$ feet

$L=\dfrac{32(2.5)^2}{4\pi^2}=5.1$ feet

$L=\dfrac{32(3)^2}{4\pi^2}=7.3$ feet

Problem Set 5.6 Merging of Exponents and Roots

1. $81^{\frac{1}{2}}=\sqrt{81}=9$

3. $27^{\frac{1}{3}}=\sqrt[3]{27}=3$

5. $(-8)^{\frac{1}{3}}=\sqrt[3]{-8}=-2$

7. $-25^{\frac{1}{2}}=-\sqrt{25}=-5$

9. $36^{-\frac{1}{2}}=\dfrac{1}{36^{\frac{1}{2}}}=\dfrac{1}{\sqrt{36}}=\dfrac{1}{6}$

11. $\left(\frac{1}{27}\right)^{-\frac{1}{3}} = \frac{1^{-\frac{1}{3}}}{27^{-\frac{1}{3}}} = \frac{27^{\frac{1}{3}}}{1^{\frac{1}{3}}} = \frac{\sqrt[3]{27}}{\sqrt[3]{1}} = \frac{3}{1} = 3$

13. $4^{\frac{3}{2}} = (\sqrt{4})^3 = 2^3 = 8$

15. $27^{\frac{4}{3}} = (\sqrt[3]{27})^4 = 3^4 = 81$

17. $(-1)^{\frac{7}{3}} = (\sqrt[3]{-1})^7 = (-1)^7 = -1$

19. $-4^{\frac{5}{2}} = -(\sqrt{4})^5 = -(2)^5 = -32$

21. $\left(\frac{27}{8}\right)^{\frac{4}{3}} = \left(\sqrt[3]{\frac{27}{8}}\right)^4 = \left(\frac{3}{2}\right)^4 = \frac{3^4}{2^4} = \frac{81}{16}$

23. $\left(\frac{1}{8}\right)^{-\frac{2}{3}} = \frac{1^{-\frac{2}{3}}}{8^{-\frac{2}{3}}} = \frac{8^{\frac{2}{3}}}{1^{\frac{2}{3}}}$

$\frac{(\sqrt[3]{8})^2}{(\sqrt[3]{1})^2} = \frac{(2)^2}{(1)^2} = \frac{4}{1} = 4$

25. $64^{-\frac{7}{6}} = \frac{1}{64^{\frac{7}{6}}} = \frac{1}{(\sqrt[6]{64})^7} = \frac{1}{2^7} = \frac{1}{128}$

27. $-25^{\frac{3}{2}} = -(\sqrt{25})^3 = -(5)^3 = -125$

29. $125^{\frac{4}{3}} = (\sqrt[3]{125})^4 = 5^4 = 625$

31. $x^{\frac{4}{3}} = \sqrt[3]{x^4}$

33. $3x^{\frac{1}{2}} = 3\sqrt{x}$

35. $(2y)^{\frac{1}{3}} = \sqrt[3]{2y}$

37. $(2x-3y)^{\frac{1}{2}} = \sqrt{2x-3y}$

39. $(2a-3b)^{\frac{2}{3}} = \sqrt[3]{(2a-3b)^2}$

41. $x^{\frac{2}{3}}y^{\frac{1}{3}} = \sqrt[3]{x^2y}$

43. $-3x^{\frac{1}{5}}y^{\frac{2}{5}} = -3\sqrt[5]{xy^2}$

45. $\sqrt{5y} = (5y)^{\frac{1}{2}} = 5^{\frac{1}{2}}y^{\frac{1}{2}}$

47. $3\sqrt{y} = 3y^{\frac{1}{2}}$

49. $\sqrt[3]{xy^2} = (xy^2)^{\frac{1}{3}} = x^{\frac{1}{3}}y^{\frac{2}{3}}$

51. $\sqrt[4]{a^2b^3} = (a^2b^3)^{\frac{1}{4}} = a^{\frac{2}{4}}b^{\frac{3}{4}} = a^{\frac{1}{2}}b^{\frac{3}{4}}$

53. $\sqrt[5]{(2x-y)^3} = (2x-y)^{\frac{3}{5}}$

55. $5x\sqrt{y} = 5xy^{\frac{1}{2}}$

57. $-\sqrt[3]{x+y} = -(x+y)^{\frac{1}{3}}$

59. $(2x^{\frac{2}{5}})(6x^{\frac{1}{4}})$

$12x^{\frac{2}{5}+\frac{1}{4}}$

$12x^{\frac{8}{20}+\frac{5}{20}}$

$12x^{\frac{13}{20}}$

61. $(y^{\frac{2}{3}})(y^{-\frac{1}{4}})$

$y^{\frac{2}{3}-\frac{1}{4}} = y^{\frac{8}{12}-\frac{3}{12}} = y^{\frac{5}{12}}$

63. $(x^{\frac{2}{5}})(4x^{-\frac{1}{2}})$

$4x^{\frac{2}{5}-\frac{1}{2}} = 4x^{\frac{4}{10}-\frac{5}{10}} = 4x^{-\frac{1}{10}} = \frac{4}{x^{\frac{1}{10}}}$

65. $(4x^{\frac{1}{2}}y)^2$

$4^2x^{\frac{1}{2}(2)}y^2$

$16xy^2$

67. $(8x^6y^3)^{\frac{1}{3}}$

$8^{\frac{1}{3}}x^{6(\frac{1}{3})}y^{3(\frac{1}{3})}$

$2x^2y$

69. $\frac{24x^{\frac{3}{5}}}{6x^{\frac{1}{3}}}$

$4x^{\frac{3}{5}-\frac{1}{3}} = 4x^{\frac{9}{15}-\frac{5}{15}} = 4x^{\frac{4}{15}}$

71. $\dfrac{48b^{\frac{1}{3}}}{12b^{\frac{3}{4}}}$

$4b^{\frac{1}{3}-\frac{3}{4}} = 4b^{\frac{4}{12}-\frac{9}{12}} = 4b^{-\frac{5}{12}} = \dfrac{4}{b^{\frac{5}{12}}}$

73. $\left(\dfrac{6x^{\frac{2}{5}}}{7y^{\frac{2}{3}}}\right)^2 = \dfrac{6^2x^{\frac{2}{5}(2)}}{7^2y^{\frac{2}{3}(2)}} = \dfrac{36x^{\frac{4}{5}}}{49y^{\frac{4}{3}}}$

75. $\left(\dfrac{x^2}{y^3}\right)^{-\frac{1}{2}} = \dfrac{x^{2(-\frac{1}{2})}}{y^{3(-\frac{1}{2})}} = \dfrac{x^{-1}}{y^{-\frac{3}{2}}} = \dfrac{y^{\frac{3}{2}}}{x}$

77. $\left(\dfrac{18x^{\frac{1}{3}}}{9x^{\frac{1}{4}}}\right)^2 = (2x^{\frac{1}{3}-\frac{1}{4}})^2 = (2x^{\frac{4}{12}-\frac{3}{12}})^2$

$(2x^{\frac{1}{12}})^2 = 2^2x^{\frac{1}{12}(2)} = 4x^{\frac{2}{12}} = 4x^{\frac{1}{6}}$

79. $\left(\dfrac{60a^{\frac{1}{5}}}{15a^{\frac{3}{4}}}\right)^2 = (4a^{\frac{1}{5}-\frac{3}{4}})^2 = (4a^{\frac{4}{20}-\frac{15}{20}})^2$

$(4a^{-\frac{11}{20}})^2 = 4^2a^{-\frac{11}{20}(2)} = 16a^{-\frac{11}{10}} = \dfrac{16}{a^{\frac{11}{10}}}$

81. $\sqrt[3]{3}\sqrt{3}$

$3^{\frac{1}{3}} \bullet 3^{\frac{1}{2}} = 3^{\frac{1}{3}+\frac{1}{2}} = 3^{\frac{2}{6}+\frac{3}{6}} = 3^{\frac{5}{6}} = \sqrt[6]{3^5} = \sqrt[6]{243}$

83. $\sqrt[4]{6}\sqrt{6}$

$6^{\frac{1}{4}} \bullet 6^{\frac{1}{2}} = 6^{\frac{1}{4}+\frac{1}{2}} = 6^{\frac{1}{4}+\frac{2}{4}} = 6^{\frac{3}{4}} = \sqrt[4]{6^3} = \sqrt[4]{216}$

85. $\dfrac{\sqrt[3]{3}}{\sqrt[4]{3}} = \dfrac{3^{\frac{1}{3}}}{3^{\frac{1}{4}}} = 3^{\frac{1}{3}-\frac{1}{4}} = 3^{\frac{4}{12}-\frac{3}{12}} = 3^{\frac{1}{12}} = \sqrt[12]{3}$

87. $\dfrac{\sqrt[3]{8}}{\sqrt[4]{4}} = \dfrac{\sqrt[3]{2^3}}{\sqrt[4]{2^2}} = \dfrac{2^{\frac{3}{3}}}{2^{\frac{2}{4}}} = \dfrac{2^1}{2^{\frac{1}{2}}}$

$2^{1-\frac{1}{2}} = 2^{\frac{1}{2}} = \sqrt{2}$

89. $\dfrac{\sqrt[4]{27}}{\sqrt{3}} = \dfrac{\sqrt[4]{3^3}}{\sqrt{3}} = \dfrac{3^{\frac{3}{4}}}{3^{\frac{1}{2}}}$

$3^{\frac{3}{4}-\frac{1}{2}} = 3^{\frac{3}{4}-\frac{2}{4}} = 3^{\frac{1}{4}} = \sqrt[4]{3}$

Problem Set 5.7 Scientific Notation

1. $89 = (8.9)(10)^1$

3. $4290 = (4.29)(10)^3$

5. $6,120,000 = (6.12)(10)^6$

7. $40,000,000 = (4)(10)^7$

9. $376.4 = (3.764)(10)^2$

11. $.347 = (3.47)(10)^{-1}$

13. $.0214 = (2.14)(10)^{-2}$

15. $.00005 = (5)(10)^{-5}$

17. $.00000000194 = (1.94)(10)^{-9}$

19. $(2.3)(10)^1 = 23$

21. $(4.19)(10)^3 = 4190$

23. $(5)(10)^8 = 500,000,000$

25. $(3.14)(10)^{10} = 31,400,000,000$

27. $(4.3)(10)^{-1} = .43$

29. $(9.14)(10)^{-4} = .000914$

31. $(5.123)(10)^{-8} = .00000005123$

33. $(0.0037)(0.00002)$

$(3.7)(10)^{-3}(2)(10)^{-5}$

$(7.4)(10)^{-8}$

$.000000074$

35. $(0.00007)(11,000)$

$(7)(10)^{-5}(1.1)(10)^4$

$(7.7)(10)^{-1}$

$.77$

37. $\frac{360,000,000}{0.0012} = \frac{(3.6)(10)^8}{(1.2)(10)^{-3}}$

$(3)(10)^{11} = 300,000,000,000$

39. $\frac{0.000064}{16,000} = \frac{(6.4)(10)^{-5}}{(1.6)(10)^4}$

$(4)(10)^{-9} = .000000004$

41. $\frac{(60,000)(0.006)}{(0.0009)(400)} = \frac{(6)(10)^4(6)(10)^{-3}}{(9)(10)^{-4}(4)(10)^2}$

$\frac{(36)(10)^1}{(36)(10)^{-2}} = (1)(10)^3 = 1000$

43. $\frac{(0.0045)(60000)}{(1800)(0.00015)}$

$\frac{(4.5)(10)^{-3}(6)(10)^4}{(1.8)(10)^3(1.5)(10)^{-4}}$

$\frac{(27)(10)^1}{(2.7)(10)^{-1}} = \frac{(2.7)(10)^1(10)^1}{(2.7)(10)^{-1}}$

$(1)(10)^3 = 1000$

45. $\sqrt{9,000,000} = \sqrt{(9)(10)^6}$

$[(9)(10)^6]^{\frac{1}{2}} = 9^{\frac{1}{2}}(10)^{6(\frac{1}{2})}$

$(3)(10)^3 = 3000$

47. $\sqrt[3]{8000} = \sqrt[3]{(8)(10)^3} = [(8)(10)^3]^{\frac{1}{3}}$

$8^{\frac{1}{3}}(10)^{3(\frac{1}{3})} = 2(10) = 20$

49. $(90,000)^{\frac{3}{2}} = [(9)(10)^4]^{\frac{3}{2}}$

$(9)^{\frac{3}{2}}(10)^{4(\frac{3}{2})} = (27)(10)^6 = 27,000,000$

Chapter 5 Review

1. $4^{-3} = \frac{1}{4^3} = \frac{1}{64}$

3. $(3^2 \bullet 3^{-3})^{-1}$

$(3^{-1})^{-1} = 3^{-1(-1)} = 3^1 = 3$

5. $\sqrt[4]{\frac{16}{81}} = \frac{2}{3}$

7. $(-1)^{-\frac{2}{3}} = \frac{1}{(-1)^{\frac{2}{3}}} = \frac{1}{(\sqrt[3]{-1})^2}$

$\frac{1}{(-1)^2} = \frac{1}{1} = 1$

9. $-16^{\frac{2}{3}} = -(\sqrt{16})^3 = -(4)^3 = -64$

11. $(4^{-2} \bullet 4^2)^{-1} = (4^0)^{-1} = (1)^{-1} = \frac{1}{1} = 1$

13. $\sqrt{54} = \sqrt{9}\sqrt{6} = 3\sqrt{6}$

15. $\frac{4\sqrt{3}}{\sqrt{6}} = 4\sqrt{\frac{3}{6}} = 4\sqrt{\frac{1}{2}} = (4)\frac{\sqrt{1}}{\sqrt{2}} = \frac{4}{\sqrt{2}}$

$\frac{4}{\sqrt{2}} \bullet \frac{\sqrt{2}}{\sqrt{2}} = \frac{4\sqrt{2}}{\sqrt{4}} = \frac{4\sqrt{2}}{2} = 2\sqrt{2}$

17. $\sqrt[3]{56} = \sqrt[3]{8}\sqrt[3]{7} = 2\sqrt[3]{7}$

19. $\sqrt{\frac{9}{5}} = \frac{\sqrt{9}}{\sqrt{5}} = \frac{3}{\sqrt{5}} = \frac{3}{\sqrt{5}} \bullet \frac{\sqrt{5}}{\sqrt{5}}$

$\frac{3\sqrt{5}}{\sqrt{25}} = \frac{3\sqrt{5}}{5}$

21. $\sqrt[3]{108x^4y^8}$

$\sqrt[3]{27x^3y^6}\ \sqrt[3]{4xy^2}$

$3xy^2\sqrt[3]{4xy^2}$

23. $\frac{2}{3}\sqrt{45xy^3}$

$\frac{2}{3}\sqrt{9y^2}\sqrt{5xy}$

$\frac{2}{3}(3y)\sqrt{5xy}$

$2y\sqrt{5xy}$

25. $(3\sqrt{8})(4\sqrt{5})$

$12\sqrt{40}$

$12\sqrt{4}\sqrt{10}$

$12(2)\sqrt{10}$

$24\sqrt{10}$

27. $3\sqrt{2}(4\sqrt{6}-2\sqrt{7})$

$12\sqrt{12}-6\sqrt{14}$

$12\sqrt{4}\sqrt{3}-6\sqrt{14}$

$12(2)\sqrt{3}-6\sqrt{14}$

$24\sqrt{3}-6\sqrt{14}$

29. $(2\sqrt{5}-\sqrt{3})(2\sqrt{5}+\sqrt{3})$

$4\sqrt{25}+2\sqrt{15}-2\sqrt{15}-\sqrt{9}$

$4(5)-3$

$20-3$

17

31. $(2\sqrt{a}+\sqrt{b})(3\sqrt{a}-4\sqrt{b})$

$6\sqrt{a^2}-8\sqrt{ab}+3\sqrt{ab}-4\sqrt{b^2}$

$6a-5\sqrt{ab}-4b$

33. $\frac{4}{\sqrt{7}-1}=\frac{4}{(\sqrt{7}-1)}\bullet\frac{(\sqrt{7}+1)}{(\sqrt{7}+1)}$

$\frac{4(\sqrt{7}+1)}{\sqrt{49}-1}=\frac{4(\sqrt{7}+1)}{7-1}$

$\frac{4(\sqrt{7}+1)}{6}=\frac{2(\sqrt{7}+1)}{3}$

35. $\frac{3}{2\sqrt{3}+3\sqrt{5}}=\frac{3}{(2\sqrt{3}+3\sqrt{5})}\bullet\frac{(2\sqrt{3}-3\sqrt{5})}{(2\sqrt{3}-3\sqrt{5})}$

$\frac{3(2\sqrt{3}-3\sqrt{5})}{4\sqrt{9}-9\sqrt{25}}=\frac{3(2\sqrt{3}-3\sqrt{5})}{4(3)-9(5)}$

$\frac{3(2\sqrt{3}-3\sqrt{5})}{12-45}=\frac{3(2\sqrt{3}-3\sqrt{5})}{-33}$

$\frac{2\sqrt{3}-3\sqrt{5}}{-11}=\frac{-2\sqrt{3}+3\sqrt{5}}{11}$

$\frac{3\sqrt{5}-2\sqrt{3}}{11}$

37. $(x^{-3}y^4)^{-2}=x^{-3(-2)}y^{4(-2)}$

$x^6y^{-8}=\frac{x^6}{y^8}$

39. $(4x^{\frac{1}{2}})(5x^{\frac{1}{5}})$

$20x^{\frac{1}{2}+\frac{1}{5}}$

$20x^{\frac{7}{10}}$

41. $\left(\frac{x^3}{y^4}\right)^{-\frac{1}{3}}=\frac{x^{3(-\frac{1}{3})}}{y^{4(-\frac{1}{3})}}=\frac{x^{-1}}{y^{-\frac{4}{3}}}=\frac{y^{\frac{4}{3}}}{x}$

43. $3\sqrt{45}-2\sqrt{20}-\sqrt{80}$

$3\sqrt{9}\sqrt{5}-2\sqrt{4}\sqrt{5}-\sqrt{16}\sqrt{5}$

$3(3)\sqrt{5}-2(2)\sqrt{5}-4\sqrt{5}$

$9\sqrt{5}-4\sqrt{5}-4\sqrt{5}$

$\sqrt{5}$

45. $3\sqrt{24}-\frac{2}{5}\sqrt{54}+\frac{1}{4}\sqrt{96}$

$3\sqrt{4}\sqrt{6}-\frac{2}{5}\sqrt{9}\sqrt{6}+\frac{1}{4}\sqrt{16}\sqrt{6}$

$3(2)\sqrt{6}-\frac{2}{5}(3)\sqrt{6}+\frac{1}{4}(4)\sqrt{6}$

$6\sqrt{6}-\frac{6}{5}\sqrt{6}+\sqrt{6}$

$7\sqrt{6}-\frac{6}{5}\sqrt{6}$

$\frac{35\sqrt{6}}{5}-\frac{6\sqrt{6}}{5}$

$\frac{29\sqrt{6}}{5}$

47. $x^{-2}+y^{-1}$

$\frac{1}{x^2}+\frac{1}{y}$; LCD $=x^2y$

$\frac{1}{x^2}\bullet\frac{y}{y}+\frac{1}{y}\bullet\frac{x^2}{x^2}$

$\frac{y}{x^2y}+\frac{x^2}{x^2y}$

$\frac{y+x^2}{x^2y}$

49. $\sqrt{7x-3}=4$

$(\sqrt{7x-3})^2=4^2$

$7x-3=16$

$7x = 19$

$x = \frac{19}{7}$

Check

$\sqrt{7\left(\frac{19}{7}\right) - 3} \stackrel{?}{=} 4$

$\sqrt{19 - 3} \stackrel{?}{=} 4$

$\sqrt{16} \stackrel{?}{=} 4$

$4 = 4$

The solution set is $\left\{\frac{19}{7}\right\}$.

51. $\sqrt{2x} = x - 4$

$(\sqrt{2x})^2 = (x - 4)^2$

$2x = x^2 - 8x + 16$

$0 = x^2 - 10x + 16$

$0 = (x - 8)(x - 2)$

$x - 8 = 0$ or $x - 2 = 0$

$x = 8$ or $x = 2$

Checking $x = 8$

$\sqrt{2(8)} \stackrel{?}{=} 8 - 4$

$\sqrt{16} \stackrel{?}{=} 4$

$4 = 4$

Checking $x = 2$

$\sqrt{2(2)} \stackrel{?}{=} 2 - 4$

$\sqrt{4} \stackrel{?}{=} -2$

$2 \neq -2$

The solution set is $\{8\}$.

53. $\sqrt[3]{2x - 1} = 3$

$(\sqrt[3]{2x - 1})^3 = 3^3$

$2x - 1 = 27$

$2x = 28$

$x = 14$

Checking

$\sqrt[3]{2(14) - 1} \stackrel{?}{=} 3$

$\sqrt[3]{27} \stackrel{?}{=} 3$

$3 = 3$

The solution set is $\{14\}$.

55. $\sqrt{x^2 + 3x - 6} = x$

$(\sqrt{x^2 + 3x - 6})^2 = x^2$

$x^2 + 3x - 6 = x^2$

$3x - 6 = 0$

$3x = 6$

$x = 2$

Check

$\sqrt{2^2 + 3(2) - 6} \stackrel{?}{=} 2$

$\sqrt{4 + 6 - 6} \stackrel{?}{=} 2$

$\sqrt{4} \stackrel{?}{=} 2$

$2 = 2$

The solution set is $\{2\}$.

57. $(.00002)(.0003)$

$(2)(10)^{-5}(3)(10)^{-4}$

$(6)(10)^{-9}$

$.000000006$

59. $(.000015)(400{,}000)$

$(1.5)(10)^{-5}(4)(10)^{5}$

$(6.0)(10)^{0}$

6

61. $\dfrac{(.00042)(.0004)}{.006}$

$\dfrac{(4.2)(10)^{-4}(4)(10)^{-4}}{(6)(10)^{-3}}$

$\dfrac{(16.8)(10)^{-8}}{(6)(10)^{-3}} = (2.8)(10)^{-5}$

$.000028$

63. $\sqrt[3]{.000000008} = \sqrt[3]{(8)(10)^{-9}}$

$[(8)(10)^{-9}]^{\frac{1}{3}} = (8)^{\frac{1}{3}}(10)^{-9\left(\frac{1}{3}\right)}$

$(2)(10)^{-3} = .002$

Chapter 5 Test

1. $(4)^{-\frac{5}{2}} = \frac{1}{4^{\frac{5}{2}}} = \frac{1}{(4^{\frac{1}{2}})^5} = \frac{1}{(2)^5} = \frac{1}{32}$

3. $\left(\frac{2}{3}\right)^{-4} = \frac{2^{-4}}{3^{-4}} = \frac{3^4}{2^4} = \frac{81}{16}$

5. $\sqrt{63} = \sqrt{9}\sqrt{7} = 3\sqrt{7}$

7. $\sqrt{52x^4y^3} = \sqrt{4x^4y^2}\sqrt{13y} = 2x^2y\sqrt{13y}$

9. $\sqrt{\frac{7}{24x^3}} = \frac{\sqrt{7}}{\sqrt{24x^3}} = \frac{\sqrt{7}}{\sqrt{4x^2}\sqrt{6x}}$

$\frac{\sqrt{7}}{2x\sqrt{6x}} = \frac{\sqrt{7}}{2x\sqrt{6x}} \bullet \frac{\sqrt{6x}}{\sqrt{6x}}$

$\frac{\sqrt{42x}}{2x\sqrt{36x^2}} = \frac{\sqrt{42x}}{2x(6x)} = \frac{\sqrt{42x}}{12x^2}$

11. $(3\sqrt{2}+\sqrt{3})(\sqrt{2}-2\sqrt{3})$

$3\sqrt{4}-6\sqrt{6}+\sqrt{6}-2\sqrt{9}$

$3(2)-5\sqrt{6}-2(3)$

$6-5\sqrt{6}-6$

$-5\sqrt{6}$

13. $\frac{3\sqrt{2}}{4\sqrt{3}-\sqrt{8}} = \frac{3\sqrt{2}}{4\sqrt{3}-\sqrt{4}\sqrt{2}} = \frac{3\sqrt{2}}{4\sqrt{3}-2\sqrt{2}}$

$\frac{3\sqrt{2}}{(4\sqrt{3}-2\sqrt{2})} \bullet \frac{(4\sqrt{3}+2\sqrt{2})}{(4\sqrt{3}+2\sqrt{2})}$

$\frac{12\sqrt{6}+6\sqrt{4}}{16\sqrt{9}-4\sqrt{4}} = \frac{12\sqrt{6}+6(2)}{16(3)-4(2)}$

$\frac{12\sqrt{6}+12}{48-8} = \frac{12\sqrt{6}+12}{40}$

$\frac{4(3\sqrt{6}+3)}{4(10)} = \frac{3\sqrt{6}+3}{10}$

15. $\frac{-84a^{\frac{1}{2}}}{7a^{\frac{4}{5}}} = -12a^{\frac{1}{2}-\frac{4}{5}} = -12a^{-\frac{3}{10}} = \frac{-12}{a^{\frac{3}{10}}}$

17. $(3x^{-\frac{1}{2}})(-4x^{\frac{3}{4}}) = -12x^{-\frac{1}{2}+\frac{3}{4}} = -12x^{\frac{1}{4}}$

19. $\frac{(.00004)(300)}{.00002} = \frac{(4)(10)^{-5}(3)(10)^2}{(2)(10)^{-5}}$

$\frac{(12)(10)^{-3}}{(2)(10)^{-5}} = (6)(10)^2 = 600$

21. $\sqrt{3x+1} = 3$

$(\sqrt{3x+1})^2 = 3^2$

$3x+1 = 9$

$3x = 8$

$x = \frac{8}{3}$

Check

$\sqrt{3\left(\frac{8}{3}\right)+1} \stackrel{?}{=} 3$

$\sqrt{8+1} \stackrel{?}{=} 3$

$\sqrt{9} \stackrel{?}{=} 3$

$3 = 3$

The solution set is $\left\{\frac{8}{3}\right\}$.

23. $\sqrt{x} = x-2$

$(\sqrt{x})^2 = (x-2)^2$

$x = x^2-4x+4$

$0 = x^2-5x+4$

$0 = (x-4)(x-1)$

$x-4 = 0$ or $x-1 = 0$

$x = 4$ or $x = 1$

Checking $x = 4$

$\sqrt{4} \stackrel{?}{=} 4-2$

$2 = 2$

Checking $x = 1$

$\sqrt{1} \stackrel{?}{=} 1-2$

$1 \neq -1$

The solution set is $\{4\}$.

25. $\sqrt{x^2-10x+28} = 2$

$(\sqrt{x^2-10x+28})^2 = 2^2$

$x^2-10x+28 = 4$

$x^2-10x+24 = 0$

$(x-6)(x-4)=0$

$x-6=0$ or $x-4=0$

$x=6$ or $x=4$

Checking $x=6$

$\sqrt{6^2-10(6)+28}\overset{?}{=}2$

$\sqrt{36-60+28}\overset{?}{=}2$

$\sqrt{4}\overset{?}{=}2$

$2=2$

Checking $x=4$

$\sqrt{4^2-10(4)+28}\overset{?}{=}2$

$\sqrt{16-40+28}\overset{?}{=}2$

$\sqrt{4}\overset{?}{=}2$

$2=2$

The solution set is $\{4, 6\}$.

Chapter 5 Cumulative Review

1. $\frac{4a^2b^3}{12a^3b}=\frac{b^2}{3a}$

$\frac{(-8)^2}{3(5)}=\frac{64}{15}$

3. $\frac{3}{n}+\frac{5}{2n}-\frac{4}{3n}$

$\frac{3}{25}+\frac{5}{2(25)}-\frac{4}{3(25)}$

$\frac{3}{25}+\frac{5}{50}-\frac{4}{75}$; LCD $=150$

$\frac{3}{25}\bullet\frac{6}{6}+\frac{5}{50}\bullet\frac{3}{3}-\frac{4}{75}\bullet\frac{2}{2}$

$\frac{18}{150}+\frac{15}{150}-\frac{8}{150}$

$\frac{25}{150}=\frac{1}{6}$

5. $2\sqrt{2x+y}-5\sqrt{3x-y}$

$2\sqrt{2(5)+6}-5\sqrt{3(5)-6}$

$2\sqrt{16}-5\sqrt{9}$

$2(4)-5(3)$

$8-15$

-7

7. $(x+3)(2x^2-x-4)$

$2x^3-x^2-4x+6x^2-3x-12$

$2x^3+5x^2-7x-12$

9. $\frac{a^2+6a-40}{a^2-4a}\div\frac{2a^2+19a-10}{a^3+a^2}$

$\frac{a^2+6a-40}{a^2-4a}\bullet\frac{a^3+a^2}{2a^2+19a-10}$

$\frac{(a+10)(a-4)}{a(a-4)}\bullet\frac{a^2(a+1)}{(2a-1)(a+10)}$

$\frac{a(a+1)}{(2a-1)}$

11. $\frac{4}{x^2+3x}+\frac{5}{x}$

$\frac{4}{x(x+3)}+\frac{5}{x}$; LCD $=x(x+3)$

$\frac{4}{x(x+3)}+\frac{5}{x}\bullet\frac{(x+3)}{(x+3)}$

$\frac{4+5(x+3)}{x(x+3)}=\frac{4+5x+15}{x(x+3)}$

$\frac{5x+19}{x(x+3)}$

13. $\frac{3}{5x^2+3x-2}-\frac{2}{5x^2-22x+8}$

$\frac{3}{(5x-2)(x+1)}-\frac{2}{(5x-2)(x-4)}$

LCD $=(5x-2)(x+1)(x-4)$

$\frac{3}{(5x-2)(x+1)}\bullet\frac{(x-4)}{(x-4)}-\frac{2}{(5x-2)(x-4)}\bullet\frac{(x+1)}{(x+1)}$

$\frac{3(x-4)-2(x+1)}{(5x-2)(x+1)(x-4)}$

$\frac{3x-12-2x-2}{(5x-2)(x+1)(x-4)}$

$\frac{x-14}{(5x-2)(x+1)(x-4)}$

15.
$$\begin{array}{r} x^2-3x-2 \\ 4x-5\,\overline{)\,4x^3-17x^2+7x+10} \\ \underline{4x^3-5x^2\qquad\qquad} \\ -12x^2+7x\quad \\ \underline{-12x^2+15x\quad} \\ -8x+10 \\ \underline{-8x+10} \\ 0 \end{array}$$

x^2-3x-2

17. $(\sqrt{x}-3\sqrt{y})(2\sqrt{x}+4\sqrt{y})$

$2\sqrt{x^2}+4\sqrt{xy}-6\sqrt{xy}-12\sqrt{y^2}$

$2x-2\sqrt{xy}-12y$

19. $\sqrt[3]{-\frac{8}{27}}=-\frac{2}{3}$

21. $32^{-\frac{1}{5}}=\frac{1}{32^{\frac{1}{5}}}=\frac{1}{2}$

23. $-9^{\frac{3}{2}}=-(9^{\frac{1}{2}})^3=-(3)^3=-27$

25. $\frac{1}{(\frac{2}{3})^{-3}}=\left(\frac{2}{3}\right)^3=\frac{2^3}{3^3}=\frac{8}{27}$

27. $6x^2+19x-20$

$(6x-5)(x+4)$

29. $9x^4+68x^2-32$

$(9x^2-4)(x^2+8)$

$(3x+2)(3x-2)(x^2+8)$

31. $27x^3-8y^3$

$(3x-2y)(9x^2+6xy+4y^2)$

33. $.06n+.08(n+50)=25$

$100[.06n+.08(n+50)]=100(25)$

$100(.06n)+100[.08(n+50)]=2500$

$6n+8(n+50)=2500$

$6n+8n+400=2500$

$14n=2100$

$n=150$

The solution set is $\{150\}$.

35. $\sqrt[3]{n^2-1}=-1$

$(\sqrt[3]{n^2-1})^3=(-1)^3$

$n^2-1=-1$

$n^2=0$

$n=0$

Check

$\sqrt[3]{0^2-1}\stackrel{?}{=}-1$

$\sqrt[3]{-1}\stackrel{?}{=}-1$

$-1=-1$

The solution set is $\{0\}$.

37. $a^2+14a+49=0$

$(a+7)(a+7)=0$

$(a+7)^2=0$

$a+7=0$

$a=-7$

The solution set is $\{-7\}$.

39. $\frac{2}{5x-2}=\frac{4}{6x+1};\ x\neq\frac{2}{5},\ x\neq-\frac{1}{6}$

$2(6x+1)=4(5x-2)$

$12x+2=20x-8$

$-8x=-10$

$x=\frac{-10}{-8}=\frac{5}{4}$

The solution set is $\left\{\frac{5}{4}\right\}$.

41. $5x-4=\sqrt{5x-4}$

$(5x-4)^2=(\sqrt{5x-4})^2$

$25x^2-40x+16=5x-4$

$25x^2-45x+20=0$

$5(5x^2-9x+4)=0$

$5(5x-4)(x-1)=0$

$5x-4=0$ or $x-1=0$

$5x=4$ or $x=1$

$x=\frac{4}{5}$ or $x=1$

Checking $x=\frac{4}{5}$

$5\left(\frac{4}{5}\right) - 4 \stackrel{?}{=} \sqrt{5\left(\frac{4}{5}\right) - 4}$

$4 - 4 \stackrel{?}{=} \sqrt{4 - 4}$

$0 \stackrel{?}{=} \sqrt{0}$

$0 = 0$

Checking $x = 1$

$5(1) - 4 \stackrel{?}{=} \sqrt{5(1) - 4}$

$5 - 4 \stackrel{?}{=} \sqrt{1}$

$1 = 1$

The solution set is $\left\{\frac{4}{5},\ 1\right\}$.

43. $(3x - 2)(4x - 1) = 0$

$3x - 2 = 0$ or $4x - 1 = 0$

$3x = 2$ or $4x = 1$

$x = \frac{2}{3}$ or $x = \frac{1}{4}$

The solution set is $\left\{\frac{1}{4},\ \frac{2}{3}\right\}$.

45. $\frac{5}{6x} - \frac{2}{3} = \frac{7}{10x};\ x \neq 0$

$30x\left(\frac{5}{6x} - \frac{2}{3}\right) = 30x\left(\frac{7}{10x}\right)$

$30x\left(\frac{5}{6x}\right) - 30x\left(\frac{2}{3}\right) = 21$

$25 - 20x = 21$

$-20x = -4$

$x = \frac{-4}{-20} = \frac{1}{5}$

The solution set is $\left\{\frac{1}{5}\right\}$.

47. $6x^4 - 23x^2 - 4 = 0$

$(6x^2 + 1)(x^2 - 4) = 0$

$(6x^2 + 1)(x + 2)(x - 2) = 0$

$6x^2 + 1 = 0$ or $x + 2 = 0$ or $x - 2 = 0$

$6x^2 = -1$ or $x = -2$ or $x = 2$

$x^2 = -\frac{1}{6}$

not a real number

The solution set is $\{-2,\ 2\}$.

49. $n^2 - 13n - 114 = 0$

$(n + 6)(n - 19) = 0$

$n + 6 = 0$ or $n - 19 = 0$

$n = -6$ or $n = 19$

The solution set is $\{-6,\ 19\}$.

51. $4(2x - 1) < 3(x + 5)$

$8x - 4 < 3x + 15$

$8x < 3x + 19$

$5x < 19$

$x < \frac{19}{5}$

The solution set is $\left(-\infty,\ \frac{19}{5}\right)$.

53. $|2x - 1| < 5$

$-5 < 2x - 1 < 5$

$-4 < 2x < 6$

$-2 < x < 3$

The solution set is $(-2,\ 3)$.

55. $\frac{1}{2}(3x - 1) - \frac{2}{3}(x + 4) \leq \frac{3}{4}(x - 1)$

$12\left[\frac{1}{2}(3x - 1) - \frac{2}{3}(x + 4)\right] \leq 12\left[\frac{3}{4}(x - 1)\right]$

$12\left[\frac{1}{2}(3x - 1)\right] - 12\left[\frac{2}{3}(x + 4)\right] \leq 9(x - 1)$

$6(3x - 1) - 8(x + 4) \leq 9x - 9$

$18x - 6 - 8x - 32 \leq 9x - 9$

$10x - 38 \leq 9x - 9$

$10x \leq 9x + 29$

$x \leq 29$

The solution set is $(-\infty,\ 29]$.

57. Let x = one part
$2250 - x$ = other part

$\frac{x}{2250 - x} = \frac{2}{3}$

$3x = 2(2250 - x)$

$3x = 4500 - 2x$

$5x = 4500$

$x = 900$

It is divided \$900 and \$1350.

59.

	Time in hours	Rate
Lolita	x	$\frac{1}{x}$
Doug	10	$\frac{1}{10}$
Together	$3\frac{20}{60} = \frac{10}{3}$	$\frac{1}{\frac{10}{3}} = \frac{3}{10}$

$\frac{1}{x} + \frac{1}{10} = \frac{3}{10}$

$10x\left(\frac{1}{x} + \frac{1}{10}\right) = 10x\left(\frac{3}{10}\right)$

$10x\left(\frac{1}{x}\right) + 10x\left(\frac{1}{10}\right) = 3x$

$10 + x = 3x$

$10 = 2x$

$5 = x$

It would take Lolita 5 hours.

61.

	rate •	time =	distance
First Jogger	$\frac{1}{8}$ mpm	x	$\frac{1}{8}x$
Second Jogger	$\frac{1}{6}$ mpm	x	$\frac{1}{6}x$

$\frac{1}{2} + \frac{1}{8}x = \frac{1}{6}x$

$24\left(\frac{1}{2} + \frac{1}{8}x\right) = 24\left(\frac{1}{6}x\right)$

$24\left(\frac{1}{2}\right) + 24\left(\frac{1}{8}x\right) = 24\left(\frac{1}{6}x\right)$

$12 + 3x = 4x$

$12 = x$

It would take 12 minutes.

Chapter 6 Quadratic Equations and Inequalities

Problem Set 6.1 Complex Numbers

1. False

3. True

5. True

7. True

9. $(6+3i)+(4+5i)$
$(6+4)+(3+5)i$
$10+8i$

11. $(-8+4i)+(2+6i)$
$(-8+2)+(4+6)i$
$-6+10i$

13. $(3+2i)-(5+7i)$
$3+2i-5-7i$
$(3-5)+(2-7)i$
$-2-5i$

15. $(-7+3i)-(5-2i)$
$-7+3i-5+2i$
$(-7-5)+(3+2)i$
$-12+5i$

17. $(-3-10i)+(2-13i)$
$(-3+2)+(-10-13)i$
$-1-23i$

19. $(4-8i)-(8-3i)$
$4-8i-8+3i$
$(4-8)+(-8+3)i$
$-4-5i$

21. $(-1-i)-(-2-4i)$
$-1-i+2+4i$
$(-1+2)+(-1+4)i$
$1+3i$

23. $\left(\frac{3}{2}+\frac{1}{3}i\right)+\left(\frac{1}{6}-\frac{3}{4}i\right)$
$\left(\frac{3}{2}+\frac{1}{6}\right)+\left(\frac{1}{3}-\frac{3}{4}\right)i$
$\frac{5}{3}-\frac{5}{12}i$

25. $\left(-\frac{5}{9}+\frac{3}{5}i\right)-\left(\frac{4}{3}-\frac{1}{6}i\right)$
$-\frac{5}{9}+\frac{3}{5}i-\frac{4}{3}+\frac{1}{6}i$
$\left(-\frac{5}{9}-\frac{4}{3}\right)+\left(\frac{3}{5}+\frac{1}{6}\right)i$
$-\frac{17}{9}+\frac{23}{30}i$

27. $\sqrt{-81}=\sqrt{-1}\sqrt{81}=i\sqrt{81}=9i$

29. $\sqrt{-14}=\sqrt{-1}\sqrt{14}=i\sqrt{14}$

31. $\sqrt{-\frac{16}{25}}=\sqrt{-1}\sqrt{\frac{16}{25}}=i\left(\frac{4}{5}\right)=\frac{4}{5}i$

33. $\sqrt{-18}=\sqrt{-1}\sqrt{18}=i\sqrt{9}\sqrt{2}$
$i(3)\sqrt{2}=3i\sqrt{2}$

35. $\sqrt{-75}=\sqrt{-1}\sqrt{75}=i\sqrt{25}\sqrt{3}$
$i(5)\sqrt{3}=5i\sqrt{3}$

37. $3\sqrt{-28}=3\sqrt{-1}\sqrt{28}=3i\sqrt{4}\sqrt{7}$
$3i(2)\sqrt{7}=6i\sqrt{7}$

39. $-2\sqrt{-80}=-2\sqrt{-1}\sqrt{80}=-2i\sqrt{16}\sqrt{5}$
$-2i(4)\sqrt{5}=-8i\sqrt{5}$

41. $12\sqrt{-90}=12\sqrt{-1}\sqrt{90}=12i\sqrt{9}\sqrt{10}$
$12i(3)\sqrt{10}=36i\sqrt{10}$

43. $\sqrt{-4}\sqrt{-16}=(i\sqrt{4})(i\sqrt{16})=(2i)(4i)$
$8i^2=8(-1)=-8$

45. $\sqrt{-3}\sqrt{-5}=(i\sqrt{3})(i\sqrt{5})$
$i^2\sqrt{15}=-\sqrt{15}$

47. $\sqrt{-9}\sqrt{-6} = (i\sqrt{9})(i\sqrt{6}) = (3i)(i\sqrt{6})$

$3i^2\sqrt{6} = 3(-1)\sqrt{6} = -3\sqrt{6}$

49. $\sqrt{-15}\sqrt{-5} = (i\sqrt{15})(i\sqrt{5}) = i^2\sqrt{75}$

$-1\sqrt{25}\sqrt{3} = -1(5)\sqrt{3} = -5\sqrt{3}$

51. $\sqrt{-2}\sqrt{-27} = (i\sqrt{2})(i\sqrt{27}) = i^2\sqrt{54}$

$-1\sqrt{9}\sqrt{6} = -1(3)\sqrt{6} = -3\sqrt{6}$

53. $\sqrt{6}\sqrt{-8} = (\sqrt{6})(i\sqrt{8}) = i\sqrt{48}$

$i\sqrt{16}\sqrt{3} = 4i\sqrt{3}$

55. $\frac{\sqrt{-25}}{\sqrt{-4}} = \frac{i\sqrt{25}}{i\sqrt{4}} = \frac{5i}{2i} = \frac{5}{2}$

57. $\frac{\sqrt{-56}}{\sqrt{-7}} = \frac{i\sqrt{56}}{i\sqrt{7}} = \sqrt{\frac{56}{7}}$

$\sqrt{8} = \sqrt{4}\sqrt{2} = 2\sqrt{2}$

59. $\frac{\sqrt{-24}}{\sqrt{6}} = \frac{i\sqrt{24}}{\sqrt{6}} = i\sqrt{\frac{24}{6}} = i\sqrt{4} = 2i$

61. $(5i)(4i) = 20i^2 = 20(-1) = -20 + 0i$

63. $(7i)(-6i) = -42i^2$

$-42(-1) = 42 + 0i$

65. $(3i)(2 - 5i)$

$6i - 15i^2$

$6i - 15(-1)$

$6i + 15$

$15 + 6i$

67. $(-6i)(-2 - 7i)$

$12i + 42i^2$

$12i + 42(-1)$

$12i - 42$

$-42 + 12i$

69. $(3 + 2i)(5 + 4i)$

$15 + 12i + 10i + 8i^2$

$15 + 22i + 8(-1)$

$15 + 22i - 8$

$7 + 22i$

71. $(6 - 2i)(7 - i)$

$42 - 6i - 14i + 2i^2$

$42 - 20i + 2(-1)$

$42 - 20i - 2$

$40 - 20i$

73. $(-3 - 2i)(5 + 6i)$

$-15 - 18i - 10i - 12i^2$

$-15 - 28i - 12(-1)$

$-15 - 28i + 12$

$-3 - 28i$

75. $(9 + 6i)(-1 - i)$

$-9 - 9i - 6i - 6i^2$

$-9 - 15i - 6(-1)$

$-9 - 15i + 6$

$-3 - 15i$

77. $(4 + 5i)^2$

$(4)^2 + 2(4)(5i) + (5i)^2$

$16 + 40i + 25i^2$

$16 + 40i + 25(-1)$

$16 + 40i - 25$

$-9 + 40i$

79. $(-2 - 4i)^2$

$(-2)^2 + 2(-2)(-4i) + (-4i)^2$

$4 + 16i + 16i^2$

$4 + 16i + 16(-1)$

$4 + 16i - 16$

$-12 + 16i$

81. $(6 + 7i)(6 - 7i)$

$36 - 42i + 42i - 49i^2$

$36 - 49(-1) + 0i$

$36 + 49 + 0i$

$85 + 0i$

83. $(-1+2i)(-1-2i)$

$1+2i-2i-4i^2$

$1-4(-1)+0i$

$1+4+0i$

$5+0i$

85. $\frac{3i}{2+4i}=\frac{3i}{(2+4i)}\bullet\frac{(2-4i)}{(2-4i)}=\frac{6i-12i^2}{4-16i^2}$

$\frac{6i-12(-1)}{4-16(-1)}=\frac{6i+12}{4+16}=\frac{12+6i}{20}$

$\frac{12}{20}+\frac{6}{20}i=\frac{3}{5}+\frac{3}{10}i$

87. $\frac{-2i}{3-5i}=\frac{-2i}{(3-5i)}\bullet\frac{(3+5i)}{(3+5i)}=\frac{-6i-10i^2}{9-25i^2}$

$\frac{-6i-10(-1)}{9-25(-1)}=\frac{-6i+10}{9+25}=\frac{10-6i}{34}$

$\frac{10}{34}-\frac{6}{34}i=\frac{5}{17}-\frac{3}{17}i$

89. $\frac{-2+6i}{3i}=\frac{(-2+6i)}{3i}\bullet\frac{(-i)}{(-i)}=\frac{2i-6i^2}{-3i^2}$

$\frac{2i-6(-1)}{-3(-1)}=\frac{2i+6}{3}=\frac{6+2i}{3}$

$\frac{6}{3}+\frac{2}{3}i=2+\frac{2}{3}i$

91. $\frac{2}{7i}=\frac{2}{7i}\bullet\frac{(-i)}{(-i)}=\frac{-2i}{-7i^2}$

$\frac{-2i}{-7(-1)}=0-\frac{2}{7}i$

93. $\frac{2+6i}{1+7i}=\frac{(2+6i)}{(1+7i)}\bullet\frac{(1-7i)}{(1-7i)}$

$\frac{2-14i+6i-42i^2}{1-49i^2}=\frac{2-8i-42(-1)}{1-49(-1)}$

$\frac{2-8i+42}{1+49}=\frac{44-8i}{50}=\frac{44}{50}-\frac{8}{50}i=\frac{22}{25}-\frac{4}{25}i$

95. $\frac{3+6i}{4-5i}=\frac{(3+6i)}{(4-5i)}\bullet\frac{(4+5i)}{(4+5i)}$

$\frac{12+15i+24i+30i^2}{16-25i^2}=\frac{12+39i+30(-1)}{16+25}$

$\frac{12+39i-30}{41}=\frac{-18+39i}{41}=-\frac{18}{41}+\frac{39}{41}i$

97. $\frac{-2+7i}{-1+i}=\frac{(-2+7i)}{(-1+i)}\bullet\frac{(-1-i)}{(-1-i)}$

$\frac{2+2i-7i-7i^2}{1-i^2}=\frac{2-5i-7(-1)}{1-(-1)}$

$\frac{2-5i+7}{1+1}=\frac{9-5i}{2}=\frac{9}{2}-\frac{5}{2}i$

99. $\frac{-1-3i}{-2-10i}=\frac{(-1-3i)}{(-2-10i)}\bullet\frac{(-2+10i)}{(-2+10i)}$

$\frac{2-10i+6i-30i^2}{4-100i^2}=\frac{2-4i-30(-1)}{4-100(-1)}$

$\frac{2-4i+30}{4+100}=\frac{32-4i}{104}$

$\frac{32}{104}-\frac{4}{104}i=\frac{4}{13}-\frac{1}{26}i$

Problem Set 6.2 Quadratic Equations

1. $x^2-9x=0$

$x(x-9)=0$

$x=0$ or $x-9=0$

$x=0$ or $x=9$

The solution set is $\{0, 9\}$.

3. $x^2=-3x$

$x^2+3x=0$

$x(x+3)=0$

$x=0$ or $x+3=0$

$x=0$ or $x=-3$

The solution set is $\{-3, 0\}$.

5. $3y^2+12y=0$

$3y(y+4)=0$

$3y=0$ or $y+4=0$

$y=0$ or $y=-4$

The solution set is $\{-4, 0\}$.

7. $5n^2-9n=0$

$n(5n-9)=0$

$n=0$ or $5n-9=0$

$n = 0$ or $5n = 9$

$n = 0$ or $n = \frac{9}{5}$

The solution set is $\left\{0, \frac{9}{5}\right\}$.

9. $x^2 + x - 30 = 0$

$(x+6)(x-5) = 0$

$x + 6 = 0$ or $x - 5 = 0$

$x = -6$ or $x = 5$

The solution set is $\{-6, 5\}$.

11. $x^2 - 19x + 84 = 0$

$(x-12)(x-7) = 0$

$x - 12 = 0$ or $x - 7 = 0$

$x = 12$ or $x = 7$

The solution set is $\{7, 12\}$.

13. $2x^2 + 19x + 24 = 0$

$(2x+3)(x+8) = 0$

$2x + 3 = 0$ or $x + 8 = 0$

$2x = -3$ or $x = -8$

$x = -\frac{3}{2}$ or $x = -8$

The solution set is $\left\{-8, -\frac{3}{2}\right\}$.

15. $15x^2 + 29x - 14 = 0$

$(3x+7)(5x-2) = 0$

$3x + 7 = 0$ or $5x - 2 = 0$

$3x = -7$ or $5x = 2$

$x = -\frac{7}{3}$ or $x = \frac{2}{5}$

The solution set is $\left\{-\frac{7}{3}, \frac{2}{5}\right\}$.

17. $25x^2 - 30x + 9 = 0$

$(5x-3)(5x-3) = 0$

$(5x-3)^2 = 0$

$5x - 3 = 0$

$5x = 3$

$x = \frac{3}{5}$

The solution set is $\left\{\frac{3}{5}\right\}$.

19. $6x^2 - 5x - 21 = 0$

$(2x+3)(3x-7) = 0$

$2x + 3 = 0$ or $3x - 7 = 0$

$2x = -3$ or $3x = 7$

$x = -\frac{3}{2}$ or $x = \frac{7}{3}$

The solution set is $\left\{-\frac{3}{2}, \frac{7}{3}\right\}$.

21. $3\sqrt{x} = x + 2$

$(3\sqrt{x})^2 = (x+2)^2$

$9x = x^2 + 4x + 4$

$0 = x^2 - 5x + 4$

$0 = (x-4)(x-1)$

$x - 4 = 0$ or $x - 1 = 0$

$x = 4$ or $x = 1$

Checking $x = 4$

$3\sqrt{4} \stackrel{?}{=} 4 + 2$

$3(2) \stackrel{?}{=} 6$

$6 = 6$

Checking $x = 1$

$3\sqrt{1} \stackrel{?}{=} 1 + 2$

$3(1) \stackrel{?}{=} 3$

$3 = 3$

The solution set is $\{1, 4\}$.

23. $\sqrt{2x} = x - 4$

$(\sqrt{2x})^2 = (x-4)^2$

$2x = x^2 - 8x + 16$

$0 = x^2 - 10x + 16$

$0 = (x-8)(x-2)$

$x - 8 = 0$ or $x - 2 = 0$

$x = 8$ or $x = 2$

Checking $x = 8$

$\sqrt{2(8)} \stackrel{?}{=} 8 - 4$

$\sqrt{16} \stackrel{?}{=} 4$

$4 = 4$

Checking $x=2$

$\sqrt{2(2)} \stackrel{?}{=} 2-4$

$\sqrt{4} \stackrel{?}{=} -2$

$2 \neq -2$

The solution set is $\{8\}$.

25. $\sqrt{3x}+6=x$

$\sqrt{3x}=x-6$

$(\sqrt{3x})^2=(x-6)^2$

$3x=x^2-12x+36$

$0=x^2-15x+36$

$0=(x-12)(x-3)$

$x-12=0$ or $x-3=0$

$x=12$ or $x=3$

Checking $x=12$

$\sqrt{3(12)}+6 \stackrel{?}{=} 12$

$\sqrt{36}+6 \stackrel{?}{=} 12$

$12=12$

Checking $x=3$

$\sqrt{3(3)}+6 \stackrel{?}{=} 3$

$\sqrt{9}+6 \stackrel{?}{=} 3$

$3+6 \stackrel{?}{=} 3$

$9 \neq 3$

The solution set is $\{12\}$.

27. $x^2-5kx=0$

$x(x-5k)=0$

$x=0$ or $x-5k=0$

$x=0$ or $x=5k$

The solution set is $\{0, 5k\}$.

29. $x^2=16k^2x$

$x^2-16k^2x=0$

$x(x-16k^2)=0$

$x=0$ or $x-16k^2=0$

$x=0$ or $x=16k^2$

The solution set is $\{0, 16k^2\}$.

31. $x^2-12kx+35k^2=0$

$(x-5k)(x-7k)=0$

$x-5k=0$ or $x-7k=0$

$x=5k$ or $x=7k$

The solution set is $\{5k, 7k\}$.

33. $2x^2+5kx-3k^2=0$

$(2x-k)(x+3k)=0$

$2x-k=0$ or $x+3k=0$

$2x=k$ or $x=-3k$

$x=\frac{k}{2}$ or $x=-3k$

The solution set is $\left\{-3k, \frac{k}{2}\right\}$.

35. $x^2=1$

$x=\pm\sqrt{1}$

$x=\pm 1$

The solution set is $\{-1, 1\}$.

37. $x^2=-36$

$x=\pm\sqrt{-36}$

$x=\pm i\sqrt{36}$

$x=\pm 6i$

The solution set is $\{-6i, 6i\}$.

39. $x^2=14$

$x=\pm\sqrt{14}$

The solution set is $\{-\sqrt{14}, \sqrt{14}\}$.

41. $n^2-28=0$

$n^2=28$

$n=\pm\sqrt{28}$

$n=\pm 2\sqrt{7}$

The solution set is $\{-2\sqrt{7}, 2\sqrt{7}\}$.

43. $3t^2=54$

$t^2=18$

$t=\pm\sqrt{18}$

$t=\pm 3\sqrt{2}$

The solution set is $\{-3\sqrt{2}, 3\sqrt{2}\}$.

45. $2t^2 = 7$

$t^2 = \frac{7}{2}$

$t = \pm\sqrt{\frac{7}{2}}$

$t = \pm\frac{\sqrt{7}}{\sqrt{2}} \bullet \frac{\sqrt{2}}{\sqrt{2}}$

$t = \pm\frac{\sqrt{14}}{2}$

The solution set is $\left\{-\frac{\sqrt{14}}{2}, \frac{\sqrt{14}}{2}\right\}$.

47. $15y^2 = 20$

$y^2 = \frac{20}{15}$

$y^2 = \frac{4}{3}$

$y = \pm\sqrt{\frac{4}{3}}$

$y = \pm\frac{2}{\sqrt{3}}$

$y = \pm\frac{2}{\sqrt{3}} \bullet \frac{\sqrt{3}}{\sqrt{3}}$

$y = \pm\frac{2\sqrt{3}}{3}$

The solution set is $\left\{-\frac{2\sqrt{3}}{3}, \frac{2\sqrt{3}}{3}\right\}$.

49. $10x^2 + 48 = 0$

$10x^2 = -48$

$x^2 = -\frac{48}{10}$

$x^2 = -\frac{24}{5}$

$x = \pm\sqrt{-\frac{24}{5}}$

$x = \pm\frac{i\sqrt{24}}{\sqrt{5}}$

$x = \pm\frac{i\sqrt{4}\sqrt{6}}{\sqrt{5}}$

$x = \pm\frac{2i\sqrt{6}}{\sqrt{5}} \bullet \frac{\sqrt{5}}{\sqrt{5}}$

$x = \pm\frac{2i\sqrt{30}}{5}$

The solution set is $\left\{-\frac{2i\sqrt{30}}{5}, \frac{2i\sqrt{30}}{5}\right\}$.

51. $24x^2 = 36$

$x^2 = \frac{36}{24}$

$x^2 = \frac{3}{2}$

$x = \pm\sqrt{\frac{3}{2}}$

$x = \pm\frac{\sqrt{3}}{\sqrt{2}} \bullet \frac{\sqrt{2}}{\sqrt{2}}$

$x = \pm\frac{\sqrt{6}}{2}$

The solution set is $\left\{-\frac{\sqrt{6}}{2}, \frac{\sqrt{6}}{2}\right\}$.

53. $(x-2)^2 = 9$

$x - 2 = \pm\sqrt{9}$

$x - 2 = \pm 3$

$x = 2 \pm 3$

$x = 2 + 3$ or $x = 2 - 3$

$x = 5$ or $x = -1$

The solution set is $\{-1, 5\}$.

55. $(x+3)^2 = 25$

$x + 3 = \pm\sqrt{25}$

$x + 3 = \pm 5$

$x = -3 \pm 5$

$x = -3 + 5$ or $x = -3 - 5$

$x = 2$ or $x = -8$

The solution set is $\{-8, 2\}$.

57. $(x+6)^2 = -4$

$x + 6 = \pm\sqrt{-4}$

$x + 6 = \pm i\sqrt{4}$

$x + 6 = \pm 2i$

$x = -6 \pm 2i$

The solution set is $\{-6-2i, -6+2i\}$.

59. $(2x-3)^2 = 1$

$2x - 3 = \pm\sqrt{1}$

$2x - 3 = \pm 1$

$2x = 3 \pm 1$

$x = \frac{3 \pm 1}{2}$

$x = \frac{3+1}{2}$ or $x = \frac{3-1}{2}$

$x = 2$ or $x = 1$

The solution set is {1, 2}.

61. $(n-4)^2 = 5$

$n - 4 = \pm\sqrt{5}$

$n = 4 \pm \sqrt{5}$

The solution set is $\{4-\sqrt{5},\ 4+\sqrt{5}\}$.

63. $(t+5)^2 = 12$

$t + 5 = \pm\sqrt{12}$

$t + 5 = \pm\sqrt{4}\sqrt{3}$

$t + 5 = \pm 2\sqrt{3}$

$t = -5 \pm 2\sqrt{3}$

The solution set is $\{-5-2\sqrt{3}, -5+2\sqrt{3}\}$.

65. $(3y-2)^2 = -27$

$3y - 2 = \pm\sqrt{-27}$

$3y - 2 = \pm i\sqrt{27}$

$3y - 2 = \pm i\sqrt{9}\sqrt{3}$

$3y - 2 = \pm 3i\sqrt{3}$

$3y = 2 \pm 3i\sqrt{3}$

$y = \frac{2 \pm 3i\sqrt{3}}{3}$

The solution set is $\left\{\frac{2 \pm 3i\sqrt{3}}{3}\right\}$.

67. $3(x+7)^2 + 4 = 79$

$3(x+7)^2 = 75$

$(x+7)^2 = 25$

$x + 7 = \pm\sqrt{25}$

$x + 7 = \pm 5$

$x = -7 \pm 5$

$x = -7 - 5$ or $x = -7 + 5$

$x = -12$ or $x = -2$

The solution set is $\{-12, -2\}$.

69. $2(5x-2)^2 + 5 = 25$

$2(5x-2)^2 = 20$

$(5x-2)^2 = 10$

$5x - 2 = \pm\sqrt{10}$

$5x = 2 \pm \sqrt{10}$

$x = \frac{2 \pm \sqrt{10}}{5}$

The solution set is $\left\{\frac{2 \pm \sqrt{10}}{5}\right\}$.

71. $4^2 + 6^2 = c^2$

$16 + 36 = c^2$

$52 = c^2$

$\sqrt{52} = c$

$\sqrt{4}\sqrt{13} = c$

$2\sqrt{13} = c$

$c = 2\sqrt{13}$ centimeters

73. $a^2 + 8^2 = 12^2$

$a^2 + 64 = 144$

$a^2 = 80$

$a = \sqrt{80}$

$a = \sqrt{16}\sqrt{5}$

$a = 4\sqrt{5}$ inches

75. $15^2 + b^2 = 17^2$

$225 + b^2 = 289$

$b^2 = 64$

$b = \sqrt{64} = 8$

$b = 8$ yards

77. $6^2 + 6^2 = c^2$

$36 + 36 = c^2$

$72 = c^2$

$\sqrt{72} = c$

$\sqrt{36}\sqrt{2} = c$

$6\sqrt{2} = c$

$c = 6\sqrt{2}$ inches

79. $a^2+a^2=8^2$

$2a^2=64$

$a^2=32$

$a=\sqrt{32}$

$a=\sqrt{16}\sqrt{2}$

$a=4\sqrt{2}$ meters

$b=4\sqrt{2}$ meters

81. $a=3$ then $c=2a=2(3)=6$

$3^2+b^2=6^2$

$9+b^2=36$

$b^2=27$

$b=\sqrt{27}$

$b=\sqrt{9}\sqrt{3}=3\sqrt{3}$ inches

$c=6$ inches and $b=3\sqrt{3}$ inches.

83. $c=14$ then $a=\frac{1}{2}c=\frac{1}{2}(14)=7$

$7^2+b^2=14^2$

$49+b^2=196$

$b^2=147$

$b=\sqrt{147}=\sqrt{49}\sqrt{3}=7\sqrt{3}$

$a=7$ centimeters and $b=7\sqrt{3}$ centimeters

85. $b=10 \quad a=\frac{1}{2}c$

$\left(\frac{1}{2}c\right)^2+10^2=c^2$

$\frac{1}{4}c^2+100=c^2$

$4\left(\frac{1}{4}c^2+100\right)=4(c^2)$

$c^2+400=4c^2$

$400=3c^2$

$\frac{400}{3}=c^2$

$\sqrt{\frac{400}{3}}=c$

$c=\frac{\sqrt{400}}{\sqrt{3}}=\frac{20}{\sqrt{3}}\bullet\frac{\sqrt{3}}{\sqrt{3}}=\frac{20\sqrt{3}}{3}$

$a=\frac{1}{2}\left(\frac{20\sqrt{3}}{3}\right)=\frac{10\sqrt{3}}{3}$

$c=\frac{20\sqrt{3}}{3}$ feet and $a=\frac{10\sqrt{3}}{3}$ feet.

87. $a^2+16^2=24^2$

$a^2+256=576$

$a^2=320$

$a=\sqrt{320}\approx 17.9$ feet

The ladder is 17.9 feet from the foundation of the house.

89. $16^2+34^2=c^2$

$256+1156=c^2$

$1412=c^2$

$c=\sqrt{1412}\approx 38$ meters

The diagonal is 38 meters.

91. Let $x=$ length of a side.

$x^2+x^2=75^2$

$2x^2=5625$

$x^2=2812.5$

$x=\sqrt{2812.5}\approx 53$ meters

The length of a side is 53 meters.

Problem Set 6.3 Completing the Square

1a. $x^2-4x-60=0$

$(x-10)(x+6)=0$

$x-10=0$ or $x-6=0$

$x=10$ or $x=-6$

b. $x^2-4x-60=0$

$x^2-4x=60$

$x^2-4x+4=60+4$

$(x-2)^2=64$

$x-2=\pm\sqrt{64}$

$x-2=\pm 8$

$x=2\pm 8$

$x=2+8$ or $x=2-8$

$x=10$ or $x=-6$

The solution set is $\{-6, 10\}$.

3a. $x^2 - 14x = -40$

$x^2 - 14x + 40 = 0$

$(x-10)(x-4) = 0$

$x - 10 = 0$ or $x - 4 = 0$

$x = 10$ or $x = 4$

b. $x^2 - 14x = -40$

$x^2 - 14x + 49 = -40 + 49$

$(x-7)^2 = 9$

$x - 7 = \pm\sqrt{9}$

$x - 7 = \pm 3$

$x = 7 \pm 3$

$x = 7 + 3$ or $x = 7 - 3$

$x = 10$ or $x = 4$

The solution set is {4, 10}.

5a. $x^2 - 5x - 50 = 0$

$(x-10)(x+5) = 0$

$x - 10 = 0$ or $x + 5 = 0$

$x = 10$ or $x = -5$

b. $x^2 - 5x - 50 = 0$

$x^2 - 5x = 50$

$x^2 - 5x + \frac{25}{4} = 50 + \frac{25}{4}$

$\left(x - \frac{5}{2}\right)^2 = \frac{225}{4}$

$x - \frac{5}{2} = \pm\sqrt{\frac{225}{4}}$

$x - \frac{5}{2} = \pm\frac{15}{2}$

$x = \frac{5}{2} \pm \frac{15}{2}$

$x = \frac{5}{2} + \frac{15}{2}$ or $x = \frac{5}{2} - \frac{15}{2}$

$x = \frac{20}{2}$ or $x = \frac{-10}{2}$

$x = 10$ or $x = -5$

The solution set is {−5, 10}.

7a. $x(x+7) = 8$

$x^2 + 7x = 8$

$x^2 + 7x - 8 = 0$

$(x+8)(x-1) = 0$

$x + 8 = 0$ or $x - 1 = 0$

$x = -8$ or $x = 1$

b. $x(x+7) = 8$

$x^2 + 7x = 8$

$x^2 + 7x + \frac{49}{4} = 8 + \frac{49}{4}$

$\left(x + \frac{7}{2}\right)^2 = \frac{81}{4}$

$x + \frac{7}{2} = \pm\sqrt{\frac{81}{4}}$

$x + \frac{7}{2} = \pm\frac{9}{2}$

$x = -\frac{7}{2} + \frac{9}{2}$ or $x = -\frac{7}{2} - \frac{9}{2}$

$x = \frac{2}{2} = 1$ or $x = -\frac{16}{2} = -8$

The solution set is {−8, 1}.

9a. $2n^2 - n - 15 = 0$

$(2n+5)(n-3) = 0$

$2n + 5 = 0$ or $n - 3 = 0$

$2n = -5$ or $n = 3$

$n = -\frac{5}{2}$ or $n = 3$

b. $2n^2 - n - 15 = 0$

$n^2 - \frac{1}{2}n - \frac{15}{2} = 0$

$n^2 - \frac{1}{2}n = \frac{15}{2}$

$n^2 - \frac{1}{2}n + \frac{1}{16} = \frac{15}{2} + \frac{1}{16}$

$\left(n - \frac{1}{4}\right)^2 = \frac{121}{16}$

$n - \frac{1}{4} = \pm\sqrt{\frac{121}{16}}$

$n - \frac{1}{4} = \pm\frac{11}{4}$

$n = \frac{1}{4} \pm \frac{11}{4}$

$n = \frac{1}{4} + \frac{11}{4}$ or $n = \frac{1}{4} - \frac{11}{4}$

$n = \frac{12}{4}$ or $n = -\frac{10}{4}$

$n = 3$ or $n = -\frac{5}{2}$

The solution set is $\left\{-\frac{5}{2}, 3\right\}$.

11a. $3n^2+7n-6=0$

$(3n-2)(n+3)=0$

$3n-2=0$ or $n+3=0$

$3n=2$ or $n=-3$

$n=\frac{2}{3}$ or $n=-3$

b. $3n^2+7n-6=0$

$n^2+\frac{7}{3}n-2=0$

$n^2+\frac{7}{3}n=2$

$n^2+\frac{7}{3}n+\frac{49}{36}=2+\frac{49}{36}$

$\left(n+\frac{7}{6}\right)^2=\frac{121}{36}$

$n+\frac{7}{6}=\pm\sqrt{\frac{121}{36}}$

$n+\frac{7}{6}=\pm\frac{11}{6}$

$n=-\frac{7}{6}\pm\frac{11}{6}$

$n=-\frac{7}{6}+\frac{11}{6}$ or $n=-\frac{7}{6}-\frac{11}{6}$

$n=\frac{4}{6}=\frac{2}{3}$ or $n=-\frac{18}{6}=-3$

The solution set is $\left\{-3, \frac{2}{3}\right\}$.

13a. $n(n+6)=160$

$n^2+6n=160$

$n^2+6n-160=0$

$(n+16)(n-10)=0$

$n+16=0$ or $n-10=0$

$n=-16$ or $n=10$

b. $n(n+6)=160$

$n^2+6n=160$

$n^2+6n+9=160+9$

$(n+3)^2=169$

$n+3=\pm\sqrt{169}$

$n+3=\pm 13$

$n=-3+13$ or $n=-3-13$

$n=10$ or $n=-16$

The solution set is $\{-16, 10\}$.

15. $x^2+4x-2=0$

$x^2+4x=2$

$x^2+4x+4=2+4$

$(x+2)^2=6$

$x+2=\pm\sqrt{6}$

$x=-2\pm\sqrt{6}$

The solution set is $\{-2\pm\sqrt{6}\}$.

17. $x^2+6x-3=0$

$x^2+6x=3$

$x^2+6x+9=3+9$

$(x+3)^2=12$

$x+3=\pm\sqrt{12}$

$x+3=\pm 2\sqrt{3}$

$x=-3\pm 2\sqrt{3}$

The solution set is $\{-3\pm 2\sqrt{3}\}$.

19. $y^2-10y=1$

$y^2-10y+25=1+25$

$(y-5)^2=26$

$y-5=\pm\sqrt{26}$

$y=5\pm\sqrt{26}$

The solution set is $\{5\pm\sqrt{26}\}$.

21. $n^2-8n+17=0$

$n^2-8n=-17$

$n^2-8n+16=-17+16$

$(n-4)^2=-1$

$n-4=\pm\sqrt{-1}$

$n-4=\pm i$

$n=4\pm i$

The solution set is $\{4\pm i\}$.

23. $n(n+12)=-9$

$n^2+12n=-9$

$n^2+12n+36=-9+36$

$(n+6)^2=27$

$n+6=\pm\sqrt{27}$

$n+6=\pm 3\sqrt{3}$

$n=-6\pm 3\sqrt{3}$

The solution set is $\{-6\pm 3\sqrt{3}\}$.

25. $n^2+2n+6=0$

$n^2+2n=-6$

$n^2+2n+1=-6+1$

$(n+1)^2=-5$

$n+1=\pm\sqrt{-5}$

$n+1=\pm i\sqrt{5}$

$n=-1\pm i\sqrt{5}$

The solution set is $\{-1\pm i\sqrt{5}\}$.

27. $x^2+3x-2=0$

$x^2+3x=2$

$x^2+3x+\frac{9}{4}=2+\frac{9}{4}$

$\left(x+\frac{3}{2}\right)^2=\frac{17}{4}$

$x+\frac{3}{2}=\pm\sqrt{\frac{17}{4}}$

$x+\frac{3}{2}=\pm\frac{\sqrt{17}}{2}$

$x=-\frac{3}{2}\pm\frac{\sqrt{17}}{2}$

$x=\frac{-3\pm\sqrt{17}}{2}$

The solution set is $\left\{\frac{-3\pm\sqrt{17}}{2}\right\}$.

29. $x^2+5x+1=0$

$x^2+5x=-1$

$x^2+5x+\frac{25}{4}=-1+\frac{25}{4}$

$\left(x+\frac{5}{2}\right)^2=\frac{21}{4}$

$x+\frac{5}{2}=\pm\sqrt{\frac{21}{4}}$

$x+\frac{5}{2}=\pm\frac{\sqrt{21}}{2}$

$x=-\frac{5}{2}\pm\frac{\sqrt{21}}{2}$

$x=\frac{-5\pm\sqrt{21}}{2}$

The solution set is $\left\{\frac{-5\pm\sqrt{21}}{2}\right\}$.

31. $y^2-7y+3=0$

$y^2-7y=-3$

$y^2-7y+\frac{49}{4}=-3+\frac{49}{4}$

$\left(y+\frac{7}{2}\right)^2=\frac{37}{4}$

$y+\frac{7}{2}=\pm\sqrt{\frac{37}{4}}$

$y+\frac{7}{2}=\pm\frac{\sqrt{37}}{2}$

$y=-\frac{7}{2}\pm\frac{\sqrt{37}}{2}$

$y=\frac{-7\pm\sqrt{37}}{2}$

The solution set is $\left\{\frac{-7\pm\sqrt{37}}{2}\right\}$.

33. $2x^2+4x-3=0$

$x^2+2x-\frac{3}{2}=0$

$x^2+2x=\frac{3}{2}$

$x^2+2x+1=\frac{3}{2}+1$

$(x+1)^2=\frac{5}{2}$

$x+1=\pm\sqrt{\frac{5}{2}}$

$x+1=\pm\frac{\sqrt{5}}{\sqrt{2}}\bullet\frac{\sqrt{2}}{\sqrt{2}}$

$x+1=\pm\frac{\sqrt{10}}{2}$

$x=-1\pm\frac{\sqrt{10}}{2}$

$x=-\frac{2}{2}\pm\frac{\sqrt{10}}{2}$

$x=\frac{-2\pm\sqrt{10}}{2}$

The solution set is $\left\{\frac{-2\pm\sqrt{10}}{2}\right\}$.

35. $3n^2 - 6n + 5 = 0$

$n^2 - 2n + \frac{5}{3} = 0$

$n^2 - 2n = -\frac{5}{3}$

$n^2 - 2n + 1 = -\frac{5}{3} + 1$

$(n-1)^2 = -\frac{2}{3}$

$n - 1 = \pm\sqrt{-\frac{2}{3}}$

$n - 1 = \pm\frac{i\sqrt{2}}{\sqrt{3}}$

$n - 1 = \pm\frac{i\sqrt{2}}{\sqrt{3}} \bullet \frac{\sqrt{3}}{\sqrt{3}}$

$n - 1 = \pm\frac{i\sqrt{6}}{3}$

$n = 1 \pm \frac{i\sqrt{6}}{3}$

$n = \frac{3}{3} \pm \frac{i\sqrt{6}}{3}$

$n = \frac{3 \pm i\sqrt{6}}{3}$

The solution set is $\left\{\frac{3 \pm i\sqrt{6}}{3}\right\}$.

37. $3x^2 + 5x - 1 = 0$

$x^2 + \frac{5}{3}x - \frac{1}{3} = 0$

$x^2 + \frac{5}{3}x = \frac{1}{3}$

$x^2 + \frac{5}{3}x + \frac{25}{36} = \frac{1}{3} + \frac{25}{36}$

$\left(x + \frac{5}{6}\right)^2 = \frac{37}{36}$

$x + \frac{5}{6} = \pm\sqrt{\frac{37}{36}}$

$x + \frac{5}{6} = \pm\frac{\sqrt{37}}{6}$

$x = -\frac{5}{6} \pm \frac{\sqrt{37}}{6}$

$x = \frac{-5 \pm \sqrt{37}}{6}$

The solution set is $\left\{\frac{-5 \pm \sqrt{37}}{6}\right\}$.

39. $x^2 + 8x - 48 = 0$

$(x+12)(x-4) = 0$

$x + 12 = 0$ or $x - 4 = 0$

$x = -12$ or $x = 4$

The solution set is $\{-12, 4\}$.

41. $2n^2 - 8n = -3$

$n^2 - 4n = -\frac{3}{2}$

$n^2 - 4n + 4 = -\frac{3}{2} + 4$

$(n-2)^2 = \frac{5}{2}$

$n - 2 = \pm\sqrt{\frac{5}{2}}$

$n - 2 = \pm\frac{\sqrt{5}}{\sqrt{2}} \bullet \frac{\sqrt{2}}{\sqrt{2}}$

$n - 2 = \pm\frac{\sqrt{10}}{2}$

$n = 2 \pm \frac{\sqrt{10}}{2}$

$n = \frac{4}{2} \pm \frac{\sqrt{10}}{2}$

$n = \frac{4 \pm \sqrt{10}}{2}$

The solution set is $\left\{\frac{4 \pm \sqrt{10}}{2}\right\}$.

43. $(3x-1)(2x+9) = 0$

$3x - 1 = 0$ or $2x + 9 = 0$

$3x = 1$ or $2x = -9$

$x = \frac{1}{3}$ or $x = -\frac{9}{2}$

The solution set is $\left\{-\frac{9}{2}, \frac{1}{3}\right\}$.

45. $(x+2)(x-7) = 10$

$x^2 - 5x - 14 = 10$

$x^2 - 5x - 24 = 0$

$(x-8)(x+3) = 0$

$x - 8 = 0$ or $x + 3 = 0$

$x = 8$ or $x = -3$

The solution set is $\{-3, 8\}$.

47. $(x-3)^2 = 12$

$x-3 = \pm\sqrt{12}$

$x-3 = \pm 2\sqrt{3}$

$x = 3 \pm 2\sqrt{3}$

The solution set is $\{3 \pm 2\sqrt{3}\}$.

49. $3n^2 - 6n + 4 = 0$

$n^2 - 2n + \frac{4}{3} = 0$

$n^2 - 2n = -\frac{4}{3}$

$n^2 - 2n + 1 = -\frac{4}{3} + 1$

$(n-1)^2 = -\frac{1}{3}$

$n - 1 = \pm\sqrt{-\frac{1}{3}}$

$n - 1 = \pm\frac{i}{\sqrt{3}}$

$n - 1 = \pm\frac{i}{\sqrt{3}} \bullet \frac{\sqrt{3}}{\sqrt{3}}$

$n - 1 = \pm\frac{i\sqrt{3}}{3}$

$n = 1 \pm \frac{i\sqrt{3}}{3}$

$n = \frac{3}{3} \pm \frac{i\sqrt{3}}{3}$

$n = \frac{3 \pm i\sqrt{3}}{3}$

The solution set is $\left\{\frac{3 \pm i\sqrt{3}}{3}\right\}$.

51. $n(n+8) = 240$

$n^2 + 8n = 240$

$n^2 + 8n - 240 = 0$

$(n+20)(n-12) = 0$

$n + 20 = 0$ or $n - 12 = 0$

$n = -20$ or $n = 12$

The solution set is $\{-20, 12\}$.

53. $3x^2 + 29x = -66$

$3x^2 + 29x + 66 = 0$

$(3x+11)(x+6) = 0$

$3x + 11 = 0$ or $x + 6 = 0$

$3x = -11$ or $x = -6$

$x = -\frac{11}{3}$ or $x = -6$

The solution set is $\left\{-6, -\frac{11}{3}\right\}$.

55. $6n^2 + 23n + 21 = 0$

$(3n+7)(2n+3) = 0$

$3n + 7 = 0$ or $2n + 3 = 0$

$3n = -7$ or $2n = -3$

$n = -\frac{7}{3}$ or $x = -\frac{3}{2}$

The solution set is $\left\{-\frac{7}{3}, -\frac{3}{2}\right\}$.

57. $x^2 + 12x = 4$

$x^2 + 12x + 36 = 4 + 36$

$(x+6)^2 = 40$

$x + 6 = \pm\sqrt{40}$

$x + 6 = \pm 2\sqrt{10}$

$x = -6 \pm 2\sqrt{10}$

The solution set is $\{-6 \pm 2\sqrt{10}\}$.

59. $12n^2 - 7n + 1 = 0$

$n^2 - \frac{7}{12}n + \frac{1}{12} = 0$

$n^2 - \frac{7}{12}n = -\frac{1}{12}$

$n^2 - \frac{7}{12}n + \frac{49}{576} = -\frac{1}{12} + \frac{49}{576}$

$\left(n - \frac{7}{24}\right)^2 = \frac{1}{576}$

$n - \frac{7}{24} = \pm\sqrt{\frac{1}{576}}$

$n - \frac{7}{24} = \pm\frac{1}{24}$

$n = \frac{7}{24} \pm \frac{1}{24}$

$n = \frac{7}{24} + \frac{1}{24}$ or $n = \frac{7}{24} - \frac{1}{24}$

$n = \frac{8}{24}$ or $n = \frac{6}{24}$

$n = \frac{1}{3}$ or $n = \frac{1}{4}$

The solution set is $\left\{\frac{1}{4}, \frac{1}{3}\right\}$.

61. $ax^2+bx+c=0$

$x^2+\frac{b}{a}x+\frac{c}{a}=0$

$x^2+\frac{b}{a}x=-\frac{c}{a}$

$x^2+\frac{b}{a}x+\frac{b^2}{4a^2}=-\frac{c}{a}+\frac{b^2}{4a^2}$

$\left(x+\frac{b}{2a}\right)^2=\frac{b^2}{4a^2}-\frac{c}{a}$

$\left(x+\frac{b}{2a}\right)^2=\frac{b^2}{4a^2}-\frac{c}{a}\bullet\frac{4a}{4a}$

$\left(x+\frac{b}{2a}\right)^2=\frac{b^2}{4a^2}-\frac{4ac}{4a^2}$

$\left(x+\frac{b}{2a}\right)^2=\frac{b^2-4ac}{4a^2}$

$x+\frac{b}{2a}=\pm\sqrt{\frac{b^2-4ac}{4a^2}}$

$x+\frac{b}{2a}=\pm\frac{\sqrt{b^2-4ac}}{2a}$

$x=-\frac{b}{2a}\pm\frac{\sqrt{b^2-4ac}}{2a}$

$x=\frac{-b\pm\sqrt{b^2-4ac}}{2a}$

The solution set is

$\left\{\frac{-b\pm\sqrt{b^2-4ac}}{2a}\right\}$.

Problem Set 6.4 Quadratic Formula

1. $x^2+4x-21=0$

b^2-4ac

$4^2-4(1)(-21)=16+84=100$

Since $b^2-4ac>0$, there will be two real solutions.

$(x+7)(x-3)=0$

$x+7=0$ or $x-3=0$

$x=-7$ or $x=3$

The solution set is $\{-7, 3\}$.

3. $9x^2-6x+1=0$

b^2-4ac

$(-6)^2-4(9)(1)=36-36=0$

Since $b^2-4ac=0$, there will be one real solution.

$(3x-1)(3x-1)=0$

$(3x-1)^2=0$

$3x-1=0$

$3x=1$

$x=\frac{1}{3}$

The solution set is $\left\{\frac{1}{3}\right\}$.

5. $x^2-7x+13=0$

b^2-4ac

$(-7)^2-4(1)(13)=49-52=-3$

Since $b^2-4ac<0$, there will be two complex solutions.

$x=\frac{-(-7)\pm\sqrt{(-7)^2-4(1)(13)}}{2(1)}$

$x=\frac{7\pm\sqrt{49-52}}{2}$

$x=\frac{7\pm\sqrt{-3}}{2}$

$x=\frac{7\pm i\sqrt{3}}{2}$

The solution set is $\left\{\frac{7\pm i\sqrt{3}}{2}\right\}$.

7. $15x^2+17x-4=0$

b^2-4ac

$(17)^2-4(15)(-4)=289+240=529$

Since $b^2-4ac>0$, there will be two real solutions.

$(3x+4)(5x-1)=0$

$3x+4=0$ or $5x-1=0$

$3x=-4$ or $5x=1$

$x=-\frac{4}{3}$ or $x=\frac{1}{5}$

The solution set is $\left\{-\frac{4}{3}, \frac{1}{5}\right\}$.

9. $3x^2+4x=2$

$3x^2+4x-2=0$

b^2-4ac

$(4)^2-4(3)(-2)=16+24=40$

Since $b^2-4ac>0$, there will be two real solutions.

$$x=\frac{-(4)\pm\sqrt{40}}{2(3)}$$

$$x=\frac{-4\pm2\sqrt{10}}{6}$$

$$x=\frac{2(-2\pm\sqrt{10})}{6}$$

$$x=\frac{-2\pm\sqrt{10}}{3}$$

The solution set is $\left\{\frac{-2\pm\sqrt{10}}{3}\right\}$.

11. $x^2+2x-1=0$

$$x=\frac{-(2)\pm\sqrt{(2)^2-4(1)(-1)}}{2(1)}$$

$$x=\frac{-2\pm\sqrt{4+4}}{2}$$

$$x=\frac{-2\pm\sqrt{8}}{2}$$

$$x=\frac{-2\pm2\sqrt{2}}{2}$$

$$x=\frac{2(-1\pm\sqrt{2})}{2}$$

$x=-1\pm\sqrt{2}$

The solution set is $\{-1\pm\sqrt{2}\}$.

13. $n^2+5n-3=0$

$$n=\frac{-(5)\pm\sqrt{(5)^2-4(1)(-3)}}{2(1)}$$

$$n=\frac{-5\pm\sqrt{25+12}}{2}$$

$$n=\frac{-5\pm\sqrt{37}}{2}$$

The solution set is $\left\{\frac{-5\pm\sqrt{37}}{2}\right\}$.

15. $a^2-8a=4$

$a^2-8a-4=0$

$$a=\frac{-(-8)\pm\sqrt{(-8)^2-4(1)(-4)}}{2(1)}$$

$$a=\frac{8\pm\sqrt{64+16}}{2}$$

$$a=\frac{8\pm\sqrt{80}}{2}$$

$$a=\frac{8\pm4\sqrt{5}}{2}$$

$$a=\frac{2(4\pm2\sqrt{5})}{2}$$

$a=4\pm2\sqrt{5}$

The solution set is $\{4\pm2\sqrt{5}\}$.

17. $n^2+5n+8=0$

$$n=\frac{-(5)\pm\sqrt{(5)^2-4(1)(8)}}{2(1)}$$

$$n=\frac{-5\pm\sqrt{25-32}}{2}$$

$$n=\frac{-5\pm\sqrt{-7}}{2}$$

$$n=\frac{-5\pm i\sqrt{7}}{2}$$

The solution set is $\left\{\frac{-5\pm i\sqrt{7}}{2}\right\}$.

19. $x^2-18x+80=0$

$$x=\frac{-(-18)\pm\sqrt{(-18)^2-4(1)(80)}}{2(1)}$$

$$x=\frac{18\pm\sqrt{324-320}}{2}$$

$$x=\frac{18\pm\sqrt{4}}{2}$$

$$x=\frac{18\pm2}{2}$$

$x=\frac{18+2}{2}$ or $x=\frac{18-2}{2}$

$x=10$ or $x=8$

The solution set is $\{8, 10\}$.

21. $-y^2 = -9y + 5$

$0 = y^2 - 9y + 5$

$y = \frac{-(-9) \pm \sqrt{(-9)^2 - 4(1)(5)}}{2(1)}$

$y = \frac{9 \pm \sqrt{81 - 20}}{2}$

$y = \frac{9 \pm \sqrt{61}}{2}$

The solution set is $\left\{\frac{9 \pm \sqrt{61}}{2}\right\}$.

23. $2x^2 + x - 4 = 0$

$x = \frac{-(1) \pm \sqrt{(1)^2 - 4(2)(-4)}}{2(2)}$

$x = \frac{-1 \pm \sqrt{1 + 32}}{4}$

$x = \frac{-1 \pm \sqrt{33}}{4}$

The solution set is $\left\{\frac{-1 \pm \sqrt{33}}{4}\right\}$.

25. $4x^2 + 2x + 1 = 0$

$x = \frac{-2 \pm \sqrt{2^2 - 4(4)(1)}}{2(4)}$

$x = \frac{-2 \pm \sqrt{4 - 16}}{8}$

$x = \frac{-2 \pm \sqrt{-12}}{8}$

$x = \frac{-2 \pm i\sqrt{12}}{8}$

$x = \frac{-2 \pm 2i\sqrt{3}}{8}$

$x = \frac{2(-1 \pm i\sqrt{3})}{8}$

$x = \frac{-1 \pm i\sqrt{3}}{4}$

The solution set is $\left\{\frac{-1 \pm i\sqrt{3}}{4}\right\}$.

27. $3a^2 - 8a + 2 = 0$

$a = \frac{-(-8) \pm \sqrt{(-8)^2 - 4(3)(2)}}{2(3)}$

$a = \frac{8 \pm \sqrt{64 - 24}}{6}$

$a = \frac{8 \pm \sqrt{40}}{6}$

$a = \frac{8 \pm 2\sqrt{10}}{6}$

$a = \frac{2(4 \pm \sqrt{10})}{6}$

$a = \frac{4 \pm \sqrt{10}}{3}$

The solution set is $\left\{\frac{4 \pm \sqrt{10}}{3}\right\}$.

29. $-2n^2 + 3n + 5 = 0$

$n = \frac{-(3) \pm \sqrt{(3)^2 - 4(-2)(5)}}{2(-2)}$

$n = \frac{-3 \pm \sqrt{9 + 40}}{-4}$

$n = \frac{-3 \pm \sqrt{49}}{-4}$

$n = \frac{-3 \pm 7}{-4}$

$n = \frac{-3 + 7}{-4}$ or $n = \frac{-3 - 7}{-4}$

$n = -1$ or $n = \frac{-10}{-4} = \frac{5}{2}$

The solution set is $\left\{-1, \frac{5}{2}\right\}$.

31. $3x^2 + 19x + 20 = 0$

$x = \frac{-19 \pm \sqrt{19^2 - 4(3)(20)}}{2(3)}$

$x = \frac{-19 \pm \sqrt{361 - 240}}{6}$

$x = \frac{-19 \pm \sqrt{121}}{6}$

$x = \frac{-19 \pm 11}{6}$

$x = \frac{-19+11}{6}$ or $x = \frac{-19-11}{6}$

$x = \frac{-8}{6} = -\frac{4}{3}$ or $x = \frac{-30}{6} = -5$

The solution set is $\left\{-5, -\frac{4}{3}\right\}$.

33. $36n^2 - 60n + 25 = 0$

$n = \frac{-(-60) \pm \sqrt{(-60)^2 - 4(36)(25)}}{2(36)}$

$n = \frac{60 \pm \sqrt{3600 - 3600}}{72}$

$n = \frac{60}{72} = \frac{5}{6}$

The solution set is $\left\{\frac{5}{6}\right\}$.

35. $4x^2 - 2x = 3$

$4x^2 - 2x - 3 = 0$

$x = \frac{-(-2) \pm \sqrt{(-2)^2 - 4(4)(-3)}}{2(4)}$

$x = \frac{2 \pm \sqrt{4+48}}{8}$

$x = \frac{2 \pm \sqrt{52}}{8}$

$x = \frac{2 \pm 2\sqrt{13}}{8}$

$x = \frac{2(1 \pm \sqrt{13})}{8}$

$x = \frac{1 \pm \sqrt{13}}{4}$

The solution set is $\left\{\frac{1 \pm \sqrt{13}}{4}\right\}$.

37. $5x^2 - 13x = 0$

$x = \frac{-(-13) \pm \sqrt{(-13)^2 - 4(5)(0)}}{2(5)}$

$x = \frac{13 \pm \sqrt{169}}{10}$

$x = \frac{13 \pm 13}{10}$

$x = \frac{13+13}{10}$ or $x = \frac{13-13}{10}$

$x = \frac{26}{10} = \frac{13}{5}$ or $x = 0$

The solution set is $\left\{0, \frac{13}{5}\right\}$.

39. $3x^2 = 5$

$3x^2 - 5 = 0$

$x = \frac{-0 \pm \sqrt{0^2 - 4(3)(-5)}}{2(3)}$

$x = \frac{\pm\sqrt{60}}{6} = \frac{\pm 2\sqrt{15}}{6} = \pm\frac{\sqrt{15}}{3}$

The solution set is $\left\{\pm\frac{\sqrt{15}}{3}\right\}$.

41. $6t^2 + t - 3 = 0$

$t = \frac{-1 \pm \sqrt{1^2 - 4(6)(-3)}}{2(6)}$

$t = \frac{-1 \pm \sqrt{1+72}}{12}$

$t = \frac{-1 \pm \sqrt{73}}{12}$

The solution set is $\left\{\frac{-1 \pm \sqrt{73}}{12}\right\}$.

43. $n^2 + 32n + 252 = 0$

$n = \frac{-32 \pm \sqrt{32^2 - 4(1)(252)}}{2(1)}$

$n = \frac{-32 \pm \sqrt{1024 - 1008}}{2}$

$n = \frac{-32 \pm \sqrt{16}}{2}$

$n = \frac{-32 \pm 4}{2}$

$n = \frac{-32+4}{2}$ or $n = \frac{-32-4}{2}$

$n = -14$ or $n = -18$

The solution set is $\{-18, -14\}$.

45. $12x^2 - 73x + 110 = 0$

$x = \frac{-(-73) \pm \sqrt{(-73)^2 - 4(12)(110)}}{2(12)}$

$x = \frac{73 \pm \sqrt{5329 - 5280}}{24}$

$x = \frac{73 \pm \sqrt{49}}{24}$

$x = \frac{73 \pm 7}{24}$

$x = \frac{73+7}{24}$ or $x = \frac{73-7}{24}$

$x = \frac{80}{24}$ or $x = \frac{66}{24}$

$x = \frac{10}{3}$ or $x = \frac{11}{4}$

The solution set is $\left\{\frac{11}{4}, \frac{10}{3}\right\}$.

47. $-2x^2 + 4x - 3 = 0$

$x = \frac{-4 \pm \sqrt{4^2 - 4(-2)(-3)}}{2(-2)}$

$x = \frac{-4 \pm \sqrt{16 - 24}}{-4}$

$x = \frac{-4 \pm \sqrt{-8}}{-4}$

$x = \frac{-4 \pm i\sqrt{8}}{-4}$

$x = \frac{-4 \pm 2i\sqrt{2}}{-4}$

$x = \frac{-2(2 \pm i\sqrt{2})}{-4}$

$x = \frac{2 \pm i\sqrt{2}}{2}$

The solution set is $\left\{\frac{2 \pm i\sqrt{2}}{2}\right\}$.

49. $-6x^2 + 2x + 1 = 0$

$x = \frac{-2 \pm \sqrt{2^2 - 4(-6)(1)}}{2(-6)}$

$x = \frac{-2 \pm \sqrt{4 + 24}}{-12}$

$x = \frac{-2 \pm \sqrt{28}}{-12}$

$x = \frac{-2 \pm 2\sqrt{7}}{-12}$

$x = \frac{-2(1 \pm \sqrt{7})}{-12}$

$x = \frac{1 \pm \sqrt{7}}{6}$

The solution set is $\left\{\frac{1 \pm \sqrt{7}}{6}\right\}$.

Problem Set 6.5 More Quadratic Equations and Applications

1. $x^2 - 4x - 6 = 0$

$x^2 - 4x = 6$

$x^2 - 4x + 4 = 6 + 4$

$(x-2)^2 = 10$

$x - 2 = \pm\sqrt{10}$

$x = 2 \pm \sqrt{10}$

The solution set is $\{2 \pm \sqrt{10}\}$.

3. $3x^2 + 23x - 36 = 0$

$(3x-4)(x+9) = 0$

$3x - 4 = 0$ or $x + 9 = 0$

$3x = 4$ or $x = -9$

$x = \frac{4}{3}$ or $x = -9$

The solution set is $\left\{-9, \frac{4}{3}\right\}$.

5. $x^2 - 18x = 9$

$x^2 - 18x + 81 = 9 + 81$

$(x-9)^2 = 90$

$x - 9 = \pm\sqrt{90}$

$x - 9 = \pm 3\sqrt{10}$

$x = 9 \pm 3\sqrt{10}$

The solution set is $\{9 \pm 3\sqrt{10}\}$.

7. $2x^2 - 3x + 4 = 0$

$x = \frac{-(-3) \pm \sqrt{(-3)^2 - 4(2)(4)}}{2(2)}$

$x = \frac{3 \pm \sqrt{9 - 32}}{4}$

$x = \frac{3 \pm \sqrt{-23}}{4}$

$x = \frac{3 \pm i\sqrt{23}}{4}$

The solution set is $\left\{\frac{3 \pm i\sqrt{23}}{4}\right\}$.

9. $135+24n+n^2=0$

$n^2+24n+135=0$

$(n+15)(n+9)=0$

$n+15=0$ or $n+9=0$

$n=-15$ or $n=-9$

The solution set is $\{-15, -9\}$.

11. $(x-2)(x+9)=-10$

$x^2+7x-18=-10$

$x^2+7x-8=0$

$(x+8)(x-1)=0$

$x+8=0$ or $x-1=0$

$x=-8$ or $x=1$

The solution set is $\{-8, 1\}$.

13. $2x^2-4x+7=0$

$$x=\frac{-(-4)\pm\sqrt{(-4)^2-4(2)(7)}}{2(2)}$$

$$x=\frac{4\pm\sqrt{16-56}}{4}$$

$$x=\frac{4\pm\sqrt{-40}}{4}$$

$$x=\frac{4\pm 2i\sqrt{10}}{4}$$

$$x=\frac{2(2\pm i\sqrt{10})}{4}$$

$$x=\frac{2\pm i\sqrt{10}}{2}$$

The solution set is $\left\{\frac{2\pm i\sqrt{10}}{2}\right\}$.

15. $x^2-18x+15=0$

$x^2-18x=-15$

$x^2-18x+81=-15+81$

$(x-9)^2=66$

$x-9=\pm\sqrt{66}$

$x=9\pm\sqrt{66}$

The solution set is $\{9\pm\sqrt{66}\}$.

17. $20y^2+17y-10=0$

$(4y+5)(5y-2)=0$

$4y+5=0$ or $5y-2=0$

$4y=-5$ or $5y=2$

$y=-\frac{5}{4}$ or $y=\frac{2}{5}$

The solution set is $\left\{-\frac{5}{4}, \frac{2}{5}\right\}$.

19. $4t^2+4t-1=0$

$$t=\frac{-4\pm\sqrt{4^2-4(4)(-1)}}{2(4)}$$

$$t=\frac{-4\pm\sqrt{16+16}}{8}$$

$$t=\frac{-4\pm\sqrt{32}}{8}$$

$$t=\frac{-4\pm 4\sqrt{2}}{8}$$

$$t=\frac{4(-1\pm\sqrt{2})}{8}$$

$$t=\frac{-1\pm\sqrt{2}}{2}$$

The solution set is $\left\{\frac{-1\pm\sqrt{2}}{2}\right\}$.

21. $n+\frac{3}{n}=\frac{19}{4}$; $n\neq 0$

$4n\left(n+\frac{3}{n}\right)=4n\left(\frac{19}{4}\right)$

$4n^2+12=19n$

$4n^2-19n+12=0$

$(4n-3)(n-4)=0$

$4n-3=0$ or $n-4=0$

$4n=3$ or $n=4$

$n=\frac{3}{4}$

The solution set is $\left\{\frac{3}{4}, 4\right\}$.

23. $\frac{3}{x}+\frac{7}{x-1}=1$; $x\neq 0$, $x\neq 1$

$x(x-1)\left(\frac{3}{x}+\frac{7}{x-1}\right)=x(x-1)(1)$

$x(x-1)\left(\frac{3}{x}\right)+x(x-1)\left(\frac{7}{x-1}\right)=x^2-x$

$3(x-1)+7x=x^2-x$

$3x-3+7x=x^2-x$

$10x-3=x^2-x$

$0=x^2-11x+3$

$x=\frac{-(-11)\pm\sqrt{(-11)^2-4(1)(3)}}{2(1)}$

$x=\frac{11\pm\sqrt{121-12}}{2}$

$x=\frac{11\pm\sqrt{109}}{2}$

The solution set is $\left\{\frac{11\pm\sqrt{109}}{2}\right\}$.

25. $\frac{12}{x-3}+\frac{8}{x}=14;\ x\neq 3,\ x\neq 0$

$x(x-3)\left(\frac{12}{x-3}+\frac{8}{x}\right)=x(x-3)(14)$

$12x+8(x-3)=14x^2-42x$

$12x+8x-24=14x^2-42x$

$20x-24=14x^2-42x$

$0=14x^2-62x+24$

$0=2(7x^2-31x+12)$

$0=2(7x-3)(x-4)$

$7x-3=0$ or $x-4=0$

$7x=3$ or $x=4$

$x=\frac{3}{7}$ or $x=4$

The solution set is $\left\{\frac{3}{7},\ 4\right\}$.

27. $\frac{3}{x-1}-\frac{2}{x}=\frac{5}{2};\ x\neq 0,\ x\neq 1$

$2x(x-1)\left(\frac{3}{x-1}-\frac{2}{x}\right)=2x(x-1)\left(\frac{5}{2}\right)$

$2x(x-1)\left(\frac{3}{x-1}\right)-2x(x-1)\left(\frac{2}{x}\right)=5x(x-1)$

$6x-4(x-1)=5x^2-5x$

$6x-4x+4=5x^2-5x$

$0=5x^2-7x-4$

$x=\frac{-(-7)\pm\sqrt{(-7)^2-4(5)(-4)}}{2(5)}$

$x=\frac{7\pm\sqrt{49+80}}{10}$

$x=\frac{7\pm\sqrt{129}}{10}$

The solution set is $\left\{\frac{7\pm\sqrt{129}}{10}\right\}$.

29. $\frac{6}{x}+\frac{40}{x+5}=7;\ x\neq 0,\ x\neq -5$

$x(x+5)\left(\frac{6}{x}+\frac{40}{x+5}\right)=x(x+5)(7)$

$6(x+5)+40x=7x(x+5)$

$6x+30+40x=7x^2+35x$

$0=7x^2-11x-30$

$0=(7x+10)(x-3)$

$7x+10=0$ or $x-3=0$

$7x=-10$ or $x=3$

$x=-\frac{10}{7}$ or $x=3$

The solution set is $\left\{-\frac{10}{7},\ 3\right\}$.

31. $\frac{5}{n-3}-\frac{3}{n+3}=1;\ n\neq 3,\ n\neq -3$

$(n-3)(n+3)\left(\frac{5}{n-3}-\frac{3}{n+3}\right)=(n-3)(n+3)(1)$

$5(n+3)-3(n-3)=n^2-9$

$5n+15-3n+9=n^2-9$

$2n+24=n^2-9$

$0=n^2-2n-33$

$n^2-2n-33=0$

$n^2-2n=33$

$n^2-2n+1=33+1$

$(n-1)^2=34$

$n-1=\pm\sqrt{34}$

$n=1\pm\sqrt{34}$

The solution set is $\{1\pm\sqrt{34}\}$.

33. $x^4-18x^2+72=0$

$(x^2-12)(x^2-6)=0$

$x^2-12=0$ or $x^2-6=0$

$x^2=12$ or $x^2=6$

$x = \pm\sqrt{12}$ or $x = \pm\sqrt{6}$

$x = \pm 2\sqrt{3}$ or $x = \pm\sqrt{6}$

The solution set is $\{\pm\sqrt{6}, \pm 2\sqrt{3}\}$.

35. $3x^4 - 35x^2 + 72 = 0$

$(3x^2 - 8)(x^2 - 9) = 0$

$3x^2 - 8 = 0$ or $x^2 - 9 = 0$

$3x^2 = 8$ or $x^2 = 9$

$x^2 = \frac{8}{3}$ or $x = \pm\sqrt{9}$

$x = \pm\sqrt{\frac{8}{3}}$ or $x = \pm 3$

$x = \pm\frac{2\sqrt{2}}{\sqrt{3}}$ or $x = \pm 3$

$x = \pm\frac{2\sqrt{2}}{\sqrt{3}} \bullet \frac{\sqrt{3}}{\sqrt{3}}$ or $x = \pm 3$

$x = \pm\frac{2\sqrt{6}}{3}$ or $x = \pm 3$

The solution set is $\left\{\pm\frac{2\sqrt{6}}{3}, \pm 3\right\}$.

37. $3x^4 + 17x^2 + 20 = 0$

$(3x^2 + 5)(x^2 + 4) = 0$

$3x^2 + 5 = 0$ or $x^2 + 4 = 0$

$3x^2 = -5$ or $x^2 = -4$

$x^2 = -\frac{5}{3}$ or $x = \pm\sqrt{-4}$

$x = \pm\sqrt{-\frac{5}{3}}$ or $x = \pm 2i$

$x = \pm\frac{i\sqrt{5}}{\sqrt{3}} \bullet \frac{\sqrt{3}}{\sqrt{3}}$ or $x = \pm 2i$

$x = \pm\frac{i\sqrt{15}}{3}$ or $x = \pm 2i$

The solution set is $\left\{\pm\frac{i\sqrt{15}}{3}, \pm 2i\right\}$.

39. $6x^2 - 29x^2 + 28 = 0$

$(3x^2 - 4)(2x^2 - 7) = 0$

$3x^2 - 4 = 0$ or $2x^2 - 7 = 0$

$3x^2 = 4$ or $2x^2 = 7$

$x^2 = \frac{4}{3}$ or $x^2 = \frac{7}{2}$

$x = \pm\sqrt{\frac{4}{3}}$ or $x = \pm\sqrt{\frac{7}{2}}$

$x = \pm\frac{2}{\sqrt{3}}$ or $x = \pm\frac{\sqrt{7}}{\sqrt{2}} \bullet \frac{\sqrt{2}}{\sqrt{2}}$

$x = \pm\frac{2}{\sqrt{3}} \bullet \frac{\sqrt{3}}{\sqrt{3}}$ or $x = \pm\frac{\sqrt{14}}{2}$

$x = \pm\frac{2\sqrt{3}}{3}$ or $x = \pm\frac{\sqrt{14}}{2}$

The solution set is $\left\{\pm\frac{2\sqrt{3}}{3}, \pm\frac{\sqrt{14}}{2}\right\}$.

41. Let $x = 1^{st}$ whole number
$x + 1 = 2^{nd}$ whole number

$x^2 + (x+1)^2 = 145$

$x^2 + x^2 + 2x + 1 = 145$

$2x^2 + 2x - 144 = 0$

$2(x^2 + x - 72) = 0$

$2(x+9)(x-8) = 0$

$x + 9 = 0$ or $x - 8 = 0$

$x = -9$ or $x = 8$

Discard the root $x = -9$.
The numbers are 8 and 9.

43. Let $x = 1^{st}$ positive integer
$x - 3 = 2^{nd}$ positive integer

$x(x-3) = 108$

$x^2 - 3x = 108$

$x^2 - 3x - 108 = 0$

$(x-12)(x+9) = 0$

$x - 12 = 0$ or $x + 9 = 0$

$x = 12$ or $x = -9$

Discard the root $x = -9$.
The numbers are 12 and 9.

45. Let $x = 1^{st}$ number
$10 - x = 2^{nd}$ number

$x(10 - x) = 22$

$10x - x^2 = 22$

$x^2 - 10x = -22$

$x^2 - 10x + 25 = -22 + 25$

$(x-5)^2 = 3$

$x-5 = \pm\sqrt{3}$

$x = 5 \pm \sqrt{3}$

$x_1 = 5+\sqrt{3}$	$x_1 = 5-\sqrt{3}$
$x_2 = 10-(5+\sqrt{3})$	$x_2 = 10-(5-\sqrt{3})$
$x_2 = 10-5-\sqrt{3}$	$x_2 = 10-5+\sqrt{3}$
$x_2 = 5-\sqrt{3}$	$x_2 = 5+\sqrt{3}$

The numbers are $5+\sqrt{3}$ and $5-\sqrt{3}$.

47.

Numbers	Reciprocals
x	$\frac{1}{x}$
$9-x$	$\frac{1}{9-x}$

$\frac{1}{x}+\frac{1}{9-x} = \frac{1}{2}$

$2x(9-x)\left(\frac{1}{x}+\frac{1}{9-x}\right) = 2x(9-x)\left(\frac{1}{2}\right)$

$2(9-x)+2x = x(9-x)$

$18-2x+2x = 9x-x^2$

$18 = 9x-x^2$

$x^2-9x+18 = 0$

$(x-6)(x-3) = 0$

$x-6=0$ or $x-3=0$

$x=6$ or $x=3$

The numbers are 6 and 3.

49. Let x = one leg
$21-x$ = other leg

$x^2+(21-x)^2 = 15^2$

$x^2+441-42x+x^2 = 225$

$2x^2-42x+216 = 0$

$2(x^2-21x+108) = 0$

$2(x-9)(x-12) = 0$

$x-9=0$ or $x-12=0$

$x=9$ or $x=12$

The legs are 9 inches and 12 inches.

51. Let x = width of walk

Area of plot and sidewalk $= (12+2x)(20+2x)$

Area of plot $= 12(20) = 240$

Area of sidewalk

$(12+2x)(20+2x)-240 = 68$

$240+24x+40x+4x^2-240 = 68$

$4x^2+64x-68 = 0$

$4(x^2+16x-17) = 0$

$4(x+17)(x-1) = 0$

$x+17=0$ or $x-1=0$

$x=-17$ or $x=1$

Discard the root $x=-17$.
The width of the walk is 1 meter.

53. Let x = width
$22-x$ = length

$x(22-x) = 112$

$22x-x^2 = 112$

$0 = x^2-22x+112$

$0 = (x-8)(x-14)$

$x-8=0$ or $x-14=0$

$x=8$ or $x=14$

The dimensions of the rectangle are 8 inches by 14 inches.

55.

	rate •	time =	distance
Charlotte	$x+5$	$\frac{250}{x+3}$	250
Lorraine	x	$\frac{180}{x}$	180

$\frac{250}{x+5} = \frac{180}{x}+1$

$x(x+5)\left(\frac{250}{x+5}\right) = x(x+5)\left(\frac{180}{x}+1\right)$

$250x = 180(x+5)+1x(x+5)$

$250x = 180x+900+x^2+5x$

$250x = x^2+185x+900$

$0 = x^2-65x+900$

$0 = (x-45)(x-20)$

$x - 45 = 0$ or $x - 20 = 0$

$x = 45$ or $x = 20$

Case 1: Lorraine drives at 45 mph and Charlotte drives at 50 mph.

Case 2: Lorraine drives at 20 mph and Charlotte drives at 25 mph.

57.

	rate •	time =	distance
1st Part	x	$\frac{330}{x}$	330
2nd Part	$x+5$	$\frac{240}{x+5}$	240

$\frac{330}{x} + \frac{240}{x+5} = 10$

$x(x+5)\left(\frac{330}{x} + \frac{240}{x+5}\right) = 10x(x+5)$

$330(x+5) + 240x = 10x^2 + 50x$

$330x + 1650 + 240x = 10x^2 + 50x$

$570x + 1650 = 10x^2 + 50x$

$0 = 10x^2 - 520x - 1650$

$0 = 10(x^2 - 52x - 165)$

$0 = 10(x - 55)(x + 3)$

$x - 55 = 0$ or $x + 3 = 0$

$x = 55$ or $x = -3$

Discard the root $x = -3$.
He drove at 55 mph.

59.

	Time for job	Rate
Terry	$x+2$	$\frac{1}{x+2}$
Tom	x	$\frac{1}{x}$

$\left(\frac{1}{x+2} + \frac{1}{x}\right)3 + \left(\frac{1}{x+2}\right)1 = 1$

$\frac{3}{x+2} + \frac{3}{x} + \frac{1}{x+2} = 1$

$\frac{4}{x+2} + \frac{3}{x} = 1$

$x(x+2)\left(\frac{4}{x+2} + \frac{3}{x}\right) = x(x+2)(1)$

$4x + 3(x+2) = x^2 + 2x$

$4x + 3x + 6 = x^2 + 2x$

$0 = x^2 - 5x - 6$

$0 = (x-6)(x+1)$

$x - 6 = 0$ or $x + 1 = 0$

$x = 6$ or $x = -1$

Discard the root $x = -1$.
It takes Tom 6 hours and Terry 8 hours.

61.

	rate pay/hour	time in hours	Pay
Anticipated	$\frac{360}{x}$	x	360
Actual	$\frac{360}{x+6}$	$x+6$	360

$\frac{360}{x} - 2 = \frac{360}{x+6}$

$x(x+6)\left(\frac{360}{x} - 2\right) = x(x+6)\left(\frac{360}{x+6}\right)$

$360(x+6) - 2x(x+6) = 360x$

$360x + 2160 - 2x^2 - 12x = 360x$

$-2x^2 + 348x + 2160 = 360x$

$-2x^2 - 12x + 2160 = 0$

$-2(x^2 + 6x - 1080) = 0$

$-2(x+36)(x-30) = 0$

$x + 36 = 0$ or $x - 30 = 0$

$x = -36$ or $x = 30$

Discard the root $x = -36$.
The time he anticipated it would take was 30 hours.

63.

	$\frac{\text{price}}{\text{person}}$	number of persons	price
Group	$\frac{100}{x+2}$	$x+2$	100
Group Less 2	$\frac{100}{x}$	x	100

$\frac{100}{x+2} + 2.50 = \frac{100}{x}$

$\frac{100}{x+2} + \frac{5}{2} = \frac{100}{x}$

$2x(x+2)\left(\frac{100}{x+2} + \frac{5}{2}\right) = 2x(x+2)\left(\frac{100}{x}\right)$

$2x(100)+5x(x+2)=200(x+2)$

$200x+5x^2+10x=200x+400$

$5x^2+210x=200x+400$

$5x^2+10x-400=0$

$5(x^2+2x-80)=0$

$5(x+10)(x-8)=0$

$x+10=0$ or $x-8=0$

$x=-10$ or $x=8$

Discard the root $x=-10$. There were 8 people that actually contributed.

65.

	$\frac{\text{price}}{\text{share}}$	number of shares	price
Purchase	$\frac{720}{x}$	x	720
Sell	$\frac{800}{x-20}$	x − 20	800

$\frac{720}{x}+8=\frac{800}{x-20}$

$x(x-20)\left(\frac{720}{x}+8\right)=x(x-20)\left(\frac{800}{x-20}\right)$

$720(x-20)+8x(x-20)=800x$

$720x-14400+8x^2-160x=800x$

$8x^2+560x-14400=800x$

$8x^2-240x-14400=0$

$8(x^2-30x-1800)=0$

$8(x-60)(x+30)=0$

$x-60=0$ or $x+30=0$

$x=60$ or $x=-30$

Discard the root $x=-30$. He sold 40 shares at \$20 per share.

67. $S=\frac{n(n+1)}{2}$

$1275=\frac{n(n+1)}{2}$

$2(1275)=2\left[\frac{n(n+1)}{2}\right]$

$2550=n(n+1)$

$2550=n^2+n$

$0=n^2+n-2550$

$0=(n+51)(n-50)$

$n+51=0$ or $n-50=0$

$n=-51$ or $n=50$

Discard the root $n=-51$. There needs to be 50 consecutive numbers.

69. $A=P(1+r)^t$

$594.05=500(1+r)^2$

$1.1881=(1+r)^2$

$(1+r)^2=1.1881$

$1+r=\pm\sqrt{1.1881}$

$r=-1\pm\sqrt{1.1881}$

$r=-1+\sqrt{1.1881}$ or $r=-1-\sqrt{1.1881}$

$r=0.09$ or $r=-2.09$

Discard the root $r=-2.09$. The rate is 9%.

Problem Set 6.6 Quadratic Inequalities

1. $(x+2)(x-1)>0$

$(x+2)(x-1)=0$

$x+2=0$ or $x-1=0$

$x=-2$ or $x=1$

Test Point:	−3	0	2
(boundaries)		−2	1
x + 2:	negative	positive	positive
x − 1:	negative	negative	positive
product:	positive	negative	positive

The solution set is $(-\infty,-2)\cup(1,\infty)$.

3. $(x+1)(x+4)<0$

$(x+1)(x+4)=0$

$x+1=0$ or $x+4=0$

$x=-1$ or $x=-4$

Test Point:	-5	-2	0
(boundaries)		-4	-1
$x+1$:	negative	negative	positive
$x+4$:	negative	positive	positive
product:	positive	negative	positive

The solution set is $(-4,-1)$.

5. $(2x-1)(3x+7)\geq 0$

$(2x-1)(3x+7)=0$

$2x-1=0$ or $3x+7=0$

$2x=1$ or $3x=-7$

$x=\frac{1}{2}$ or $x=-\frac{7}{3}$

Test Point:	-3	0	1
(boundaries)		$-\frac{7}{3}$	$\frac{1}{2}$
$2x-1$:	negative	negative	positive
$3x+7$:	negative	positive	positive
product:	positive	negative	positive

The solution set is $\left(-\infty,-\frac{7}{3}\right]\cup\left[\frac{1}{2},\infty\right)$.

7. $(x+2)(4x-3)\leq 0$

$(x+2)(4x-3)=0$

$x+2=0$ or $4x-3=0$

$x=-2$ or $4x=3$

$x=-2$ or $x=\frac{3}{4}$

Test Point:	-3	0	1
(boundaries)		-2	$\frac{3}{4}$
$x+2$:	negative	positive	positive
$4x-3$:	negative	negative	positive
product:	positive	negative	positive

The solution set is $\left[-2,\frac{3}{4}\right]$.

9. $(x+1)(x-1)(x-3)>0$

$(x+1)(x-1)(x-3)=0$

$x+1=0$ or $x-1=0$ or $x-3=0$

$x=-1$ or $x=1$ or $x=3$

Test Point:	-2	0	2	4
(boundaries)		-1	1	3
$x+1$:	neg	pos	pos	pos
$x-1$:	neg	neg	pos	pos
$x-3$:	neg	neg	neg	pos
product:	neg	pos	neg	pos

The solution set is $(-1,1)\cup(3,\infty)$.

11. $x(x+2)(x-4)\leq 0$

$x(x+2)(x-4)=0$

$x=0$ or $x+2=0$ or $x-4=0$

$x=0$ or $x=-2$ or $x=4$

Test Point:	-3	-1	2	5
(boundaries)		-2	0	4
x:	neg	neg	pos	pos
$x+2$:	neg	pos	pos	pos
$x-4$:	neg	neg	neg	pos
product:	neg	pos	neg	pos

The solution set is $(-\infty,-2]\cup[0,4]$.

13. $\frac{x+1}{x-2}>0$; $x\neq 2$

$x+1=0$ or $x-2=0$

$x=-1$ or $x=2$

Test Point:	-2	0	3
(boundaries)		-1	2
$x+1$:	negative	positive	positive
$x-2$:	negative	negative	positive
quotient:	positive	negative	positive

The solution set is $(-\infty,-1)\cup(2,\infty)$.

15. $\frac{x-3}{x+2}<0$; $x\neq -2$

$x-3=0$ or $x+2=0$

$x=3$ or $x=-2$

Test Point:	-3	0	4
(boundaries)		-2	3
$x-3$:	negative	negative	positive
$x+2$:	negative	positive	positive
quotient:	positive	negative	positive

The solution set is $(-2,3)$.

17. $\frac{2x-1}{x} \geq 0;\ x \neq 0$

$2x-1=0$ or $x=0$

$2x=1$ or $x=0$

$x=\frac{1}{2}$ or $x=0$

	$(-\infty, 0)$	$(0, \frac{1}{2})$	$(\frac{1}{2}, \infty)$
Test Point:	-1	$\frac{1}{4}$	1
$2x-1$:	negative	negative	positive
x:	negative	positive	positive
quotient:	positive	negative	positive

The solution set is $(-\infty, 0) \cup \left[\frac{1}{2}, \infty\right)$.

19. $\frac{-x+2}{x-1} \leq 0;\ x \neq 1$

$-x+2=0$ or $x-1=0$

$2=x$ or $x=1$

	$(-\infty, 1)$	$(1, 2)$	$(2, \infty)$
Test Point:	0	$\frac{3}{2}$	3
$-x+2$:	positive	positive	negative
$x-1$:	negative	positive	positive
quotient:	negative	positive	negative

The solution set is $(-\infty, 1) \cup [2, \infty)$.

21. $x^2+2x-35<0$

$(x+7)(x-5)<0$

$(x+7)(x-5)=0$

$x+7=0$ or $x-5=0$

$x=-7$ or $x=5$

	$(-\infty, -7)$	$(-7, 5)$	$(5, \infty)$
Test Point:	-8	0	6
$x+7$:	negative	positive	positive
$x-5$:	negative	negative	positive
product:	positive	negative	positive

The solution set is $(-7, 5)$.

23. $x^2-11x+28>0$

$(x-7)(x-4)>0$

$(x-7)(x-4)=0$

$x-7=0$ or $x-4=0$

$x=7$ or $x=4$

	$(-\infty, 4)$	$(4, 7)$	$(7, \infty)$
Test Point:	0	6	8
$x-7$:	negative	negative	positive
$x-4$:	negative	positive	positive
product:	positive	negative	positive

The solution set is $(-\infty, 4) \cup (7, \infty)$.

25. $3x^2+13x-10 \leq 0$

$(3x-2)(x+5) \leq 0$

$(3x-2)(x+5)=0$

$3x-2=0$ or $x+5=0$

$3x=2$ or $x=-5$

$x=\frac{2}{3}$ or $x=-5$

	$(-\infty, -5)$	$(-5, \frac{2}{3})$	$(\frac{2}{3}, \infty)$
Test Point:	-6	0	1
$3x-2$:	negative	negative	positive
$x+5$:	negative	positive	positive
product:	positive	negative	positive

The solution set is $\left[-5, \frac{2}{3}\right]$.

27. $8x^2+22x+5 \geq 0$

$(2x+5)(4x+1) \geq 0$

$(2x+5)(4x+1)=0$

$2x+5=0$ or $4x+1=0$

$2x=-5$ or $4x=-1$

$x=-\frac{5}{2}$ or $x=-\frac{1}{4}$

	$(-\infty, -\frac{5}{2})$	$(-\frac{5}{2}, -\frac{1}{4})$	$(-\frac{1}{4}, \infty)$
Test Point:	-3	-1	0
$2x+5$:	negative	positive	positive
$4x+1$:	negative	negative	positive
product:	positive	negative	positive

The solution set is $\left(-\infty, -\frac{5}{2}\right] \cup \left[-\frac{1}{4}, \infty\right)$.

29. $x(5x-36) > 32$

$5x^2 - 36x > 32$

$5x^2 - 36x - 32 > 0$

$(5x+4)(x-8) > 0$

$(5x+4)(x-8) = 0$

$5x+4=0$ or $x-8=0$

$5x=-4$ or $x=8$

$x=-\frac{4}{5}$ or $x=8$

Test Point:	-1	0	10
	($-\frac{4}{5}$)		(8)
$5x+4$:	negative	positive	positive
$x-8$:	negative	negative	positive
product:	positive	negative	positive

The solution set is $\left(-\infty, -\frac{4}{5}\right) \cup (8, \infty)$.

31. $x^2 - 14x + 49 \geq 0$

$(x-7)(x-7) \geq 0$

$(x-7)(x-7) = 0$

$(x-7)^2 = 0$

$x-7=0$

$x=7$

Test Point:	5	9
	(7)	
$x-7$:	negative	positive
$x-7$:	negative	positive
product:	positive	positive

The solution set is $(-\infty, \infty)$.

33. $4x^2 + 20x + 25 \leq 0$

$(2x+5)(2x+5) \leq 0$

$(2x+5)(2x+5) = 0$

$(2x+5)^2 = 0$

$2x+5=0$

$2x=-5$

$x=-\frac{5}{2}$

Test Point:	-3	0
	($-\frac{5}{2}$)	
$2x+5$:	negative	positive
$2x+5$:	negative	positive
product:	positive	positive

The solution set is $\left\{-\frac{5}{2}\right\}$.

35. $(x+1)(x-3)^2 > 0$

$(x+1)(x-3)(x-3) > 0$

$(x+1)(x-3)^2 = 0$

$x+1=0$ or $x-3=0$

$x=-1$ or $x=3$

Test Point:	-2	0	4
	(-1)		(3)
$x+1$:	negative	positive	positive
$x-3$:	negative	negative	positive
$x-3$:	negative	negative	positive
product:	negative	positive	positive

The solution set is $(-1, 3) \cup (3, \infty)$.

37. $\frac{2x}{x+3} > 4$

$\frac{2x}{x+3} - 4 > 0$

$\frac{2x}{x+3} - \frac{4(x+3)}{x+3} > 0$

$\frac{2x-4x-12}{x+3} > 0$

$\frac{-2x-12}{x+3} > 0$; $x \neq -3$

$-2x-12=0$ or $x+3=0$

$-2x=12$ or $x=-3$

$x=-6$ or $x=-3$

Test Point:	-7	-5	0
	(-6)		(-3)
$-2x-12$:	positive	negative	negative
$x+3$:	negative	negative	positive
quotient:	negative	positive	negative

The solution set is $(-6, -3)$.

39. $\frac{x-1}{x-5} \le 2$

$\frac{x-1}{x-5} - 2 \le 0$

$\frac{x-1}{x-5} - \frac{2(x-5)}{x-5} \le 0$

$\frac{x-1-2(x-5)}{x-5} \le 0$

$\frac{x-1-2x+10}{x-5} \le 0$

$\frac{-x+9}{x-5} \le 0; \ x \ne 5$

$-x+9=0$ or $x-5=0$

$9=x$ or $x=5$

Test Point:	4	6	10
(critical values)	5	9	
$-x+9$:	positive	positive	negative
$x-5$:	negative	positive	positive
quotient:	negative	positive	negative

The solution set is $(-\infty, 5) \cup [9, \infty)$.

41. $\frac{x+2}{x-3} > -2$

$\frac{x+2}{x-3} + 2 > 0$

$\frac{x+2}{x-3} + \frac{2(x-3)}{x-3} > 0$

$\frac{x+2+2(x-3)}{x-3} > 0$

$\frac{x+2+2x-6}{x-3} > 0$

$\frac{3x-4}{x-3} > 0; \ x \ne 3$

$3x-4=0$ or $x-3=0$

$3x=4$ or $x=3$

$x=\frac{4}{3}$ or $x=3$

Test Point:	0	2	4
(critical values)	$\frac{4}{3}$	3	
$3x-4$:	negative	positive	positive
$x-3$:	negative	negative	positive
quotient:	positive	negative	positive

The solution set is $\left(-\infty, \frac{4}{3}\right) \cup (3, \infty)$.

43. $\frac{3x+2}{x+4} \le 2$

$\frac{3x+2}{x+4} - 2 \le 0$

$\frac{3x+2}{x+4} - \frac{2(x+4)}{x+4} \le 0$

$\frac{3x+2-2(x+4)}{x+4} \le 0$

$\frac{3x+2-2x-8}{x+4} \le 0$

$\frac{x-6}{x+4} \le 0; \ x \ne -4$

$x-6=0$ or $x+4=0$

$x=6$ or $x=-4$

Test Point:	-5	0	7
(critical values)	-4	6	
$x-6$:	negative	negative	positive
$x+4$:	negative	positive	positive
quotient:	positive	negative	positive

The solution set is $(-4, 6]$.

45. $\frac{x+1}{x-2} < 1$

$\frac{x+1}{x-2} - 1 < 0$

$\frac{x+1}{x-2} - \frac{1(x-2)}{x-2} < 0$

$\frac{x+1-1(x-2)}{x-2} < 0$

$\frac{x+1-x+2}{x-2} < 0$

$\frac{3}{x-2} < 0; \ x \ne 2$

$x-2=0$

$x=2$

Test Point:	1	3
(critical value)	2	
3:	positive	positive
$x-2$:	negative	positive
quotient:	negative	positive

The solution set is $(-\infty, 2)$.

Chapter 6 Review

1. $(-7+3i)+(9-5i)$

   $(-7+9)+(3-5)i$

   $2-2i$

3. $5i(3-6i)$

   $15i-30i^2$

   $15i-30(-1)$

   $15i+30$

   $30+15i$

5. $(-2-3i)(4-8i)$

   $-8+16i-12i+24i^2$

   $-8+4i-24$

   $-32+4i$

7. $\frac{4+3i}{6-2i}=\frac{(4+3i)}{(6-2i)}\bullet\frac{(6+2i)}{(6+2i)}$

   $\frac{24+8i+18i+6i^2}{36-4i^2}=\frac{24+26i-6}{36+4}$

   $\frac{18+26i}{40}=\frac{18}{40}+\frac{26}{40}i=\frac{9}{20}+\frac{13}{20}i$

9. $4x^2-20x+25=0$

   b^2-4ac

   $(-20)^2-4(4)(25)$

   $400-400=0$

   Since $b^2-4ac=0$, there will be one real solution with multiplicity of two.

11. $7x^2-2x-14=0$

    b^2-4ac

    $(-2)^2-4(7)(-14)$

    $4+392=396$

    Since $b^2-4ac>0$, there will be two real solutions.

13. $x^2-17x=0$

    $x(x-17)=0$

    $x=0 \quad \text{or} \quad x-17=0$

    $x=0 \quad \text{or} \quad x=17$

    The solution set is $\{0, 17\}$.

15. $(2x-1)^2=-64$

    $2x-1=\pm\sqrt{-64}$

    $2x-1=\pm 8i$

    $2x=1\pm 8i$

    $x=\frac{1\pm 8i}{2}$

    The solution set is $\left\{\frac{1\pm 8i}{2}\right\}$.

17. $x^2+2x-9=0$

    $x^2+2x=9$

    $x^2+2x+1=9+1$

    $(x+1)^2=10$

    $x+1=\pm\sqrt{10}$

    $x=-1\pm\sqrt{10}$

    The solution set is $\{-1\pm\sqrt{10}\}$.

19. $4\sqrt{x}=x-5$

    $(4\sqrt{x})^2=(x-5)^2$

    $16x=x^2-10x+25$

    $0=x^2-26x+25$

    $0=(x-25)(x-1)$

    $x-25=0 \quad \text{or} \quad x-1=0$

    $x=25 \quad \text{or} \quad x=1$

    Checking $x=25$

    $4\sqrt{25}\stackrel{?}{=}25-5$

    $4(5)\stackrel{?}{=}20$

    $20=20$

    Checking $x=1$

    $4\sqrt{1}\stackrel{?}{=}1-5$

    $4(1)\stackrel{?}{=}-4$

    $4\neq -4$

    The solution set is $\{25\}$.

21. $n^2 - 10n = 200$

$n^2 - 10n - 200 = 0$

$(n-20)(n+10) = 0$

$n - 20 = 0$ or $n + 10 = 0$

$n = 20$ or $n = -10$

The solution set is $\{-10, 20\}$.

23. $x^2 - x + 3 = 0$

$x = \dfrac{-1(-1) \pm \sqrt{(-1)^2 - 4(1)(3)}}{2(1)}$

$x = \dfrac{1 \pm \sqrt{1-12}}{2}$

$x = \dfrac{1 \pm \sqrt{-11}}{2}$

$x = \dfrac{1 \pm i\sqrt{11}}{2}$

The solution set is $\left\{\dfrac{-1 \pm i\sqrt{11}}{2}\right\}$.

25. $2a^2 + 4a - 5 = 0$

$a = \dfrac{-4 \pm \sqrt{4^2 - 4(2)(-5)}}{2(2)}$

$a = \dfrac{-4 \pm \sqrt{16+40}}{4}$

$a = \dfrac{-4 \pm \sqrt{56}}{4}$

$a = \dfrac{-4 \pm 2\sqrt{14}}{4}$

$a = \dfrac{2(-2 \pm \sqrt{14})}{4}$

$a = \dfrac{-2 \pm \sqrt{14}}{2}$

The solution set is $\left\{\dfrac{-2 \pm \sqrt{14}}{2}\right\}$.

27. $x^2 + 4x + 9 = 0$

$x^2 + 4x = -9$

$x^2 + 4x + 4 = -9 + 4$

$(x+2)^2 = -5$

$x + 2 = \pm\sqrt{-5}$

$x + 2 = \pm i\sqrt{5}$

$x = -2 \pm i\sqrt{5}$

The solution set is $\{-2 \pm i\sqrt{5}\}$.

29. $\frac{3}{x} + \frac{2}{x+3} = 1;\ x \neq 0,\ x \neq -3$

$x(x+3)\left(\frac{3}{x} + \frac{2}{x+3}\right) = x(x+3)(1)$

$3(x+3) + 2x = x(x+3)$

$3x + 9 + 2x = x^2 + 3x$

$0 = x^2 - 2x - 9$

$x^2 - 2x = 9$

$x^2 - 2x + 1 = 9 + 1$

$(x-1)^2 = 10$

$x - 1 = \pm\sqrt{10}$

$x = 1 \pm \sqrt{10}$

The solution set is $\{1 \pm \sqrt{10}\}$.

31. $\frac{3}{n-2} = \frac{n+5}{4};\ n \neq 2$

$3(4) = (n-2)(n+5)$

$12 = n^2 + 3n - 10$

$0 = n^2 + 3n - 22$

$n = \dfrac{-3 \pm \sqrt{3^2 - 4(1)(-22)}}{2(1)}$

$n = \dfrac{-3 \pm \sqrt{9+88}}{2}$

$n = \dfrac{-3 \pm \sqrt{97}}{2}$

The solution set is $\left\{\dfrac{-3 \pm \sqrt{97}}{2}\right\}$.

33. $2x^2 + x - 21 \leq 0$

$(2x+7)(x-3) \leq 0$

$(2x+7)(x-3) = 0$

$2x + 7 = 0$ or $x - 3 = 0$

$2x = -7$ or $x = 3$

$x = -\frac{7}{2}$ or $x = 3$

		$-\frac{7}{2}$	3
Test Point:	-4	0	4
$2x + 7$:	negative	positive	positive
$x - 3$:	negative	negative	positive
product:	positive	negative	positive

The solution set is $\left[-\frac{7}{2}, 3\right]$.

35. $\frac{2x-1}{x+1} > 4$

$\frac{2x-1}{x+1} - 4 > 0$

$\frac{2x-1}{x+1} - \frac{4(x+1)}{x+1} > 0$

$\frac{2x-1-4(x+1)}{x+1} > 0$

$\frac{2x-1-4x-4}{x+1} > 0$

$\frac{-2x-5}{x+1} > 0, \ x \neq -1$

$-2x-5=0$ or $x+1=0$

$-2x=5$ or $x=-1$

$x=-\frac{5}{2}$ or $x=-1$

Critical points: $-\frac{5}{2}$ and -1

Test Point:	-3	-2	0
$-2x-5$:	positive	negative	negative
$x+1$:	negative	negative	positive
quotient:	negative	positive	negative

The solution set is $\left(-\frac{5}{2}, -1\right)$.

37.

	price/share	number of shares	Cost
Purchase	$\frac{250}{x}$	x	250
Sell	$\frac{300}{x-5}$	$x-5$	300

$\frac{250}{x} + 5 = \frac{300}{x-5}$

$x(x-5)\left(\frac{250}{x}+5\right) = x(x-5)\left(\frac{300}{x-5}\right)$

$250(x-5) + 5x(x-5) = 300x$

$250x - 1250 + 5x^2 - 25x = 300x$

$5x^2 + 225x - 1250 = 300x$

$5x^2 - 75x - 1250 = 0$

$5(x^2 - 15x - 250) = 0$

$5(x-25)(x+10) = 0$

$x-25=0$ or $x+10=0$

$x=25$ or $x=-10$

Discard the root $x=-10$.
She sold 20 shares at
\$15 per share.

39. Let s = length of a side
Area of square $= s^2$
Perimeter of square $= 4s$

$s^2 = 2(4s)$

$s^2 = 8s$

$s^2 - 8s = 0$

$s(s-8) = 0$

$s=0$ or $s-8=0$

$s=0$ or $s=8$

Discard the root $s=0$.
The length of the side
of the square is 8 units.

41. Let x = width
$19-x$ = length

$x(19-x) = 84$

$19x - x^2 = 84$

$0 = x^2 - 19x + 84$

$0 = (x-12)(x-7)$

$x-12=0$ or $x-7=0$

$x=12$ or $x=7$

The dimensions of the rectangle
are 7 inches by 12 inches.

43. Let x = width of strip
Area of the 40×60 lot $= 2400$ sq. m.
width of new lot $= x+40$
length of new lot $= x+60$

$(x+40)(x+60) = 2400 + 1100$

$x^2 + 100x + 2400 = 3500$

$x^2 + 100x - 1100 = 0$

$(x-10)(x+110) = 0$

$x-10=0$ or $x+110=0$

$x=10$ or $x=-110$

Discard the root $x=-110$.
The width of the strip
should be 10 meters.

Chapter 6 Test

1. $(3-4i)(5+6i)$

$15+18i-20i-24i^2$

$15-2i+24$

$39-2i$

3. $x^2=7x$

$x^2-7x=0$

$x(x-7)=0$

$x=0$ or $x-7=0$

$x=0$ or $x=7$

The solution set is $\{0, 7\}$.

5. $x^2+3x-18=0$

$(x+6)(x-3)=0$

$x+6=0$ or $x-3=0$

$x=-6$ or $x=3$

The solution set is $\{-6, 3\}$.

7. $5x^2-2x+1=0$

$x=\frac{-(-2)\pm\sqrt{(-2)^2-4(5)(1)}}{2(5)}$

$x=\frac{2\pm\sqrt{4-20}}{10}$

$x=\frac{2\pm\sqrt{-16}}{10}$

$x=\frac{2\pm 4i}{10}$

$x=\frac{2(1\pm 2i)}{10}$

$x=\frac{1\pm 2i}{5}$

The solution set is $\left\{\frac{1\pm 2i}{5}\right\}$.

9. $(3x-1)^2+36=0$

$(3x-1)^2=-36$

$3x-1=\pm\sqrt{-36}$

$3x-1=\pm 6i$

$3x=1\pm 6i$

$x=\frac{1\pm 6i}{3}$

The solution set is $\left\{\frac{1\pm 6i}{3}\right\}$.

11. $(2x+1)(3x-2)=55$

$6x^2-x-2=55$

$6x^2-x-57=0$

$(6x-19)(x+3)=0$

$6x-19=0$ or $x+3=0$

$6x=19$ or $x=-3$

$x=\frac{19}{6}$ or $x=-3$

The solution set is $\left\{-3, \frac{19}{6}\right\}$.

13. $x^4+12x-64=0$

$(x^2+16)(x^2-4)=0$

$x^2+16=0$ or $x^2-4=0$

$x^2=-16$ or $x^2=4$

$x=\pm\sqrt{-16}$ or $x=\pm\sqrt{4}$

$x=\pm 4i$ or $x=\pm 2$

The solution set is $\{\pm 2, \pm 4i\}$.

15. $3x^2-2x-3=0$

$x=\frac{-(-2)\pm\sqrt{(-2)^2-4(3)(-3)}}{2(3)}$

$x=\frac{2\pm\sqrt{4+36}}{6}$

$x=\frac{2\pm\sqrt{40}}{6}$

$x=\frac{2\pm 2\sqrt{10}}{6}$

$x=\frac{2(1\pm\sqrt{10})}{6}$

$x=\frac{1\pm\sqrt{10}}{3}$

The solution set is $\left\{\frac{1\pm\sqrt{10}}{3}\right\}$.

17. $4x^2-3x=-5$

$4x^2-3x+5=0$

b^2-4ac

$(-3)^2-4(4)(5)$

$9-80=-71$

Since $b^2-4ac<0$, there will be two nonreal complex solutions.

19. $\frac{3x-1}{x+2} > 0$; $x \neq -2$

$3x - 1 = 0$ or $x + 2 = 0$

$3x = 1$ or $x = -2$

$x = \frac{1}{3}$ or $x = -2$

	-2		$\frac{1}{3}$
Test Point:	-3	0	1
$3x-1$:	negative	negative	positive
$x+2$:	negative	positive	positive
quotient:	positive	negative	positive

The solution set is

$(-\infty, -2) \cup \left(\frac{1}{3}, \infty\right)$.

21. $c = 24$

Let a be the side opposite the $30°$ angle, then $a = \frac{1}{2}(24) = 12$.

$12^2 + b^2 = 24^2$

$144 + b^2 = 576$

$b^2 = 432$

$b = \sqrt{432}$

$b = 20.8$ feet

The ladder reaches 20.8 feet up the building.

23.

	price/share	number of shares	Cost
Purchase	$\frac{3000}{x}$	x	3000
Sell	$\frac{3000}{x-50}$	$x-50$	3000

$\frac{3000}{x} + 5 = \frac{3000}{x-50}$

$x(x-50)\left(\frac{3000}{x} + 5\right) = x(x-50)\left(\frac{3000}{x-50}\right)$

$3000(x-50) + 5x(x-50) = 3000x$

$3000x - 150{,}000 + 5x^2 - 250x = 3000x$

$5x^2 + 2750x - 150{,}000 = 3000x$

$5x^2 - 250x - 150{,}000 = 0$

$5(x^2 - 50x - 30{,}000) = 0$

$5(x-200)(x+150) = 0$

$x - 200 = 0$ or $x + 150 = 0$

$x = 200$ or $x = -150$

Discard the root $x = -150$.
Dana sold 150 shares.

25. Let x = one number
$6 - x$ = other number

$x(6-x) = 4$

$6x - x^2 = 4$

$-x^2 + 6x = 4$

$x^2 - 6x = -4$

$x^2 - 6x + 9 = -4 + 9$

$(x-3)^2 = 5$

$x - 3 = \pm\sqrt{5}$

$x = 3 \pm \sqrt{5}$

The larger number is $3 + \sqrt{5}$.

Chapter 7 Coordinate Geometry and Graphing Techniques

Problem Set 7.1 Coordinate Geometry

1. $d = 12 - (-3) = 15$

3. $d = 2 - 7 = -5$

5. $d = -13 - (-4) = -9$

7. $d = -3 - (-14) = 11$

9. $d = |4 - (-6)| = 10$

11. $d = |-8 - (-3)| = 5$

13. (3, 2) and (9, 10)

$d = \sqrt{(9-3)^2 + (10-2)^2}$

$d = \sqrt{6^2 + 8^2}$

$d = \sqrt{36 + 64}$

$d = \sqrt{100} = 10$

15. (−1, 3) and (4, 2)

$d = \sqrt{[4-(-1)]^2 + (2-3)^2}$

$d = \sqrt{5^2 + (-1)^2}$

$d = \sqrt{25 + 1} = \sqrt{26}$

17. (−2, −4) and (2, 2)

$d = \sqrt{[2-(-2)]^2 + [2-(-4)]^2}$

$d = \sqrt{4^2 + 6^2}$

$d = \sqrt{16 + 36}$

$d = \sqrt{52} = 2\sqrt{13}$

19. (5, −3) and (−7, −19)

$d = \sqrt{(-7-5)^2 + [-19-(-3)]^2}$

$d = \sqrt{(-12)^2 + (-16)^2}$

$d = \sqrt{144 + 256}$

$d = \sqrt{400} = 20$

21. (−2, −1) and (−6, −9)

$d = \sqrt{[-6-(-2)]^2 + [-9-(-1)]^2}$

$d = \sqrt{(-4)^2 + (-8)^2}$

$d = \sqrt{16 + 64}$

$d = \sqrt{80} = 4\sqrt{5}$

23. (6, −4) and (9, −7)

$d = \sqrt{(9-6)^2 + [-7-(-4)]^2}$

$d = \sqrt{3^2 + (-3)^2}$

$d = \sqrt{9 + 9}$

$d = \sqrt{18} = 3\sqrt{2}$

25. (3, 4), (5, −2), and (−9, 0)

The distance between (3, 4) and (5, −2),

$d = \sqrt{(5-3)^2 + (-2-4)^2}$

$d = \sqrt{2^2 + (-6)^2}$

$d = \sqrt{4 + 36}$

$d = \sqrt{40}$

The distance between (3, 4) and (−9, 0),

$d = \sqrt{(-9-3)^2 + (0-4)^2}$

$d = \sqrt{(-12)^2 + (-4)^2}$

$d = \sqrt{144 + 16}$

$d = \sqrt{160}$

The distance between (5, −2) and (−9, 0),

$d = \sqrt{(-9-5)^2 + [0-(-2)]^2}$

$d=\sqrt{(-14)^2+(2)^2}$

$d=\sqrt{196+4}$

$d=\sqrt{200}$

Verifying the right triangle

$(\sqrt{40})^2+(\sqrt{160})^2 \stackrel{?}{=} (\sqrt{200})^2$

$40+160 \stackrel{?}{=} 200$

$200=200$

Yes, the points are vertices of a right triangle.

27. The distance between $(-2,-4)$ and $(2,-1)$.

$d=\sqrt{[2-(-2)]^2+[-1-(-4)]^2}$

$d=\sqrt{4^2+3^2}$

$d=\sqrt{16+9}$

$d=\sqrt{25}=5$

The distance between $(2,-1)$ and $(6, 2)$,

$d=\sqrt{(6-2)^2+[2-(-1)]^2}$

$d=\sqrt{4^2+3^2}$

$d=\sqrt{16+9}$

$d=\sqrt{25}=5$

Since the distances are equal, $(2,-1)$ is the midpoint.

29. $(4, 1)$, $(1,-3)$, and $(-5, 5)$

The distance between $(4, 1)$ and $(1,-3)$,

$d=\sqrt{(1-4)^2+(-3-1)^2}$

$d=\sqrt{(-3)^2+(-4)^2}$

$d=\sqrt{9+16}$

$d=\sqrt{25}=5$

The distance between $(4, 1)$ and $(-5, 5)$,

$d=\sqrt{(-5-4)^2+(5-1)^2}$

$d=\sqrt{(-9)^2+(4)^2}$

$d=\sqrt{81+16}=\sqrt{97}$

The distance between $(1,-3)$ and $(-5, 5)$,

$d=\sqrt{(-5-1)^2+[5-(-3)]^2}$

$d=\sqrt{(-6)^2+(8)^2}$

$d=\sqrt{36+64}$

$d=\sqrt{100}=10$

Perimeter $=5+\sqrt{97}+10$

Perimeter $=15+\sqrt{97}$

31. $(-4, 9)$, $(8, 4)$, $(3,-8)$, and $(-9,-3)$

The distance between $(-4, 9)$ and $(8, 4)$,

$d=\sqrt{[8-(-4)]^2+(4-9)^2}$

$d=\sqrt{12^2+(-5)^2}$

$d=\sqrt{144+25}$

$d=\sqrt{169}=13$

The distance between $(-4, 9)$ and $(-9,-3)$,

$d=\sqrt{[-9-(-4)]^2+(-3-9)^2}$

$d=\sqrt{(-5)^2+(-12)^2}$

$d=\sqrt{25+144}$

$d=\sqrt{169}=13$

The distance between $(8, 4)$ and $(3,-8)$,

$d=\sqrt{(3-8)^2+(-8-4)^2}$

$d=\sqrt{(-5)^2+(-12)^2}$

$d=\sqrt{25+144}$

$d=\sqrt{169}=13$

The distance from $(-9,-3)$ and $(3,-8)$,

$d=\sqrt{[3-(-9)]^2+[-8-(-3)]^2}$

$d=\sqrt{12^2+(-5)^2}$

$d=\sqrt{144+25}$

$d=\sqrt{169}=13$

All the sides are equal to 13.

The diagonal of a square with sides equal to 13.

$\text{diagonal} = 13^2 + 13^2$

$\text{diagonal} = 169 + 169$

$\text{diagonal} = \sqrt{2(169)}$

$\text{diagonal} = 13\sqrt{2}$

The diagonal is from $(-4, 9)$ to $(3, -8)$.

$d = \sqrt{[3-(-4)]^2 + (-8-9)^2}$

$d = \sqrt{7^2 + 17^2}$

$d = \sqrt{49 + 289}$

$d = \sqrt{338}$

$d = \sqrt{169(2)}$

$d = 13\sqrt{2}$

So the vertices do represent a square.

33. $(4, 2)$ and $(7, 4)$

$m = \frac{4-2}{7-4} = \frac{2}{3}$

35. $(-1, 3)$ and $(3, -3)$

$m = \frac{-3-3}{3-(-1)} = \frac{-6}{4} = -\frac{3}{2}$

37. $(-1, -2)$ and $(-7, -10)$

$m = \frac{-10-(-2)}{-7-(-1)} = \frac{-8}{-6} = \frac{4}{3}$

39. $(-2, 4)$ and $(7, 4)$

$m = \frac{4-4}{7-(-2)} = \frac{0}{9} = 0$

41. (a, b) and (c, d)

$m = \frac{d-b}{c-a}$

43. $(-2, 4)$ and $(4, 10)$

$m = \frac{10-4}{4-(-2)} = \frac{6}{6} = 1$

45. $(-1, 5)$ and $(-1, -7)$

$m = \frac{-7-5}{-1-(-1)} = \frac{-12}{0}$

The slope is undefined.

47. $(-2, 4)$ and $(x, 6)$

$m = \frac{6-4}{x-(-2)}$

$\frac{2}{9} = \frac{2}{x+2}$

$2(x+2) = 18$

$2x + 4 = 18$

$2x = 14$

$x = 7$

49. $(5, 2)$ and $(-3, y)$

$m = \frac{y-2}{-3-5}$

$-\frac{7}{8} = \frac{y-2}{-8}$

$-7(-8) = 8(y-2)$

$56 = 8y - 16$

$72 = 8y$

$9 = y$

51. $(2, 4)$ $m = \frac{5}{6}$

$(2+6,\ 4+5) = (8, 9)$

$(8+6,\ 9+5) = (14, 14)$

$(14+6,\ 14+5) = (20, 19)$

or other various answers.

53. $(-2, 1)$ $m = 3 = \frac{3}{1}$

$(-2+1,\ 1+3) = (-1, 4)$

$(-1+1,\ 4+3) = (0, 7)$

$(0+1,\ 7+3) = (1, 10)$

or other various answers.

55. $(-3, -2)$ $m = \frac{-2}{3}$

$(-3+3, -2-2) = (0, -4)$

$(0+3, -4-2) = (3, -6)$

$(3+3, -6-2) = (6, -8)$

or other various answers.

57. $(4, -1)$ $m = \frac{-3}{2}$

$(4+2, -1-3) = (6, -4)$

$(6+2, -4-3) = (8, -7)$

$(8+2, -7-3) = (10, -10)$

or other various answers.

59a. $\frac{2}{100} = \frac{y}{5280}$

$100y = 2(5280)$

$100y = 10560$

$y = 105.6$ ft

b. $\frac{30}{100} = \frac{75}{x}$

$30x = 100(75)$

$30x = 7500$

$x = 250$ feet

61. $2\frac{1}{4}\% = 2.25\% = .0225 = \frac{225}{10000}$

$\frac{225}{10000} = \frac{y}{45}$

$10000y = 225(45)$

$10000y = 10125$

$y = 1.0$ feet

Problem Set 7.2 Graphing Techniques Linear Equations and Inequalities

1 - 35. See back of textbook for graphs.

37. $4x + 5y = 20$

Let $x = 0$

$4(0) + 5y = 20$

$5y = 20$

$y = 4$

$(0, 4)$

Let $y = 0$

$4x + 5(0) = 20$

$4x = 20$

$x = 5$

$(5, 0)$

$m = \frac{0-4}{5-0} = -\frac{4}{5}$

39. $3x - y = 15$

Let $x = 0$

$3(0) - y = 15$

$-y = 15$

$y = -15$

$(0, -15)$

Let $y = 0$

$3x - 0 = 15$

$3x = 15$

$x = 5$

$(5, 0)$

$m = \frac{0-(-15)}{5-0} = \frac{15}{5} = 3$

41. $2x + 3y = 7$

Let $x = 0$

$2(0) + 3y = 7$

$3y = 7$

$y = \frac{7}{3}$

$\left(0, \frac{7}{3}\right)$

Let $y = 0$

$2x + 3(0) = 7$

$2x = 7$

$x = \frac{7}{2}$

$\left(\frac{7}{2}, 0\right)$

$$m = \frac{0 - \frac{7}{3}}{\frac{7}{2} - 0} = \frac{-\frac{7}{3}}{\frac{7}{2}} = \frac{2}{7} \bullet \frac{-7}{3} = -\frac{2}{3}$$

43. $-3x + 4y = 12$

Let $x = 0$

$-3(0) + 4y = 12$

$4y = 12$

$y = 3$

$(0, 3)$

Let $y = 0$

$-3x + 4(0) = 12$

$-3x = 12$

$x = -4$

$(-4, 0)$

$$m = \frac{0 - 3}{-4 - 0} = \frac{-3}{-4} = \frac{3}{4}$$

45. $y = -3x - 1$

Let $x = 0$

$y = -3(0) - 1$

$y = -1$

$(0, -1)$

Let $x = 2$

$y = -3(2) - 1$

$y = -6 - 1$

$y = -7$

$(2, -7)$

$$m = \frac{-7 - (-1)}{2 - 0} = \frac{-6}{2} = -3$$

47. $x = 2y - 1$

Let $y = 0$

$x = 2(0) - 1$

$x = -1$

$(-1, 0)$

Let $y = 1$

$x = 2(1) - 1$

$x = 2 - 1$

$x = 1$

$(1, 1)$

$$m = \frac{1 - 0}{1 - (-1)} = \frac{1}{2}$$

Problem Set 7.3 Determining the Equation of a Line

1. $m = \frac{1}{2}$, $(3, 4)$

$y - 4 = \frac{1}{2}(x - 3)$

$2(y - 4) = 2\left[\frac{1}{2}(x - 3)\right]$

$2y - 8 = x - 3$

$-8 = x - 2y - 3$

$-5 = x - 2y$

$x - 2y = -5$

3. $m = -\frac{5}{6}$, $(-2, 7)$

$y - 7 = -\frac{5}{6}[x - (-2)]$

$y - 7 = -\frac{5}{6}(x + 2)$

$6(y - 7) = 6\left[-\frac{5}{6}(x + 2)\right]$

$6y - 42 = -5(x + 2)$

$6y - 42 = -5x - 10$

$5x + 6y - 42 = -10$

$5x + 6y = 32$

5. $m = 3$, $(-1, -4)$

$y - (-4) = 3[x - (-1)]$

$y + 4 = 3(x + 1)$

$y + 4 = 3x + 3$

$4 = 3x - y + 3$

$1 = 3x - y$

$3x - y = 1$

7. $m = -\frac{3}{2}$, $(2, -8)$

$y - (-8) = -\frac{3}{2}(x - 2)$

$y + 8 = -\frac{3}{2}(x - 2)$

$2(y + 8) = 2\left[-\frac{3}{2}(x - 2)\right]$

$2y + 16 = -3(x - 2)$

$2y + 16 = -3x + 6$

$3x + 2y + 16 = 6$

$3x + 2y = -10$

9. $(1, 3)$, $(5, 7)$

$m = \frac{7 - 3}{5 - 1} = \frac{4}{4} = 1$

$y - 3 = 1(x - 1)$

$y - 3 = x - 1$

$-3 = x - y - 1$

$-2 = x - y$

$x - y = -2$

11. $(-2, 2)$, $(3, -1)$

$m = \frac{-1 - 2}{3 - (-2)} = -\frac{3}{5}$

$y - 2 = -\frac{3}{5}[x - (-2)]$

$y - 2 = -\frac{3}{5}(x + 2)$

$5(y - 2) = 5\left[-\frac{3}{5}(x + 2)\right]$

$5y - 10 = -3(x + 2)$

$5y - 10 = -3x - 6$

$3x + 5y - 10 = -6$

$3x + 5y = 4$

13. $(-7, -3)$, $(-2, 4)$

$m = \frac{4 - (-3)}{-2 - (-7)} = \frac{7}{5}$

$y - 4 = \frac{7}{5}[x - (-2)]$

$y - 4 = \frac{7}{5}(x + 2)$

$5(y - 4) = 5\left[\frac{7}{5}(x + 2)\right]$

$5y - 20 = 7(x + 2)$

$5y - 20 = 7x + 14$

$-20 = 7x - 5y + 14$

$-34 = 7x - 5y$

$7x - 5y = -34$

15. $(2, -4)$, $(7, -4)$

$m = \frac{-4 - (-4)}{7 - 2} = \frac{0}{5} = 0$

$y - (-4) = 0(x - 2)$

$y + 4 = 0$

$y = -4$

17. $(-3, 0)$, $(0, 6)$

$m = \frac{6 - 0}{0 - (-3)} = \frac{6}{3} = 2$

$y - 6 = 2(x - 0)$

$y - 6 = 2x$

$-6 = 2x - y$

$2x - y = -6$

19. $m = \frac{2}{7}$, $b = 3$

$y = \frac{2}{7}x + 3$

21. $m = 3$, $b = -2$

$y = 3x - 2$

23. $m = -\frac{1}{4}$, $b = -5$

$y = -\frac{1}{4}x - 5$

25. $m = -\frac{1}{2}$, $b = \frac{2}{3}$

$y = -\frac{1}{2}x + \frac{2}{3}$

27. $(-3, 0)$, $(0, 4)$

$m = \frac{4 - 0}{0 - (-3)} = \frac{4}{3}$

$y = \frac{4}{3}x + 4$

$3(y) = 3\left(\frac{4}{3}x + 4\right)$

$3y = 4x + 12$

$0 = 4x - 3y + 12$

$-12 = 4x - 3y$

$4x - 3y = -12$

29. (3, −5), parallel to y-axis

$x = 3$

31. (4, 7), perpendicular to y-axis

$y = 7$

33. $y = \frac{5}{6}x - 2$

$m = \frac{5}{6}$

$y = \frac{5}{6}x + 8$

$m = \frac{5}{6}$

The lines are parallel.

35. $5x - 7y = 8$

$-7y = -5x + 8$

$y = \frac{5}{7}x - \frac{8}{7}$

$m = \frac{5}{7}$

$7x + 5y = 9$

$5y = -7x + 9$

$y = -\frac{7}{5}x + \frac{9}{5}$

$m = -\frac{7}{5}$

The lines are perpendicular.

37. $2x + 3y = 9$

$3y = -2x + 9$

$y = -\frac{2}{3}x + 3$

$m = -\frac{2}{3}$

$3x + 2y = 4$

$2y = -3x + 4$

$y = -\frac{3}{2}x + 2$

$m = -\frac{3}{2}$

The lines are a pair of intersecting lines that are not perpendicular.

39. $y = -2x - 5$

$m = -2$

$6x + 3y = 14$

$3y = -6x + 14$

$y = -2x + \frac{14}{3}$

$m = -2$

The lines are parallel.

41. $y = 3x$

$m = 3$

$y = -3x$

$m = -3$

The lines are a pair of intersecting lines that are not perpendicular.

43. $x = 2y + 4$

$x - 4 = 2y$

$\frac{1}{2}x - 2 = y$

$y = \frac{1}{2}x - 2$

$m = \frac{1}{2}$

$y = -2x + 7$

$m = -2$

The lines are perpendicular.

45. (2, 4), parallel to $5x - y = 12$

$5x - y = 12$

$-y = -5x + 12$

$y = 5x - 12$

$m = 5$

Parallel lines have the same slope.

$(2, 4), \ m = 5$

$y - 4 = 5(x - 2)$

$y - 4 = 5x - 10$

$-4 = 5x - y - 10$

$6 = 5x - y$

$5x - y = 6$

47. $(0, 0)$, parallel to $3x + 2y = 8$

$3x + 2y = 8$

$2y = -3x + 8$

$y = -\frac{3}{2}x + 4$

$m = -\frac{3}{2}$

Parallel lines have the same slope.

$(0, 0), \ m = -\frac{3}{2}$

$y - 0 = -\frac{3}{2}(x - 0)$

$y = -\frac{3}{2}x$

$2y = 2\left(-\frac{3}{2}x\right)$

$2y = -3x$

$3x + 2y = 0$

49. $(2, -2)$, perpendicular to $2x + y = 6$.

$y = -2x + 6$

$m = -2$

Perpendicular lines have slopes that are negative reciprocals.

$(2, -2), \ m = \frac{1}{2}$

$y - (-2) = \frac{1}{2}(x - 2)$

$y + 2 = \frac{1}{2}(x - 2)$

$2(y + 2) = 2\left[\frac{1}{2}(x - 2)\right]$

$2y + 4 = x - 2$

$4 = x - 2y - 2$

$6 = x - 2y$

$x - 2y = 6$

51. $(-4, -1)$, perpendicular to $y = -3x$

$y = -3x$

$m = -3$

Perpendicular lines have slopes that are negative reciprocals.

$(-4, -1), \ m = \frac{1}{3}$

$y - (-1) = \frac{1}{3}[x - (-4)]$

$y + 1 = \frac{1}{3}(x + 4)$

$3(y + 1) = 3\left[\frac{1}{3}(x + 4)\right]$

$3y + 3 = x + 4$

$3 = x - 3y + 4$

$-1 = x - 3y$

$x - 3y = -1$

53. $m = \frac{1}{5}, \ (2, 6)$

$y = mx + b$

$6 = \frac{1}{5}(2) + b$

$6 = \frac{2}{5} + b$

$\frac{30}{5} - \frac{2}{5} = b$

$\frac{28}{5} = b$

$y = \frac{1}{5}x + \frac{28}{5}$

55. $m = -\frac{5}{4}, \ (-2, -3)$

$y = mx + b$

$-3 = -\frac{5}{4}(-2) + b$

$-3 = \frac{5}{2} + b$

$-\frac{6}{2} - \frac{5}{2} = b$

$-\frac{11}{2} = b$

$y = -\frac{5}{4}x - \frac{11}{2}$

57. See back of textbook for graph.

59. (0, 7), (− 2, − 1), (2, − 2), (4, 6)

Slope between (0, 7) and (− 2, − 1),

$m = \frac{-1-7}{-2-0} = \frac{-8}{-2} = 4$

Slope between (− 2, − 1) and (2, − 2),

$m = \frac{-2-(-1)}{2-(-2)} = \frac{-1}{4} = -\frac{1}{4}$

Slope between (2, − 2) and (4, 6),

$m = \frac{6-(-2)}{4-2} = \frac{8}{2} = 4$

Slope between (4, 6) and (0, 7),

$m = \frac{7-6}{0-4} = \frac{1}{-4} = -\frac{1}{4}$

From the slopes we can tell that opposite sides are parallel and adjacent sides are perpendicular.

61. (1, 1), (3, 7), and (− 2, − 8)

Slope between (1, 1) and (3, 7),

$m = \frac{7-1}{3-1} = \frac{6}{2} = 3$

Slope between (1, 1) and (− 2, − 8),

$m = \frac{-8-1}{-2-1} = \frac{-9}{-3} = 3$

The slopes are the same so the points are on a straight line.

Problem Set 7.4 Graphing Parabolas

1 - 31. See back of textbook for graphs.

Problem Set 7.5 More Parabolas and Some Circles

1 - 29. See back of textbook for graphs.

31. $y = x^2 - 8x + 15$

$0 = x^2 - 8x + 15$

$0 = (x-5)(x-3)$

$x - 5 = 0$ or $x - 3 = 0$

$x = 5$ or $x = 3$

The x-intercepts are 3 and 5.

$y = x^2 - 8x + 15$

$y = x^2 - 8x + 16 - 16 + 15$

$y = (x-4)^2 - 1$

Vertex (4, − 1)

33. $y = 2x^2 - 28x + 96$

$0 = 2(x^2 - 14x + 48)$

$0 = 2(x-6)(x-8)$

$x - 6 = 0$ or $x - 8 = 0$

$x = 6$ or $x = 8$

The x-intercepts are 6 and 8.

$y = 2(x^2 - 14x) + 96$

$y = 2(x^2 - 14x + 49) - 98 + 96$

$y = 2(x-7)^2 - 2$

Vertex (7, − 2)

35. $y = -x^2 + 10x - 24$

$0 = -(x^2 - 10x + 24)$

$0 = -(x-6)(x-4)$

$x - 6 = 0$ or $x - 4 = 0$

$x = 6$ or $x = 4$

The x-intercepts are 6 and 4.

$y = -(x^2 - 10x) - 24$

$y = -(x^2 - 10x + 25) + 25 - 24$

$y = -(x-5)^2 + 1$

Vertex (5, 1)

37. $y = x^2 - 14x + 44$

$0 = x^2 - 14x + 44$

$$x = \frac{-(-14) \pm \sqrt{(-14)^2 - 4(1)(44)}}{2(1)}$$

$$x = \frac{14 \pm \sqrt{196 - 176}}{2}$$

$$x = \frac{14 \pm \sqrt{20}}{2}$$

$$x = \frac{14 \pm 2\sqrt{5}}{2}$$

$$x = \frac{2(7 \pm \sqrt{5})}{2}$$

$$x = 7 \pm \sqrt{5}$$

The x-intercepts are $7 + \sqrt{5}$ and $7 - \sqrt{5}$.

$$y = x^2 - 14x + 44$$

$$y = x^2 - 14x + 49 - 49 + 44$$

$$y = (x-7)^2 - 5$$

Vertex $(7, -5)$

39. $y = -x^2 + 9x - 21$

$$0 = -x^2 + 9x - 21$$

$$x = \frac{-(9) \pm \sqrt{9^2 - 4(-1)(-21)}}{2(-1)}$$

$$x = \frac{-9 \pm \sqrt{81 - 84}}{-2}$$

$$x = \frac{-9 \pm \sqrt{-3}}{-2}$$

Since these are complex roots, there are no x-intercepts.

$$y = -x^2 + 9x - 21$$

$$y = -(x^2 - 9x) - 21$$

$$y = -\left(x^2 - 9x + \frac{81}{4}\right) + \frac{81}{4} - 21$$

$$y = -\left(x - \frac{9}{2}\right)^2 - \frac{3}{4}$$

Vertex $\left(\frac{9}{2}, -\frac{3}{4}\right)$

41. $y = -4x^2 + 4x + 4$

$$0 = -4x^2 + 4x + 4$$

$$x = \frac{-4 \pm \sqrt{4^2 - 4(-4)(4)}}{2(-4)}$$

$$x = \frac{-4 \pm \sqrt{16 + 64}}{-8}$$

$$x = \frac{-4 \pm \sqrt{80}}{-8}$$

$$x = \frac{-4 \pm 4\sqrt{5}}{-8}$$

$$x = \frac{-4(1 \pm \sqrt{5})}{-8}$$

$$x = \frac{1 \pm \sqrt{5}}{2}$$

The x-intercepts are $\frac{1+\sqrt{5}}{2}$ and $\frac{1-\sqrt{5}}{2}$.

$$y = -4x^2 + 4x + 4$$

$$y = -4(x^2 - x) + 4$$

$$y = -4\left(x^2 - x + \frac{1}{4}\right) + 1 + 4$$

$$y = -4\left(x - \frac{1}{2}\right)^2 + 5$$

Vertex $\left(\frac{1}{2}, 5\right)$

43. center $(3, 5)$, $r = 2$

$$(x-3)^2 + (y-5)^2 = 2^2$$

$$x^2 - 6x + 9 + y^2 - 10y + 25 = 4$$

$$x^2 + y^2 - 6x - 10y + 30 = 0$$

45. center $(-2, 1)$, $r = 3$

$$[x - (-2)]^2 + (y-1)^2 = 3^2$$

$$(x+2)^2 + (y-1)^2 = 9$$

$$x^2 + 4x + 4 + y^2 - 2y + 1 = 9$$

$$x^2 + y^2 + 4x - 2y - 4 = 0$$

47. center $(0, 0)$, $r = \sqrt{3}$

$$(x-0)^2 + (y-0)^2 = (\sqrt{3})^2$$

$$x^2 + y^2 = 3$$

$$x^2 + y^2 - 3 = 0$$

49. center $(-2,-5)$, $r=2\sqrt{3}$

$[x-(-2)]^2+[y-(-5)]^2=(2\sqrt{3})^2$

$(x+2)^2+(y+5)^2=4(3)$

$x^2+4x+4+y^2+10y+25=12$

$x^2+y^2+4x+10y+17=0$

51. center $(0,-3)$, $r=6$

$(x-0)^2+[y-(-3)]^2=6^2$

$x^2+(y+3)^2=36$

$x^2+y^2+6y+9=36$

$x^2+y^2+6y-27=0$

53. $x^2+y^2-2x-8y+8=0$

$x^2-2x+y^2-8y=-8$

$x^2-2x+1+y^2-8y+16=-8+1+16$

$(x-1)^2+(y-4)^2=9$

center $(1,\ 4)$

$r=\sqrt{9}=3$

55. $x^2+y^2-6x+4y-23=0$

$x^2-6x+y^2+4y=23$

$x^2-6x+9+y^2+4y+4=23+9+4$

$(x-3)^2+(y+2)^2=36$

center $(3,-2)$

$r=\sqrt{36}=6$

57. $x^2+y^2=24$

center $(0,\ 0)$

$r=\sqrt{24}=2\sqrt{6}$

59. $x^2+y^2+4x+12y+15=0$

$x^2+4x+y^2+12y=-15$

$x^2+4x+4+y^2+12y+36=-15+4+36$

$(x+2)^2+(y+6)^2=25$

center $(-2,-6)$

$r=\sqrt{25}=5$

61. $x^2+y^2+4x-3=0$

$x^2+4x+y^2=3$

$x^2+4x+4+y^2=3+4$

$(x+2)^2+y^2=7$

center $(-2,\ 0)$

$r=\sqrt{7}$

63. $x^2+y^2+6x-8y=0$

$x^2+6x+y^2-8y=0$

$x^2+6x+9+y^2-8y+16=0+9+16$

$(x+3)^2+(y-4)^2=25$

center $(-3,\ 4)$

$r=\sqrt{25}=5$

65. Center $(0,-5)$ passes through $(0,\ 0)$.

$r=\sqrt{(0-0)^2+(-5-0)^2}$

$r=\sqrt{25}=5$

$(x-0)^2+[y-(-5)]^2=5^2$

$x^2+(y+5)^2=5^2$

$x^2+y^2+10y+25=25$

$x^2+y^2+10y=0$

67. Center $(-4,-3)$ passes through $(0,\ 0)$.

$r=\sqrt{(-4-0)^2+(-3-0)^2}$

$r=\sqrt{(-4)^2+(-3)^2}$

$r=\sqrt{16+9}$

$r=\sqrt{25}=5$

$[x-(-4)]^2+[y-(-3)]^2=5^2$

$(x+4)^2+(y+3)^2=25$

$x^2+8x+16+y^2+6y+9=25$

$x^2+y^2+8x+6y=0$

Problem Set 7.6 Ellipses and Hyperbolas

1 - 3. See back of textbook for graphs.

5. $x^2 - y^2 = 1$

Finding asymptotes

$x^2 - y^2 = 0$

$y^2 = x^2$

$y = \pm\sqrt{x}$

$y = x$ or $y = -x$

7. $y^2 - 4x^2 = 9$

Finding asymptotes

$y^2 - 4x^2 = 0$

$y^2 = 4x^2$

$y = \pm\sqrt{4x^2}$

$y = \pm 2x$

$y = 2x$ or $y = -2x$

9. See back of textbook for graph.

11. $5x^2 - 2y^2 = 20$

Finding asymptotes

$5x^2 - 2y^2 = 0$

$2y^2 = 5x^2$

$y^2 = \frac{5}{2}x^2$

$y = \pm\sqrt{\frac{5x^2}{2}}$

$y = \pm\frac{x\sqrt{5}}{\sqrt{2}} \bullet \frac{\sqrt{2}}{\sqrt{2}}$

$y = \pm\frac{\sqrt{10}}{2}x$

$y = \frac{\sqrt{10}}{2}x$ or $y = -\frac{\sqrt{10}}{2}x$

13. See back of textbook for graph.

15. $y^2 - 16x^2 = 4$

Finding asymptotes

$y^2 = 16x^2$

$y = \pm\sqrt{16x^2}$

$y = \pm 4x$

$y = 4x$ or $y = -4x$

17. $-4x^2 + y^2 = -4$

$4x^2 - y^2 = 4$

Finding asymptotes

$4x^2 - y^2 = 0$

$y^2 = 4x^2$

$y = \pm\sqrt{4x^2}$

$y = \pm 2x$

$y = 2x$ or $y = -2x$

19. $25y^2 - 3x^2 = 75$

Finding asymptotes

$25y^2 - 3x^2 = 0$

$25y^2 = 3x^2$

$y^2 = \frac{3}{25}x^2$

$y = \pm\sqrt{\frac{3}{25}x^2}$

$y = \pm\frac{\sqrt{3}}{5}x$

$y = \frac{\sqrt{3}}{5}x$ or $y = \frac{-\sqrt{3}}{5}x$

21 - 23. See back of textbook for graphs.

25. $4x^2 + 8x + 16y^2 - 64y + 4 = 0$

$4x^2 + 8x + 16y^2 - 64y = -4$

$4(x^2 + 2x) + 16(y^2 - 4y) = -4$

$4(x^2+2x+1)+16(y^2 - 4y+4) = -4+4+64$

$4(x+1)^2 + 16(y-2)^2 = 64$

center $(-1, 2)$

To find the endpoints of minor axis let $x = -1$.

$4(-1+1)^2 + 16(y-2)^2 = 64$

$16(y-2)^2 = 64$

$(y-2)^2 = 4$

$y - 2 = \pm\sqrt{4}$

$y-2=2$ or $y-2=-2$

$y=4$ or $y=0$

$(-1, 4)$ and $(-1, 0)$

To find endpoints of major axis let $y=2$.

$4(x+1)^2+16(2-2)^2=64$

$4(x+1)^2=64$

$(x+1)^2=16$

$x+1=\pm\sqrt{16}$

$x+1=4$ or $x+1=-4$

$x=3$ or $x=-5$

$(3, 2)$ and $(-5, 2)$

27. $-4x^2+32x+9y^2-18y-91=0$

$-4(x^2-8x)+9(y^2-2y)=91$

$-4(x^2-8x+16)+9(y^2-2y+1)=91-64+9$

$-4(x-4)^2+9(y-1)^2=36$

Finding asymptotes

$-4(x-4)^2+9(y-1)^2=0$

$9(y-1)^2=4(x-4)^2$

$(y-1)^2=\frac{4}{9}(x-1)^2$

$y-1=\pm\sqrt{\frac{4}{9}(x-1)^2}$

$y-1=\pm\frac{2}{3}(x-1)$

$y-1=\frac{2}{3}(x-1)$ or $y-1=-\frac{2}{3}(x-1)$

$y-1=\frac{2}{3}x-\frac{2}{3}$ or $y-1=-\frac{2}{3}x+\frac{2}{3}$

$y=\frac{2}{3}x+\frac{1}{3}$ or $y=-\frac{2}{3}x+\frac{5}{3}$

29. $x^2+8x+9y^2+36y+16=0$

$x^2+8x+9(y^2+4y)=-16$

$x^2+8x+16+9(y^2+4y+4)=-16+16+36$

$(x+4)^2+9(y+2)^2=36$

Center $(-4, -2)$

To find the endpoints of the minor axis let $x=-4$.

$(-4+4)^2+9(y+2)^2=36$

$9(y+2)^2=36$

$(y+2)^2=4$

$y+2=\pm\sqrt{4}$

$y+2=2$ or $y+2=-2$

$y=0$ or $y=-4$

$(-4, 0)$ or $(-4, -4)$

To find the endpoints of the major axis let $y=-2$.

$(x+4)^2+9(-2+2)^2=36$

$(x+4)^2=36$

$x+4=6$ or $x+4=-6$

$x=2$ or $x=-10$

$(2, -2)$ or $(-10, -2)$

31. $-4x^2+24x+16y^2+64y-36=0$

$-4(x^2-6x)+16(y^2+4y)=36$

$-4(x^2-6x+9)+16(y^2+4y+4)=36-36+64$

$-4(x-3)^2+16(y+2)^2=64$

Finding asymptotes

$-4(x-3)^2+16(y+2)^2=0$

$16(y+2)^2=4(x-3)^2$

$(y+2)^2=\frac{1}{4}(x-3)^2$

$y+2=\pm\sqrt{\frac{1}{4}(x-3)^2}$

$y+2=\pm\frac{1}{2}(x-3)$

$y+2=\frac{1}{2}(x-3)$ or $y+2=-\frac{1}{2}(x-3)$

$y+2=\frac{1}{2}x-\frac{3}{2}$ or $y+2=-\frac{1}{2}x+\frac{3}{2}$

$y=\frac{1}{2}x-\frac{7}{2}$ or $y=-\frac{1}{2}x-\frac{1}{2}$

Problem Set 7.7 More on Graphing

1 $(3, 5)$ original point

a. $(3, -5)$ symmetric with respect to x-axis.

b. $(-3, 5)$ symmetric with respect to y-axis.

c. $(-3, -5)$ symmetric with respect to the origin.

3. $(-2,-4)$ original point

a. $(-2, 4)$ symmetric with respect to x-axis.

b. $(2,-4)$ symmetric with respect to y-axis.

c. $(2, 4)$ symmetric with respect to the origin.

5. $(0,-1)$ original point

a. $(0, 1)$ symmetric with respect to x-axis.

b. $(0,-1)$ symmetric with respect to y-axis.

c. $(0, 1)$ symmetric with respect to origin.

7. $x=y^2+4$

To test for symmetry for the:

x-axis replace y with $-y$

$x=(-y)^2+4$

$x=y^2+4$

It is symmetric to the x-axis.

y-axis replace x with $-x$

$-x=y^2+4$

It is not symmetric to the y-axis.

origin replace x with $-x$ and y with $-y$

$-x=(-y)^2+4$

$-x=y^2+4$

It is not symmetric to the origin.

9. $x^2=y^3$

To test for symmetry for the:

x-axis replace y with $-y$

$x^2=(-y)^3$

$x^2=-y^3$

It is not symmetric to the x axis.

y-axis replace x with $-x$

$(-x)^2=y^3$

$x^2=y^3$

It is symmetric to the y-axis.

origin replace x with $-x$ and y with $-y$

$(-x)^2=(-y)^3$

$x^2=-y^3$

It is not symmetric to the origin.

11. $x^2-2x+3y^2=4$

To test for symmetry for the:

x-axis replace y with $-y$

$x^2-2x+3(-y)^2=4$

$x^2-2x+3y^2=4$

It is symmetric to the x-axis.

y-axis replace x with $-x$

$(-x)^2-2(-x)+3y^2=4$

$x^2+2x+3y^2=4$

It is not symmetric to the y-axis.

origin replace x with $-x$ and y with $-y$

$(-x)^2-2(-x)+3(-y)^2=4$

$x^2+2y-3y^2=4$

It is not symmetric to the origin.

13. $y=x$

To test for symmetry for the:

x-axis replace y with $-y$

$-y=x$

It is not symmetric to the x-axis.

y-axis replace x with $-x$

$y=-x$

It is not symmetric to the y-axis.

origin replace x with $-x$ and y with $-y$

$-y=-x$

$y=x$

It is symmetric to the origin.

15. $xy = 4$

To test for symmetry for the:

x-axis replace y with $-y$

$x(-y) = 4$

$-xy = 4$

It is not symmetric to the x-axis.

y-axis replace x with $-x$

$(-x)(y) = 4$

$-xy = 4$

It is not symmetric to the y-axis.

origin replace x with $-x$ and y with $-y$

$(-x)(-y) = 4$

$xy = 4$

It is symmetric to the origin.

17. $x^2 + 5y^2 = 9$

To test for symmetry for the:

x-axis replace y with $-y$

$x^2 + 5(-y)^2 = 9$

$x^2 + 5y^2 = 9$

It is symmetric to the x-axis.

y-axis replace x with $-x$

$(-x)^2 + 5y^2 = 9$

$x^2 + 5y^2 = 9$

It is symmetric to the y-axis.

origin replace x with $-x$ and y with $-y$

$(-x)^2 + 5(-y)^2 = 9$

$x^2 + 5y^2 = 9$

It is symmetric to the origin.

19. $y = x^4 + x^2$

To test for symmetry for the:

x-axis replace y with $-y$

$-y = x^4 + x^2$

It is not symmetric to the x-axis.

y-axis replace x with $-x$

$y = (-x)^4 + (-x)^2$

$y = x^4 + x^2$

It is symmetric to the y-axis.

origin replace x with $-x$ and y with $-y$

$-y = (-x)^4 + (-x)^2$

$-y = x^4 + x^2$

It is not symmetric to the origin.

21 - 49. See back of textbook for graphs.

Chapter 7 Review

1a. $(3, 4), (-2, -2)$

$$m = \frac{-2-4}{-2-3} = \frac{-6}{-5} = \frac{6}{5}$$

b. $(-2, 3), (4, -1)$

$$m = \frac{-1-3}{4-(-2)} = \frac{-4}{6} = -\frac{2}{3}$$

3. $(2, 3), (5, -1), (-4, -5)$

Distance between $(2, 3)$ and $(5, -1)$,

$d = \sqrt{(5-2)^2 + (-1-3)^2}$

$d = \sqrt{3^2 + (-4)^2}$

$d = \sqrt{9 + 16}$

$d = \sqrt{25} = 5$

Distance between $(2, 3)$ and $(-4, -5)$,

$d = \sqrt{(-4-2)^2 + (-5-3)^2}$

$d = \sqrt{(-6)^2 + (-8)^2}$

$d = \sqrt{36 + 64}$

$d = \sqrt{100} = 10$

Distance between $(5, -1)$ and $(-4, -5)$,

$d = \sqrt{(-4-5)^2 + [-5-(-1)]^2}$

$d = \sqrt{(-9)^2 + (-4)^2}$

$d = \sqrt{81 + 16}$

$d = \sqrt{97}$

5. $m = -\frac{3}{7}$, $b = 4$

$y = -\frac{3}{7}x + 4$

$7(y) = 7\left(-\frac{3}{7}x + 4\right)$

$7y = -3x + 28$

$3x + 7y = 28$

7. (2, 5), parallel to $x - 2y = 4$

$x - 2y = 4$

$-2y = -x + 4$

$y = \frac{1}{2}x - 2$

$m = \frac{1}{2}$

Parallel lines have the same slope.

(2, 5), $m = \frac{1}{2}$

$y - 5 = \frac{1}{2}(x - 2)$

$2(y - 5) = 2\left[\frac{1}{2}(x - 2)\right]$

$2y - 10 = x - 2$

$-10 = x - 2y - 2$

$-8 = x - 2y$

$x - 2y = -8$

9 - 21. See back of textbook for graphs.

23. $x^2 + y^2 + 6x - 8y + 16 = 0$

$x^2 + 6x + y^2 - 8y = -16$

$x^2 + 6x + 9 + y^2 - 8y + 16 = -16 + 9 + 16$

$(x + 3)^2 + (y - 4)^2 = 9$

Center $(-3, 4)$

$r = \sqrt{9} = 3$

25. $9x^2 - 4y^2 = 72$

Finding asymptotes

$9x^2 - 4y^2 = 0$

$4y^2 = 9x^2$

$y^2 = \frac{9}{4}x^2$

$y = \pm\sqrt{\frac{9}{4}x^2}$

$y = \pm\frac{3}{2}x$

$y = \frac{3}{2}x$ and $y = -\frac{3}{2}x$

Chapter 7 Test

1. $(-2, 4)$ and $(3, -2)$

$m = \frac{-2 - 4}{3 - (-2)} = \frac{-6}{5}$

3. $(4, 2)$ and $(-3, -1)$

$d = \sqrt{(-1 - 2)^2 + (-3 - 4)^2}$

$d = \sqrt{(-3)^2 + (-7)^2}$

$d = \sqrt{9 + 49}$

$d = \sqrt{58}$

5. $(-4, 2)$ and $(2, 1)$

$m = \frac{1 - 2}{2 - (-4)} = -\frac{1}{6}$

$y - 1 = -\frac{1}{6}(x - 2)$

$y - 1 = -\frac{1}{6}x + \frac{1}{3}$

$y = -\frac{1}{6}x + \frac{4}{3}$

7. (4, 7) perpendicular to $x - 6y = 9$.

$x - 6y = 9$

$-6y = -x + 9$

$y = \frac{1}{6}x - \frac{3}{2}$

$m = \frac{1}{6}$

Perpendicular lines have slopes that are negative reciprocals.

$(4,\ 7),\ m = -6$

$y - 7 = -6(x - 4)$

$y - 7 = -6x + 24$

$6x + y - 7 = 24$

$6x + y = 31$

9. $x^2 + y^2 + 6x - 8y - 11 = 0$

$x^2 + 6x + y^2 - 8y = 11$

$x^2 + 6x + 9 + y^2 - 8y + 16 = 11 + 9 + 16$

$(x+3)^2 + (y-4)^2 = 36$

$r = \sqrt{36} = 6$

11. $y = 3x^2 - 18x + 23$

$y = 3(x^2 - 6x) + 23$

$y = 3(x^2 - 6x + 9) - 27 + 23$

$y = 3(x-3)^2 - 4$

Vertex $(3, -4)$

13. $16x^2 - 9y^2 = 12$

Finding asymptotes

$16x^2 - 9y^2 = 0$

$9y^2 = 16x^2$

$y^2 = \frac{16}{9}x^2$

$y = \pm\sqrt{\frac{16}{9}x^2}$

$y = \pm\frac{4}{3}x$

$y = \frac{4}{3}x$ or $y = -\frac{4}{3}x$

15. $y = -5x^2 - 2$

parabola

17. $2x^2 + 5x + 2y^2 - 9y - 4 = 0$

circle

19 - 25. See back of textbook for graphs.

Chapter 7 Cumulative Review

1. $\sqrt{32xy^3}$

$\sqrt{16y^2}\sqrt{2xy}$

$4y\sqrt{2xy}$

3. $\frac{3\sqrt{2}}{2\sqrt{3}} = \frac{3\sqrt{2}}{2\sqrt{3}} \bullet \frac{\sqrt{3}}{\sqrt{3}} = \frac{3\sqrt{6}}{2(3)} = \frac{\sqrt{6}}{2}$

5. $\sqrt[3]{48x^5}$

$\sqrt[3]{8x^3}\ \sqrt[3]{6x^2}$

$2x\ \sqrt[3]{6x^2}$

7. $\frac{3}{2\sqrt{5}-1} = \frac{3}{(2\sqrt{5}-1)} \bullet \frac{(2\sqrt{5}+1)}{(2\sqrt{5}+1)}$

$\frac{3(2\sqrt{5}+1)}{4\sqrt{25}-1} = \frac{6\sqrt{5}+3}{4(5)-1} = \frac{6\sqrt{5}+3}{19}$

9. $(3\sqrt{6})(2\sqrt{8})$

$6\sqrt{48}$

$6\sqrt{16}\sqrt{3}$

$6(4)\sqrt{3}$

$24\sqrt{3}$

11. $(2\sqrt{3} + \sqrt{5})(\sqrt{3} - 2\sqrt{5})$

$2\sqrt{9} - 4\sqrt{15} + \sqrt{15} - 2\sqrt{25}$

$2(3) - 3\sqrt{15} - 2(5)$

$6 - 3\sqrt{15} - 10$

$-4 - 3\sqrt{15}$

13. $\frac{9y^2}{x^2 + 12x + 36} \div \frac{12y}{x^2 + 6x}$

$\frac{9y^2}{(x+6)(x+6)} \bullet \frac{x(x+6)}{12y}$

$\frac{3xy}{4(x+6)}$

15. $\frac{3}{5x} - \frac{2}{3x} + \frac{5x}{6}$; LCD = 30x

$\frac{3}{5x} \bullet \frac{6}{6} - \frac{2}{3x} \bullet \frac{10}{10} + \frac{5x}{6} \bullet \frac{5x}{5x}$

$\frac{18}{30x} - \frac{20}{30x} + \frac{25x^2}{30x}$

$\frac{25x^2 - 2}{30x}$

17.

$$\begin{array}{r|l} & \quad\quad\quad 4x - 3 \\ 5x - 6 & 20x^2 - 39x + 18 \\ & \underline{20x^2 - 24x} \\ & \quad\quad - 15x + 18 \\ & \quad\quad \underline{- 15x + 18} \\ & \quad\quad\quad\quad 0 \end{array}$$

$4x - 3$

19. $\frac{(.00063)(960000)}{(3200)(.0000021)}$

$\frac{(6.3)(10)^{-4}(9.6)(10)^5}{(3.2)(10)^3(2.1)(10)^{-6}}$

$\frac{(3)(3)(10)^1}{(10)^{-3}} = 9(10)^4 = 90{,}000$

21. $\sqrt{.000009}$

$\sqrt{9(10^{-6})}$

$3(10)^{-3}$

$.003$

23. $(-2 + 5i)(4 - 7i)$

$-8 + 14i + 20i - 35i^2$

$-8 + 34i + 35$

$27 + 34i$

25. $\frac{2}{5i} = \frac{2}{5i} \bullet \frac{(-i)}{(-i)} = \frac{-2i}{-5i^2}$

$\frac{-2i}{5} = 0 - \frac{2}{5}i$

27. $\frac{-1 - 4i}{-2 + 8i}$

$\frac{(-1 - 4i)}{(-2 + 8i)} \bullet \frac{(-2 - 8i)}{(-2 - 8i)}$

$\frac{2 + 8i + 8i + 32i^2}{4 - 64i^2}$

$\frac{2 + 16i - 32}{4 + 64}$

$\frac{-30 + 16i}{68}$

$\frac{-30}{68} + \frac{16}{68}i$

$-\frac{15}{34} + \frac{4}{17}i$

29. $(-8)^{\frac{1}{3}} = -2$

31. $\sqrt[4]{16} = 2$

33. $9x^2 + 12xy + 4y^2$

$(3x + 2y)(3x + 2y)$

$(3x + 2y)^2$

35. $4x^4 - 13x^2 + 9$

$(4x^2 - 9)(x^2 - 1)$

$(2x + 3)(2x - 3)(x + 1)(x - 1)$

37. $x^2 - 4xy + 4y^2 - 4$

$(x - 2y)^2 - 4$

$[(x - 2y) + 2][(x - 2y) - 2]$

$(x - 2y + 2)(x - 2y - 2)$

39. $x^3 = 8x$

$x^3 - 8x = 0$

$x(x^2 - 8) = 0$

$x = 0$ or $x^2 - 8 = 0$

$x^2 = 8$

$x = \pm\sqrt{8}$

$x = \pm 2\sqrt{2}$

The solution set is $\{-2\sqrt{2}, 0, 2\sqrt{2}\}$.

41. $(y+2)^2 = -24$

$y+2 = \pm\sqrt{-24}$

$y+2 = \pm 2i\sqrt{6}$

$y = -2 \pm 2i\sqrt{6}$

The solution set is

$\{-2-2i\sqrt{6}, -2+2i\sqrt{6}\}$.

43. $x^3 = 8$

$x^3 - 8 = 0$

$(x-2)(x^2+2x+4) = 0$

$x-2=0$ or $x^2+2x+4=0$

$x=2$ or $x = \dfrac{-2 \pm \sqrt{2^2-4(1)(4)}}{2(1)}$

$x = \dfrac{-2 \pm \sqrt{4-16}}{2}$

$x = \dfrac{-2 \pm \sqrt{-12}}{2}$

$x = \dfrac{-2 \pm 2i\sqrt{3}}{2}$

$x = \dfrac{2(-1 \pm i\sqrt{3})}{2}$

$x = -1 \pm i\sqrt{3}$

The solution set is

$\{2, -1-i\sqrt{3}, -1+i\sqrt{3}\}$.

45. $(x-5)(4x-1) = 23$

$4x^2 - x - 20x + 5 = 23$

$4x^2 - 21x - 18 = 0$

$(4x+3)(x-6) = 0$

$4x+3=0$ or $x-6=0$

$4x = -3$ or $x = 6$

$x = -\frac{3}{4}$ or $x = 6$

The solution set is $\left\{-\frac{3}{4}, 6\right\}$.

47. $(x-1)(x+4) = (x+1)(2x-5)$

$x^2+3x-4 = 2x^2-3x-5$

$0 = x^2-6x-1$

$x = \frac{-(-6) \pm \sqrt{(-6)^2 - 4(1)(-1)}}{2(1)}$

$x = \frac{6 \pm \sqrt{36+4}}{2}$

$x = \frac{6 \pm \sqrt{40}}{2}$

$x = \frac{6 \pm 2\sqrt{10}}{2}$

$x = \frac{2(3 \pm \sqrt{10})}{2}$

$x = 3 \pm \sqrt{10}$

The solution set is
$\{3 - \sqrt{10},\ 3 + \sqrt{10}\}$.

49. $t^2 - 2t = -4$

$t^2 - 2t + 1 = -4 + 1$

$(t-1)^2 = -3$

$t - 1 = \pm\sqrt{-3}$

$t - 1 = \pm i\sqrt{3}$

$t = 1 \pm i\sqrt{3}$

The solution set is
$\{1 - i\sqrt{3},\ 1 + i\sqrt{3}\}$.

51. $x^2 - 4x = 192$

$x^2 - 4x - 192 = 0$

$(x+12)(x-16) = 0$

$x + 12 = 0$ or $x - 16 = 0$

$x = -12$ or $x = 16$

The solution set is $\{-12, 16\}$.

53. $\frac{3n}{n^2+n-6} + \frac{2}{n^2+4n+3} = \frac{n}{n^2-n-2}$

$\frac{3n}{(n+3)(n-2)} + \frac{2}{(n+1)(n+3)} = \frac{n}{(n-2)(n+1)};\ n \neq -3,\ n \neq -1,\ n \neq 2$

$(n+3)(n-2)(n+1)\left[\frac{3n}{(n+3)(n-2)} + \frac{2}{(n+1)(n+3)}\right] = (n+3)(n-2)(n+1)\left[\frac{n}{(n-2)(n+1)}\right]$

$3n(n+1) + 2(n-2) = n(n+3)$

$3n^2 + 3n + 2n - 4 = n^2 + 3n$

$2n^2 + 2n - 4 = 0$

$2(n^2 + n - 2) = 0$

$2(n+2)(n-1) = 0$

$n + 2 = 0$ or $n - 1 = 0$

$n = -2$ or $n = 1$

The solution set is $\{-2, 1\}$.

55. $\sqrt{x+4} = \sqrt{x-1} + 1$

$(\sqrt{x+4})^2 = (\sqrt{x-1} + 1)^2$

$x + 4 = (\sqrt{x-1})^2 + 2(1)(\sqrt{x-1}) + 1^2$

$x + 4 = x - 1 + 2\sqrt{x-1} + 1$

$4 = 2\sqrt{x-1}$

$2 = \sqrt{x-1}$

$2^2 = (\sqrt{x-1})^2$

$4 = x - 1$

$5 = x$

check

$\sqrt{5+4} \stackrel{?}{=} \sqrt{5-1} + 1$

$\sqrt{9} \stackrel{?}{=} \sqrt{4} + 1$

$3 \stackrel{?}{=} 2 + 1$

$3 = 3$

The solution set is $\{5\}$.

57. $\frac{2}{x-1} = \frac{x+4}{3}$; $x \neq 1$

$2(3) = (x-1)(x+4)$

$0 = x^2 + 3x - 10$

$0 = (x+5)(x-2)$

$x + 5 = 0$ or $x - 2 = 0$

$x = -5$ or $x = 2$

The solution set is $\{-5, 2\}$.

59. $\frac{3}{4} = \frac{y-1}{x-2}$ for y

$3(x-2) = 4(y-1)$

$3x - 6 = 4y - 4$

$3x - 2 = 4y$

$\frac{3x-2}{4} = y$

61. $V = C\left(1 - \frac{T}{N}\right)$ for T

$V = C - \frac{CT}{N}$

$V - C = -\frac{CT}{N}$

$-\frac{N}{C}(V - C) = -\frac{N}{C}\left(-\frac{CT}{N}\right)$

$\frac{-NV + NC}{C} = T$

$T = \frac{NC - NV}{C}$

63. $|3x + 5| < 2$

$-2 < 3x + 5 < 2$

$-7 < 3x < -3$

$-\frac{7}{3} < x < -1$

$\left(-\frac{7}{3}, -1\right)$

65. $(x+1)(2x-3) < 0$

$(x+1)(2x-3) = 0$

$x + 1 = 0$ or $2x - 3 = 0$

$x = -1$ or $2x = 3$

$x = -1$ or $x = \frac{3}{2}$

Critical points: -1, $\frac{3}{2}$

Test Point:	-2	0	2
$x + 1$:	negative	positive	positive
$2x - 3$:	negative	negative	positive
product:	positive	negative	positive

The solution set is $\left(-1, \frac{3}{2}\right)$.

67. $\frac{x-3}{x-5} \geq 0$; $x \neq 5$

$x - 3 = 0$ or $x - 5 = 0$

$x = 3$ or $x = 5$

Critical points: 3, 5

Test Point:	0	4	6
$x - 3$:	negative	positive	positive
$x - 5$:	negative	negative	positive
product:	positive	negative	positive

The solution set is $(-\infty, 3] \cup (5, \infty)$.

69. $\frac{2x}{x+3} > 4$

$\frac{2x}{x+3} - 4 > 0$

$\frac{2x}{x+3} - \frac{4(x+3)}{x+3} > 0$

$\frac{2x - 4(x+3)}{x+3} > 0$

$\frac{2x - 4x - 12}{x+3} > 0$

$\frac{-2x - 12}{x+3} > 0;\ x \neq -3$

$-2x - 12 = 0$ or $x + 3 = 0$

$-2x = 12$ or $x = -3$

$x = -6$ or $x = -3$

Test Point:	-7	-4	0
	(-6)	(-3)	
$-2x - 12$:	positive	negative	negative
$x + 3$:	negative	negative	positive
quotient:	negative	positive	negative

The solution set is $(-6, -3)$.

71. Let $x =$ one number
$-2 - x =$ other number

$x(-2 - x) = -35$

$-2x - x^2 = -35$

$0 = x^2 + 2x - 35$

$0 = (x+7)(x-5)$

$x + 7 = 0$ or $x - 5 = 0$

$x = -7$ or $x = 5$

The numbers are -7 and 5.

73. Let $x =$ width of frame
Dimensions of picture and frame.
width $= 2x + 3$
length $= 2x + 5$

$A = lw$

$24 = (2x+5)(2x+3)$

$24 = 4x^2 + 16x + 15$

$0 = 4x^2 + 16x - 9$

$0 = (2x-1)(2x+9)$

$2x - 1 = 0$ or $2x + 9 = 0$

$2x = 1$ or $2x = -9$

$x = \frac{1}{2}$ or $x = -\frac{9}{2}$

Discard the root $x = -\frac{9}{2}$.
The width of the frame is $\frac{1}{2}$ inch.

Chapter 8 Functions

Problem Set 8.1 Relations and Functions

1. Domain $= \{1, 2, 3, 4\}$.
   Range $= \{5, 8, 11, 14\}$.
   It is a function.

3. Domain $= \{0, 1\}$.
   Range $= \{-2\sqrt{6}, -5, 5, 2\sqrt{6}\}$.
   It is not a function.

5. Domain $= \{1, 2, 3, 4, 5\}$.
   Range $= \{2, 5, 10, 17, 26\}$.
   It is a function.

7. $\{(x, y) \mid 5x - 2y = 6\}$
   Domain $= \{$all reals$\}$.
   Range $= \{$all reals$\}$.
   It is a function.

9. $\{(x, y) \mid x^2 = y^3\}$
   Domain $= \{$all reals$\}$.
   To find range solve for y.
   $y^3 = x^2$
   $y = \sqrt[3]{x^2}$
   Since $x^2 \geq 0$, then $\sqrt[3]{x^2} \geq 0$.
   Range $= \{y \mid y \geq 0\}$.
   It is a function.

11. $f(x) = 7x - 2$
    Domain $= \{$all reals$\}$.

13. $f(x) = \frac{1}{x-1}$
    $x - 1 = 0$
    $x = 1$
    $\{x \mid x \neq 1\}$

15. $g(x) = \frac{3x}{4x-3}$
    $4x - 3 = 0$
    $4x = 3$
    $x = \frac{3}{4}$
    $\left\{x \mid x \neq \frac{3}{4}\right\}$

17. $h(x) = \frac{2}{(x+1)(x-4)}$
    $x + 1 = 0$ or $x - 4 = 0$
    $x = -1$ or $x = 4$
    $\{x \mid x \neq -1$ and $x \neq 4\}$

19. $f(x) = \frac{14}{x^2 + 3x - 40}$
    $f(x) = \frac{14}{(x+8)(x-5)}$
    $x + 8 = 0$ or $x - 5 = 0$
    $x = -8$ or $x = 5$
    $\{x \mid x \neq -8$ and $x \neq 5\}$

21. $f(x) = \frac{-4}{x^2 + 6x}$
    $f(x) = \frac{-4}{x(x+6)}$
    $x = 0$ or $x + 6 = 0$
    $x = -6$
    $\{x \mid x \neq 0$ and $x \neq -6\}$

23. $f(t) = \frac{4}{t^2 + 9}$
    $\{$all reals$\}$

25. $f(t) = \frac{3t}{t^2 - 4}$
    $f(t) = \frac{3t}{(t+2)(t-2)}$
    $t + 2 = 0$ or $t - 2 = 0$
    $t = -2$ or $t = 2$
    $\{t \mid t \neq -2$ and $t \neq 2\}$

27. $h(x) = \sqrt{x+4}$

$x+4 \geq 0$

$x \geq -4$

$\{x \mid x \geq -4\}$

29. $f(s) = \sqrt{4s-5}$

$4s-5 \geq 0$

$4s \geq 5$

$s \geq \frac{5}{4}$

$\left\{s \mid s \geq \frac{5}{4}\right\}$

31. $f(x) = \sqrt{x^2-16}$

$x^2-16 \geq 0$

$(x+4)(x-4) \geq 0$

$x+4=0 \quad \text{or} \quad x-4=0$

$x=-4 \quad \text{or} \quad x=4$

	-4		4
Test Point:	-5	0	5
$x+4$:	negative	positive	positive
$x-4$:	negative	negative	positive
product:	positive	negative	positive

$\{x \mid x \leq -4 \text{ or } x \geq 4\}$

33. $f(x) = \sqrt{x^2-3x-18}$

$f(x) = \sqrt{(x-6)(x+3)}$

$x-6=0 \quad \text{or} \quad x+3=0$

$x=6 \quad \text{or} \quad x=-3$

$(x-6)(x+3) \geq 0$

	-3		6
Test Point:	-4	0	7
$x-6$:	negative	negative	positive
$x+3$:	negative	positive	positive
product:	positive	negative	positive

$\{x \mid x \leq -3 \text{ or } x \geq 6\}$

35. $f(x) = \sqrt{1-x^2}$

$f(x) = \sqrt{(1-x)(1+x)}$

$(1-x)(1+x) \geq 0$

$1-x=0 \quad \text{or} \quad 1+x=0$

$1=x \quad \text{or} \quad x=-1$

	-1		1
Test Point:	-2	0	2
$1-x$:	positive	positive	negative
$1+x$:	negative	positive	positive
product:	negative	positive	negative

$\{x \mid -1 \leq x \leq 1\}$

37. $f(x) = 5x-2$

$f(0) = 5(0)-2 = -2$

$f(2) = 5(2)-2 = 10-2 = 8$

$f(-1) = 5(-1)-2 = -5-2 = -7$

$f(-4) = 5(-4)-2 = -20-2 = -22$

39. $f(x) = \frac{1}{2}x - \frac{3}{4}$

$f(-2) = \frac{1}{2}(-2) - \frac{3}{4} = -1 - \frac{3}{4} = -\frac{7}{4}$

$f(0) = \frac{1}{2}(0) - \frac{3}{4} = -\frac{3}{4}$

$f\left(\frac{1}{2}\right) = \frac{1}{2}\left(\frac{1}{2}\right) - \frac{3}{4} = \frac{1}{4} - \frac{3}{4} = -\frac{1}{2}$

$f\left(\frac{2}{3}\right) = \frac{1}{2}\left(\frac{2}{3}\right) - \frac{3}{4} = \frac{1}{3} - \frac{3}{4} = -\frac{5}{12}$

41. $g(x) = 2x^2-5x-7$

$g(-1) = 2(-1)^2-5(-1)-7$

$g(-1) = 2+5-7 = 0$

$g(2) = 2(2)^2-5(2)-7$

$g(2) = 8-10-7 = -9$

$g(-3) = 2(-3)^2-5(-3)-7$

$g(-3) = 18+15-7 = 26$

$g(4) = 2(4)^2-5(4)-7$

$g(4) = 32-20-7 = 5$

43. $h(x) = -2x^2-x+4$

$h(-2) = -2(-2)^2-(-2)+4$

$h(-2) = -8+2+4 = -2$

$h(-3) = -2(-3)^2-(-3)+4$

$h(-3) = -18+3+4 = -11$

$h(4) = -2(4)^2 - (4) + 4$
$h(4) = -32 - 4 + 4 = -32$

$h(5) = -2(5)^2 - (5) + 4$
$h(5) = -50 - 5 + 4 = -51$

45. $f(x) = \sqrt{2x+1}$
$f(3) = \sqrt{2(3)+1} = \sqrt{7}$
$f(4) = \sqrt{2(4)+1} = \sqrt{9} = 3$
$f(10) = \sqrt{2(10)+1} = \sqrt{21}$
$f(12) = \sqrt{2(12)+1} = \sqrt{25} = 5$

47. $f(x) = \frac{-4}{x+3}$
$f(1) = \frac{-4}{1+3} = \frac{-4}{4} = -1$
$f(-1) = \frac{-4}{-1+3} = \frac{-4}{2} = -2$
$f(3) = \frac{-4}{3+3} = \frac{-4}{6} = -\frac{2}{3}$
$f(-6) = \frac{-4}{-6+3} = \frac{-4}{-3} = \frac{4}{3}$

49. $f(x) = 5x^2 - 2x + 3$ and
$g(x) = -x^2 + 4x - 5$

$f(-2) = 5(-2)^2 - 2(-2) + 3$
$f(-2) = 20 + 4 + 3 = 27$

$f(3) = 5(3)^2 - 2(3) + 3$
$f(3) = 45 - 6 + 3 = 42$

$g(-4) = -(-4)^2 + 4(-4) - 5$
$g(-4) = -16 - 16 - 5 = -37$

$g(6) = -(6)^2 + 4(6) - 5$
$g(6) = -36 + 24 - 5 = -17$

51. $f(x) = 3|x| - 1$ and
$g(x) = -|x| + 1$

$f(-2) = 3|-2| - 1$
$f(-2) = 3(2) - 1 = 5$

$f(3) = 3|3| - 1$
$f(3) = 3(3) - 1 = 8$

$g(-4) = -|-4| + 1$
$g(-4) = -(4) + 1 = -3$

$g(5) = -|5| + 1$
$g(5) = -5 + 1 = -4$

53. $f(x) = -3x + 6$

$f(a) = -3a + 6$

$f(a+h) = -3(a+h) + 6$
$f(a+h) = -3a - 3h + 6$

$$\frac{f(a+h) - f(a)}{h} = \frac{-3a - 3h + 6 - (-3a + 6)}{h}$$
$$= \frac{-3a - 3h + 6 + 3a - 6}{h}$$
$$= \frac{-3h}{h}$$
$$= -3$$

55. $f(x) = -x^2 - 1$

$f(a) = -a^2 - 1$

$f(a+h) = -(a+h)^2 - 1$
$f(a+h) = -(a^2 + 2ah + h^2) - 1$
$f(a+h) = -a^2 - 2ah - h^2 - 1$

$$\frac{f(a+h) - f(a)}{h} = \frac{-a^2 - 2ah - h^2 - 1 - (-a^2 - 1)}{h}$$
$$= \frac{-a^2 - 2ah - h^2 - 1 + a^2 + 1}{h}$$
$$= \frac{-2ah - h^2}{h}$$
$$= \frac{h(-2a - h)}{h}$$
$$= -2a - h$$

57. $f(x) = 2x^2 - x + 8$

$f(a) = 2a^2 - a + 8$

$f(a+h) = 2(a+h)^2 - (a+h) + 8$
$f(a+h) = 2(a^2 + 2ah + h^2) - a - h + 8$
$f(a+h) = 2a^2 + 4ah + 2h^2 - a - h + 8$

$$\frac{f(a+h)-f(a)}{h}=\frac{2a^2+4ah+2h^2-a-h+8-(2a^2-a+8)}{h}$$
$$=\frac{2a^2+4ah+2h^2-a-h+8-2a^2+a-8}{h}$$
$$=\frac{4ah-h+2h^2}{h}$$
$$=\frac{h(4a-1+2h)}{h}$$
$$=4a-1+2h$$

59. $f(x)=-4x^2-7x-9$

$f(a)=-4a^2-7a-9$

$f(a+h)=-4(a+h)^2-7(a+h)-9$

$f(a+h)=-4(a^2+2ah+h^2)-7a-7h-9$

$f(a+h)=-4a^2-8ah-4h^2-7a-7h-9$

$$\frac{f(a+h)-f(a)}{h}=\frac{-4a^2-8ah-4h^2-7a-7h-9-(-4a^2-7a-9)}{h}$$
$$=\frac{-4a^2-8ah-4h^2-7a-7h-9+4a^2+7a+9}{h}$$
$$=\frac{-8ah-7h-4h^2}{h}$$
$$=\frac{h(-8a-7-4h)}{h}$$
$$=-8a-7-4h$$

61. $h(t)=64t-16t^2$

$h(1)=64(1)-16(1)^2$

$h(1)=64-16=48$

$h(2)=64(2)-16(2)^2$

$h(2)=128-64=64$

$h(3)=64(3)-16(3)^2$

$h(3)=192-144=48$

$h(4)=64(4)-16(4)^2$

$h(4)=256-256=0$

63. $C(m)=50+0.32m$

$C(75)=50+0.32(75)$

$C(75)=50+24=\$74$

$C(150) = 50 + 0.32(150)$

$C(150) = 50 + 48 = \$98$

$C(225) = 50 + 0.32(225)$

$C(225) = 50 + 72 = \$122$

$C(650) = 50 + 0.32(650)$

$C(650) = 50 + 208 = \$258$

65. $I(r) = 500r$

$I(0.11) = 500(0.11) = \$55$

$I(0.12) = 500(0.12) = \$60$

$I(0.135) = 500(0.135) = \$67.50$

$I(0.15) = 500(0.15) = \$75$

Problem Set 8.2 Functions: Graphs and Applications

1 - 29. See back of textbook for graphs.

31. $C(x) = 2x^2 - 320x + 12,920$

$C(x) = 2(x^2 - 160x) + 12,920$

$C(x) = 2(x^2 - 160x + 6400) - 12,800 + 12,920$

$C(x) = 2(x - 80)^2 + 120$

The minimum cost is when 80 items are produced.

33. Let $x =$ one number
$30 - x =$ other number

$F(x) = x^2 + 10(30 - x)$

$F(x) = x^2 + 300 - 10x$

$F(x) = x^2 - 10x + 25 - 25 + 300$

$F(x) = (x - 5)^2 + 275$

The numbers are 5 and 25.

35. Let $x =$ width
$120 - x =$ length

$A(x) = x(120 - x)$

$A(x) = 120x - x^2$

$A(x) = -x^2 + 120x$

$A(x) = -(x^2 - 120x)$

$A(x) = -(x^2 - 120x + 3600) + 3600$

$A(x) = -(x - 60)^2 + 3600$

The dimensions should be 60 meters by 60 meters.

37. Let $x =$ number of decreases

$R(x) = \left(15 - \frac{1}{4}x\right)(1000 + 20x)$

$R(x) = 15000 + 300x - 250x - 5x^2$

$R(x) = -5x^2 + 50x + 15000$

$R(x) = -5(x^2 - 10x) + 15000$

$R(x) = -5(x^2 - 10x + 25) + 125 + 15000$

$R(x) = -5(x - 5)^2 + 15125$

There should be 5 rate decreases.

$\text{Rate} = 15 - 5(.25) = \$13.75$

$\text{Subscribers} = 1000 + 20(5) = 1100$

39 - 47. See back of textbook for graphs.

Problem Set 8.3 Graphing Made Easy Via Transformations

See back of textbook for graphs.

Problem Set 8.4 Combining Functions

1. $f(x) = 3x - 4$ and $g(x) = 5x + 2$

$(f + g)(x) = 3x - 4 + 5x + 2$

$(f + g)(x) = 8x - 2$

$(f - g)(x) = 3x - 4 - (5x + 2)$

$(f - g)(x) = 3x - 4 - 5x - 2$

$(f - g)(x) = -2x - 6$

$(f \bullet g)(x) = (3x - 4)(5x + 2)$

$(f \bullet g)(x) = 15x^2 - 14x - 8$

$\left(\frac{f}{g}\right)(x) = \frac{3x - 4}{5x + 2};\ x \neq -\frac{2}{5}$

3. $f(x) = x^2 - 6x + 4$ and $g(x) = -x - 1$

$(f+g)(x) = x^2 - 6x + 4 - x - 1$
$(f+g)(x) = x^2 - 7x + 3$

$(f-g)(x) = x^2 - 6x + 4 - (-x - 1)$
$(f-g)(x) = x^2 - 6x + 4 + x + 1$
$(f-g)(x) = x^2 - 5x + 5$

$(f \bullet g)(x) = (x^2 - 6x + 4)(-x - 1)$
$(f \bullet g)(x) = -x^3 + 6x^2 - 4x - x^2 + 6x - 4$
$(f \bullet g)(x) = -x^3 + 5x^2 + 2x - 4$
$\left(\frac{f}{g}\right)(x) = \frac{x^2 - 6x + 4}{-x - 1};\ x \neq -1$

5. $f(x) = x^2 - x - 1$ and $g(x) = x^2 + 4x - 5$

$(f+g)(x) = x^2 - x - 1 + x^2 + 4x - 5$
$(f+g)(x) = 2x^2 + 3x - 6$

$(f-g)(x) = x^2 - x - 1 - (x^2 + 4x - 5)$
$(f-g)(x) = x^2 - x - 1 - x^2 - 4x + 5$
$(f-g)(x) = -5x + 4$

$(f \bullet g)(x) = (x^2 - x - 1)(x^2 + 4x - 5)$
$(f \bullet g)(x) = x^4 - x^3 - x^2 + 4x^3 - 4x^2 - 4x - 5x^2 + 5x + 5$
$(f \bullet g)(x) = x^4 + 3x^3 - 10x^2 + x + 5$
$\left(\frac{f}{g}\right)(x) = \frac{x^2 - x - 1}{x^2 + 4x - 5};\ x \neq -5,\ 1$

7. $f(x) = \sqrt{x - 1}$ and $g(x) = \sqrt{x}$

$(f+g)(x) = \sqrt{x-1} + \sqrt{x}$

$(f-g)(x) = \sqrt{x-1} - \sqrt{x}$

$(f \bullet g)(x) = (\sqrt{x-1})(\sqrt{x})$
$(f \bullet g)(x) = \sqrt{x^2 - x}$

$\left(\frac{f}{g}\right)(x) = \frac{\sqrt{x-1}}{\sqrt{x}}$

$\left(\frac{f}{g}\right)(x) = \frac{\sqrt{x-1}}{\sqrt{x}} \bullet \frac{\sqrt{x}}{\sqrt{x}}$

$\left(\frac{f}{g}\right)(x) = \frac{\sqrt{x^2 - x}}{x};\ x \neq 0$

9. $f(x) = 9x - 2$ and $g(x) = -4x + 6$

$(f \circ g)(x) = 9(-4x + 6) - 2$
$(f \circ g)(x) = -36x + 54 - 2$
$(f \circ g)(x) = -36x + 52$

$(f \circ g)(-2) = -36(-2) + 52$
$(f \circ g)(-2) = 72 + 52 = 124$

$(g \circ f)(x) = -4(9x - 2) + 6$
$(g \circ f)(x) = -36x + 8 + 6$
$(g \circ f)(x) = -36x + 14$

$(g \circ f)(4) = -36(4) + 14$
$(g \circ f)(4) = -130$

11. $f(x) = 4x^2 - 1$ and $g(x) = 4x + 5$

$(f \circ g)(x) = 4(4x + 5)^2 - 1$
$(f \circ g)(x) = 4(16x^2 + 40x + 25) - 1$
$(f \circ g)(x) = 64x^2 + 160x + 100 - 1$
$(f \circ g)(x) = 64x^2 + 160x + 99$

$(f \circ g)(1) = 64(1)^2 + 160(1) + 99$
$(f \circ g)(1) = 323$

$(g \circ f)(x) = 4(4x^2 - 1) + 5$
$(g \circ f)(x) = 16x^2 - 4 + 5$
$(g \circ f)(x) = 16x^2 + 1$

$(g \circ f)(4) = 16(4)^2 + 1$
$(g \circ f)(4) = 257$

13. $f(x) = \frac{1}{x}$ and $g(x) = \frac{2}{x - 1}$

$(f \circ g)(x) = \frac{1}{\frac{2}{x-1}} = \frac{x-1}{2}$

$(f \circ g)(2) = \frac{2-1}{2} = \frac{1}{2}$

$(g \circ f)(x) = \frac{2}{\frac{1}{x} - 1}$

$(g \circ f)(x) = \frac{x}{x} \bullet \frac{2}{\left(\frac{1}{x} - 1\right)}$

$(g \circ f)(x) = \frac{2x}{1-x}$

$(g \circ f)(-1) = \frac{2(-1)}{1-(-1)} = \frac{-2}{2} = -1$

15. $f(x) = \frac{1}{x-2}$ and $g(x) = \frac{4}{x-1}$

$(f \circ g)(x) = \frac{1}{\frac{4}{x-1} - 2}$

$(f \circ g)(x) = \frac{(x-1)}{(x-1)} \bullet \frac{1}{\left(\frac{4}{x-1} - 2\right)}$

$(f \circ g)(x) = \frac{x-1}{4-2(x-1)}$

$(f \circ g)(x) = \frac{x-1}{4-2x+2} = \frac{x-1}{-2x+6}$

$(f \circ g)(3) = \frac{3-1}{-2(3)+6} = \frac{2}{0}$

$(f \circ g)(3)$ is undefined.

The domain of f is $\{x | x \neq 2\}$. Since 2 is not in the domain of f, then 2 cannot be in the domain of $g \circ f$. So $(g \circ f)(2)$ is undefined.

17. $f(x) = \sqrt{3x-2}$ and $g(x) = -x+4$

$(f \circ g)(x) = \sqrt{3(-x+4)-2}$

$(f \circ g)(x) = \sqrt{-3x+12-2}$

$(f \circ g)(x) = \sqrt{-3x+10}$

$(f \circ g)(1) = \sqrt{-3(1)+10} = \sqrt{7}$

$(g \circ f)(x) = -\sqrt{3x-2}+4$

$(g \circ f)(6) = -\sqrt{3(6)-2}+4$

$(g \circ f)(6) = -\sqrt{18-2}+4$

$(g \circ f)(6) = -\sqrt{16}+4 = 0$

19. $f(x) = |4x-5|$ and $g(x) = x^3$

$(f \circ g)(x) = |4(x)^3 - 5|$

$(f \circ g)(x) = |4x^3 - 5|$

$(f \circ g)(-2) = |4(-2)^3 - 5|$

$(f \circ g)(-2) = |-37| = 37$

$(g \circ f)(x) = (|4x-5|)^3$

$(g \circ f)(2) = (|4(2)-5|)^3$

$(g \circ f)(2) = (3)^3 = 27$

21. $f(x) = 3x$ and $g(x) = 5x-1$

$(f \circ g)(x) = 3(5x-1)$

$(f \circ g)(x) = 15x-3$

Domain = {all reals}.

$(g \circ f)(x) = 5(3x)-1$

$(g \circ f)(x) = 15x-1$

Domain = {all reals}.

23. $f(x) = -2x+1$ and $g(x) = 7x+4$

$(f \circ g)(x) = -2(7x+4)+1$

$(f \circ g)(x) = -14x-8+1$

$(f \circ g)(x) = -14x-7$

Domain = {all reals}.

$(g \circ f)(x) = 7(-2x+1)+4$

$(g \circ f)(x) = -14x+7+4$

$(g \circ f)(x) = -14x+11$

Domain = {all reals}.

25. $f(x) = 3x+2$ and $g(x) = x^2+3$

$(f \circ g)(x) = 3(x^2+3)+2$

$(f \circ g)(x) = 3x^2+9+2$

$(f \circ g)(x) = 3x^2+11$

Domain = {all reals}.

$(g \circ f)(x) = (3x+2)^2+3$

$(g \circ f)(x) = 9x^2+12x+4+3$

$(g\circ f)(x) = 9x^2 + 12x + 7$

Domain = {all reals}.

27. $f(x) = 2x^2 - x + 2$ and $g(x) = -x + 3$

$(f\circ g)(x) = 2(-x+3)^2 - (-x+3) + 2$

$(f\circ g)(x) = 2(x^2 - 6x + 9) + x - 3 + 2$

$(f\circ g)(x) = 2x^2 - 12x + 18 + x - 1$

$(f\circ g)(x) = 2x^2 - 11x + 17$

Domain = {all reals}.

$(g\circ f)(x) = -(2x^2 - x + 2) + 3$

$(g\circ f)(x) = -2x^2 + x - 2 + 3$

$(g\circ f)(x) = -2x^2 + x + 1$

Domain = {all reals}.

29. $f(x) = \frac{3}{x}$ and $g(x) = 4x - 9$

$(f\circ g)(x) = \frac{3}{4x-9}$

The domain of f is $\{x|x \neq 0\}$.
Therefore $g(x) \neq 0$.

$4x - 9 = 0$

$4x = 9$

$x = \frac{9}{4}$

The domain of g is {all reals}.

So the domain of $f\circ g$ is $\left\{x|x \neq \frac{9}{4}\right\}$.

$(g\circ f)(x) = 4\left(\frac{3}{x}\right) - 9$

$(g\circ f)(x) = \frac{12}{x} - 9$

$(g\circ f)(x) = \frac{12}{x} - \frac{9x}{x} = \frac{12-9x}{x}$

The domain of g is {all reals}.
The domain of f is $\{x|x \neq 0\}$.
so, the domain of $g\circ f$ is
$\{x|x \neq 0\}$.

31. $f(x) = \sqrt{x+1}$ and $g(x) = 5x + 3$

$(f\circ g)(x) = \sqrt{5x+3+1}$

$(f\circ g)(x) = \sqrt{5x+4}$

Domain of f

$x + 1 \geq 0$

$x \geq -1$

The domain of f is $\{x|x \geq -1\}$.

Therefore $g(x) \geq -1$

$5x + 3 \geq -1$

$5x \geq -4$

$x \geq -\frac{4}{5}$

The domain of g is {all reals}.
So, the domain of $f\circ g$ is
$\left\{x|x \geq -\frac{4}{5}\right\}$.

$(g\circ f)(x) = 5\sqrt{x+1} + 3$

The domain of g is {all reals}.
The domain of f is $\{x|x \geq -1\}$.
So, the domain of $g\circ f$ is
$\{x|x \geq -1\}$.

33. $f(x) = \frac{1}{x}$ and $g(x) = \frac{1}{x-4}$

$(f\circ g)(x) = \frac{1}{\frac{1}{x-4}}$

$(f\circ g) = x - 4$

The domain of f is $\{x|x \neq 0\}$.
So, $g(x) \neq 0$, however $g(x)$
will never be zero.
The domain of g is $\{x|x \neq 4\}$.
So, the domain of $f\circ g$ is
$\{x|x \neq 4\}$.

$(g\circ f)(x) = \frac{1}{\frac{1}{x} - 4}$

$(g\circ f)(x) = \frac{x}{x} \bullet \frac{1}{\left(\frac{1}{x} - 4\right)}$

$(g\circ f)(x) = \frac{x}{1-4x}$

The domain of g is $\{x|x \neq 4\}$.
So, $f(x) \neq 4$.

$\frac{1}{x} = 4$

$x\left(\frac{1}{x}\right) = 4x$

$1 = 4x$

$\frac{1}{4} = x$

The domain of f is $\{x|x \neq 0\}$.

So, the domain of $g \circ f$ is

$\left\{x | x \neq \frac{1}{4},\ x \neq 0\right\}$.

35. $f(x) = \sqrt{x}$ and $g(x) = \frac{4}{x}$

$(f \circ g)(x) = \sqrt{\frac{4}{x}}$

$(f \circ g)(x) = \frac{2}{\sqrt{x}} \bullet \frac{\sqrt{x}}{\sqrt{x}}$

$(f \circ g)(x) = \frac{2\sqrt{x}}{x}$

The domain of f is $\{x | x \geq 0\}$.
So, $g(x) \geq 0$.
$\frac{4}{x} \geq 0$
This is a rational inequality.

	0	
Test Point:	-2	2
4:	positive	positive
x:	negative	positive
quotient:	negative	positive

$x \geq 0$
The domain of g is $\{x | x \neq 0\}$.
So, the domain of $f \circ g$ is
$\{x | x > 0\}$.

$(g \circ f)(x) = \frac{4}{\sqrt{x}}$

$(g \circ f)(x) = \frac{4}{\sqrt{x}} \bullet \frac{\sqrt{x}}{\sqrt{x}}$

$(g \circ f)(x) = \frac{4\sqrt{x}}{x}$

The domain of g is $\{x | x \neq 0\}$.

$f(x) \neq 0$

$\sqrt{x} \neq 0$

$\sqrt{x} = 0$

$(\sqrt{x})^2 = 0^2$

$x = 0$

The domain of f is $\{x | x \geq 0\}$.
So, the domain of $g \circ f$ is
$\{x | x > 0\}$.

37. $f(x) = \frac{3}{2x}$ and $g(x) = \frac{1}{x+1}$

$(f \circ g)(x) = \frac{3}{2\left(\frac{1}{x+1}\right)}$

$(f \circ g)(x) = \frac{3}{\frac{2}{x+1}}$

$(f \circ g)(x) = \frac{3(x+1)}{2}$

$(f \circ g)(x) = \frac{3x+3}{2}$

The domain of f is $\{x | x \neq 0\}$.
So, $g(x) \neq 0$. However $g(x)$
will never be zero.
The domain of g is $\{x | x \neq -1\}$.
So, the domain of $f \circ g$ is
$\{x | x \neq -1\}$.

$(g \circ f)(x) = \frac{1}{\frac{3}{2x}+1}$

$(g \circ f)(x) = \frac{2x}{2x} \bullet \frac{1}{\left(\frac{3}{2x}+1\right)}$

$(g \circ f)(x) = \frac{2x}{3+2x}$

The domain of g is $\{x | x \neq -1\}$.
So, $f(x) \neq -1$.

$\frac{3}{2x} = -1$

$3 = -2x$

$-\frac{3}{2} = x$

The domain of f is $\{x | x \neq 0\}$.
So, the domain of $g \circ f$ is
$\left\{x | x \neq -\frac{3}{2} \text{ and } x \neq 0\right\}$.

39. $f(x) = 3x$ and $g(x) = \frac{1}{3}x$

$(f \circ g)(x) = 3\left(\frac{1}{3}x\right) = x$

$(g \circ f)(x) = \frac{1}{3}(3x) = x$

41. $f(x) = 4x+2$ and $g(x) = \frac{x-2}{4}$

$(f \circ g)(x) = 4\left(\frac{x-2}{4}\right)+2$

$(f \circ g)(x) = x-2+2$

$(f \circ g)(x) = x$

$(g \circ f)(x) = \frac{4x+2-2}{4}$

$(g \circ f)(x) = \frac{4x}{4}$

$(g \circ f)(x) = x$

43. $f(x) = \frac{1}{2}x + \frac{3}{4}$ and $g(x) = \frac{4x-3}{2}$

$(f \circ g)(x) = \frac{1}{2}\left(\frac{4x-3}{2}\right) + \frac{3}{4}$

$(f \circ g)(x) = \frac{4x-3}{4} + \frac{3}{4}$

$(f \circ g)(x) = \frac{4x}{4} - \frac{3}{4} + \frac{3}{4}$

$(f \circ g)(x) = \frac{4x}{4}$

$(f \circ g)(x) = x$

$(g \circ f)(x) = \frac{4\left(\frac{1}{2}x + \frac{3}{4}\right) - 3}{2}$

$(g \circ f)(x) = \frac{2x+3-3}{2}$

$(g \circ f)(x) = \frac{2x}{2}$

$(g \circ f)(x) = x$

45. $f(x) = -\frac{1}{4}x - \frac{1}{2}$ and $g(x) = -4x - 2$

$(f \circ g)(x) = -\frac{1}{4}(-4x-2) - \frac{1}{2}$

$(f \circ g)(x) = x + \frac{1}{2} - \frac{1}{2}$

$(f \circ g)(x) = x$

$(g \circ f)(x) = -4\left(-\frac{1}{4}x - \frac{1}{2}\right) - 2$

$(g \circ f)(x) = x + 2 - 2$

$(g \circ f)(x) = x$

Problem Set 8.5 Inverse Functions

1. Not a function

3. Function

5. Function

7. Function

9. One-to-one function

11. Not one-to-one function

13. Not one-to-one function

15. One-to-one function

17. $f(x) = 7x - 2$

Assume

$f(x_1) = f(x_2)$

$7x_1 - 2 = 7x_2 - 2$

$7x_1 = 7x_2$

$x_1 = x_2$

So, $f(x) = 7x - 2$ is one-to-one.

19. $f(x) = x^4$

Assume

$f(x_1) = f(x_2)$

${x_1}^4 = {x_2}^4$

${x_1}^4 - {x_2}^4 = 0$

$({x_1}^2 - {x_2}^2)({x_1}^2 + {x_2}^2) = 0$

$(x_1 - x_2)(x_1 + x_2)({x_1}^2 + {x_2}^2) = 0$

$x_1 - x_2 = 0$ or $x_1 + x_2 = 0$ or ${x_1}^2 + {x_2}^2 = 0$

$x_1 = x_2$ or $x_1 = -x_2$

So, $f(x) = x^4$ is not one-to-one.

21. $f(x) = |x|$

Assume

$f(x_1) = f(x_2)$

$|x_1| = |x_2|$

$x_1 = x_2$ or $x_1 = -x_2$

So, $f(x) = |x|$ is not one-to-one.

23. $f(x) = -2x$

Assume

$f(x_1) = f(x_2)$

$-2x_1 = -2x_2$

$x_1 = x_2$

So, $f(x) = -2x$ is one-to-one.

25. $f = \{(1, 3), (2, 6), (3, 11), (4, 18)\}$

a. $D = \{1, 2, 3, 4\}$ $R = \{3, 6, 11, 18\}$

b. $f^{-1} = \{(3, 1), (6, 2), (11, 3), (18, 4)\}$

c. Domain of $f^{-1} = \{3, 6, 11, 18\}$
Range of $f^{-1} = \{1, 2, 3, 4\}$

27. $f = \{(-2, -1), (-1, 1), (0, 5), (5, 10)\}$

a. $D = \{-2, -1, 0, 5\}$ $R = \{-1, 1, 5, 10\}$

b. $f^{-1} = \{(-1, -2), (1, -1), (5, 0), (10, 5)\}$

c. Domain of $f^{-1} = \{-1, 1, 5, 10\}$
Range of $f^{-1} = \{-2, -1, 0, 5\}$

29. $f(x) = 5x - 4$

$f^{-1}(x) = \frac{x+4}{5}$

$(f \circ f^{-1})(x) = 5\left(\frac{x+4}{5}\right) - 4$

$(f \circ f^{-1})(x) = x + 4 - 4$

$(f \circ f^{-1})(x) = x$

$(f^{-1} \circ f)(x) = \frac{5x - 4 + 4}{5}$

$(f^{-1} \circ f)(x) = \frac{5x}{5}$

$(f^{-1} \circ f)(x) = x$

31. $f(x) = -2x + 1$

$f^{-1}(x) = -\frac{x-1}{2}$

$(f \circ f^{-1})(x) = -2\left(-\frac{x-1}{2}\right) + 1$

$(f \circ f^{-1})(x) = x - 1 + 1$

$(f \circ f^{-1})(x) = x$

$(f^{-1} \circ f)(x) = -\frac{-2x + 1 - 1}{2}$

$(f^{-1} \circ f)(x) = -\frac{-2x}{2}$

$(f^{-1} \circ f)(x) = x$

33. $f(x) = \frac{4}{5}x$

$f^{-1}(x) = \frac{5}{4}x$

$(f \circ f^{-1})(x) = \frac{4}{5}\left(\frac{5}{4}x\right)$

$(f \circ f^{-1})(x) = x$

$(f^{-1} \circ f)(x) = \frac{5}{4}\left(\frac{4}{5}x\right)$

$(f^{-1} \circ f)(x) = x$

35. $f(x) = \frac{1}{2}x + 4$

$f^{-1}(x) = 2(x - 4)$

$f^{-1}(x) = 2x - 8$

$(f \circ f^{-1})(x) = \frac{1}{2}(2x - 8) + 4$

$(f \circ f^{-1})(x) = x - 4 + 4$

$(f \circ f^{-1})(x) = x$

$(f^{-1} \circ f)(x) = 2\left(\frac{1}{2}x + 4\right) - 8$

$(f^{-1} \circ f)(x) = x + 8 - 8$

$(f^{-1} \circ f)(x) = x$

37. $f(x) = \frac{1}{3}x - \frac{2}{5}$

$f^{-1}(x) = 3\left(x + \frac{2}{5}\right)$

$f^{-1}(x) = 3x + \frac{6}{5}$

$(f \circ f^{-1})(x) = \frac{1}{3}\left(3x + \frac{6}{5}\right) - \frac{2}{5}$

$(f \circ f^{-1})(x) = x + \frac{2}{5} - \frac{2}{5}$

$(f \circ f^{-1})(x) = x$

$(f^{-1} \circ f)(x) = 3\left(\frac{1}{3}x - \frac{2}{5}\right) + \frac{6}{5}$

$(f^{-1} \circ f)(x) = x - \frac{6}{5} + \frac{6}{5}$

$(f^{-1} \circ f)(x) = x$

39. $f(x) = 9x + 4$

$y = 9x + 4$

Interchange x and y.

$x = 9y + 4$

$x-4=9y$

$\frac{x-4}{9}=y$

$y=\frac{x-4}{9}$

$f^{-1}(x)=\frac{x-4}{9}$

$(f\circ f^{-1})(x)=9\left(\frac{x-4}{9}\right)+4$

$(f\circ f^{-1})(x)=x-4+4$

$(f\circ f^{-1})(x)=x$

$(f^{-1}\circ f)(x)=\frac{9x+4-4}{9}$

$(f^{-1}\circ f)(x)=\frac{9x}{9}$

$(f^{-1}\circ f)(x)=x$

41. $f(x)=-5x-4$

$y=-5x-4$

Interchange x and y.

$x=-5y-4$

$x+4=-5y$

$\frac{x+4}{-5}=y$

$y=\frac{-x-4}{5}$

$f^{-1}(x)=\frac{-x-4}{5}$

$(f\circ f^{-1})(x)=-5\left(\frac{-x-4}{5}\right)-4$

$(f\circ f^{-1})(x)=x+4-4$

$(f\circ f^{-1})(x)=x$

$(f^{-1}\circ f)(x)=\frac{-(-5x-4)-4}{5}$

$(f^{-1}\circ f)(x)=\frac{5x+4-4}{5}$

$(f^{-1}\circ f)(x)=\frac{5x}{5}$

$(f^{-1}\circ f)(x)=x$

43. $f(x)=-\frac{2}{3}x+7$

$y=-\frac{2}{3}x+7$

Interchange x and y.

$x=-\frac{2}{3}y+7$

$x-7=-\frac{2}{3}y$

$-\frac{3}{2}(x-7)=-\frac{3}{2}\left(-\frac{2}{3}y\right)$

$\frac{-3(x-7)}{2}=y$

$y=\frac{-3x+21}{2}$

$f^{-1}(x)=\frac{-3x+21}{2}$

$(f\circ f^{-1})(x)=-\frac{2}{3}\left(\frac{-3x+21}{2}\right)+7$

$(f\circ f^{-1})(x)=\frac{-3x+21}{-3}+7$

$(f\circ f^{-1})(x)=x-7+7$

$(f\circ f^{-1})(x)=x$

$(f^{-1}\circ f)(x)=\frac{-3\left(-\frac{2}{3}x+7\right)+21}{2}$

$(f^{-1}\circ f)(x)=\frac{2x-21+21}{2}$

$(f^{-1}\circ f)(x)=\frac{2x}{2}$

$(f^{-1}\circ f)(x)=x$

45. $f(x)=\frac{4}{3}x-\frac{1}{4}$

$y=\frac{4}{3}x-\frac{1}{4}$

Interchange x and y.

$x=\frac{4}{3}y-\frac{1}{4}$

$x+\frac{1}{4}=\frac{4}{3}y$

$\frac{3}{4}\left(x+\frac{1}{4}\right)=\frac{3}{4}\left(\frac{4}{3}y\right)$

$\frac{3}{4}x+\frac{3}{16}=y$

$y=\frac{3}{4}x+\frac{3}{16}$

$f^{-1}(x)=\frac{3}{4}x+\frac{3}{16}$

$(f\circ f^{-1})(x)=\frac{4}{3}\left(\frac{3}{4}x+\frac{3}{16}\right)-\frac{1}{4}$

$(f\circ f^{-1})(x)=x+\frac{1}{4}-\frac{1}{4}$

$(f\circ f^{-1})(x)=x$

$(f^{-1} \circ f)(x) = \frac{3}{4}\left(\frac{4}{3}x - \frac{1}{4}\right) + \frac{3}{16}$

$(f^{-1} \circ f)(x) = x - \frac{3}{16} + \frac{3}{16}$

$(f^{-1} \circ f)(x) = x$

47. $f(x) = -\frac{3}{7}x - \frac{2}{3}$

$y = -\frac{3}{7}x - \frac{2}{3}$

Interchange x and y.

$x = -\frac{3}{7}y - \frac{2}{3}$

$x + \frac{2}{3} = -\frac{3}{7}y$

$-\frac{7}{3}\left(x + \frac{2}{3}\right) = -\frac{7}{3}\left(-\frac{3}{7}y\right)$

$-\frac{7}{3}x - \frac{14}{9} = y$

$y = -\frac{7}{3}x - \frac{14}{9}$

$f^{-1}(x) = -\frac{7}{3}x - \frac{14}{9}$

$(f \circ f^{-1})(x) = -\frac{3}{7}\left(-\frac{7}{3}x - \frac{14}{9}\right) - \frac{2}{3}$

$(f \circ f^{-1})(x) = x + \frac{2}{3} - \frac{2}{3}$

$(f \circ f^{-1})(x) = x$

$(f^{-1} \circ f)(x) = -\frac{7}{3}\left(-\frac{3}{7}x - \frac{2}{3}\right) - \frac{14}{9}$

$(f^{-1} \circ f)(x) = x + \frac{14}{9} - \frac{14}{9}$

$(f^{-1} \circ f)(x) = x$

49. $f(x) = \sqrt{x}$, for $x \geq 0$

$y = \sqrt{x}$

Interchange x and y.

$x = \sqrt{y}$

$(x)^2 = (\sqrt{y})^2$

$x^2 = y$

$f^{-1}(x) = x^2$, for $x \geq 0$

$(f \circ f^{-1})(x) = \sqrt{x^2} = x$

$(f^{-1} \circ f)(x) = (\sqrt{x})^2 = x$

51. $f(x) = x^2 + 4$, for $x \geq 0$

$y = x^2 + 4$

Interchange x and y.

$x = y^2 + 4$

$x - 4 = y^2$

$\sqrt{x-4} = y$

$f^{-1}(x) = \sqrt{x-4}$, for $x \geq 4$

$(f \circ f^{-1})(x) = (\sqrt{x-4})^2 + 4$

$(f \circ f^{-1})(x) = x - 4 + 4 = x$

$(f^{-1} \circ f)(x) = \sqrt{(x^2+4) - 4}$

$(f^{-1} \circ f)(x) = \sqrt{x^2} = x$

53a. $f(x) = 4x$

$y = 4x$

Interchange x and y.

$x = 4y$

$\frac{1}{4}x = y$

$f^{-1}(x) = \frac{1}{4}x$

55a. $f(x) = -\frac{1}{3}x$

$y = -\frac{1}{3}x$

Interchange x and y.

$x = -\frac{1}{3}y$

$-3x = y$

$y = -3x$

$f^{-1}(x) = -3x$

57a. $f(x) = 3x - 3$

$y = 3x - 3$

Interchange x and y.

$x = 3y - 3$

$x + 3 = 3y$

$\frac{x+3}{3} = y$

$y = \frac{x+3}{3}$

$f^{-1}(x) = \frac{x+3}{3}$

59a. $f(x) = -2x-4$

$y = -2x-4$

Interchange x and y.

$x = -2y-4$

$x+4 = -2y$

$\frac{x+4}{-2} = y$

$y = \frac{-x-4}{2}$

$f^{-1}(x) = \frac{-x-4}{2}$

61a. $f(x) = x^2;\ x \geq 0$

$y = x^2$

Interchange x and y.

$x = y^2$

$\pm\sqrt{x} = y$

Since the domain of f is $x \geq 0$ the range of f^{-1} is restricted to the positive root.

$y = \sqrt{x}$

$f^{-1}(x) = \sqrt{x};\ x \geq 0$

63. $f(x) = \sqrt{x-1}$ for $x \geq 1$

$y = \sqrt{x-1}$

Interchange x and y.

$x = \sqrt{y-1}$

$x^2 = y-1$

$x^2+1 = y$

$f^{-1}(x) = x^2+1$, for $x \geq 0$

Problem Set 8.6 Direct and Inverse Variations

1. $y = \frac{k}{x^2}$

3. $C = \frac{kg}{t^3}$

5. $V = kr^3$

7. $S = ke^2$

9. $V = khr^2$

11. $y = kx$

$8 = k(12)$

$\frac{8}{12} = k$

$\frac{2}{3} = k$

13. $y = kx^2$

$-144 = k(6)^2$

$-144 = 36k$

$-4 = k$

15. $V = kBh$

$96 = k(24)(12)$

$96 = 288k$

$\frac{96}{288} = k$

$\frac{1}{3} = k$

17. $y = \frac{k}{x}$

$-4 = \frac{k}{\frac{1}{2}}$

$-4 = 2k$

$-2 = k$

19. $r = \frac{k}{t^2}$

$\frac{1}{8} = \frac{k}{4^2}$

$\frac{1}{8} = \frac{k}{16}$

$16\left(\frac{1}{8}\right) = 16\left(\frac{k}{16}\right)$

$2 = k$

21. $y = \frac{kx}{z}$

$45 = \frac{k(18)}{2}$

$45 = 9k$

$5 = k$

23. $y = \frac{kx}{z^2}$

$81 = \frac{k(36)}{(2)^2}$

$81 = \frac{k(36)}{4}$

$81 = 9k$

$9 = k$

25. $y = kx$

$36 = k(48)$

$\frac{36}{48} = k$

$\frac{3}{4} = k$

$y = \frac{3}{4}x$

$y = \frac{3}{4}(12)$

$y = 9$

27. $y = \frac{k}{x}$

$\frac{1}{9} = \frac{k}{12}$

$12\left(\frac{1}{9}\right) = 12\left(\frac{k}{12}\right)$

$\frac{4}{3} = k$

$y = \frac{\frac{4}{3}}{x}$

$y = \frac{4}{3x}$

$y = \frac{4}{3(8)}$

$y = \frac{1}{6}$

29. $A = kbh$

$60 = k(12)(10)$

$60 = 120k$

$\frac{1}{2} = k$

$A = \frac{1}{2}bh$

$A = \frac{1}{2}(16)(14)$

$A = 112$

31. $V = \frac{k}{P}$

$15 = \frac{k}{20}$

$20(15) = 20\left(\frac{k}{20}\right)$

$300 = k$

$V = \frac{300}{P}$

$V = \frac{300}{25}$

$V = 12$ cubic centimeters

33. $V = \frac{kT}{P}$

$48 = \frac{k(320)}{20}$

$48 = 16k$

$3 = k$

$V = \frac{3T}{P}$

$V = \frac{3(280)}{30}$

$V = 28$

35. $P = k\sqrt{l}$

$4 = k\sqrt{12}$

$4 = k(2\sqrt{3})$

$\frac{4}{2\sqrt{3}} = k$

$k = \frac{2}{\sqrt{3}}$

$k = \frac{2}{\sqrt{3}} \bullet \frac{\sqrt{3}}{\sqrt{3}} = \frac{2\sqrt{3}}{3}$

$P = \frac{2\sqrt{3}}{3}\sqrt{l} = \frac{2\sqrt{3l}}{3}$

$P = \frac{2\sqrt{3(3)}}{3} = \frac{2\sqrt{9}}{3} = \frac{2(3)}{3} = 2$ sec.

37. $R = \frac{kl}{d^2}$

$1.5 = \frac{k(200)}{(.5)^2}$

$1.5 = \frac{k(200)}{.25}$

$1.5 = 800k$

$\frac{1.5}{800} = k$

$.001875 = k$

$R = \frac{.0018751}{d^2}$

$R = \frac{.001875(400)}{(.25)^2}$

$R = \frac{.001875(400)}{.0625}$

$R = 12$ ohms

39. $I = krt$

a. $385 = k(.11)(2)$

$385 = .22k$

$1750 = k$

$I = 1750rt$

$I = 1750(.12)(1)$

$I = \$210$

b. $819 = k(.12)(3)$

$819 = .36k$

$2275 = k$

$I = 2275rt$

$I = 2275(.14)(2)$

$I = \$637$

c. $1960 = k(.14)(4)$

$1960 = .56k$

$3500 = k$

$I = 3500rt$

$I = 3500(.15)(2)$

$I = \$1050$

41. $V = khr^2$

$549.5 = k(7)(5)^2$

$549.5 = 175k$

$3.14 = k$

$V = 3.14hr^2$

$V = 3.14(14)(9)^2$

$V = 3560.76$ cubic meters

43. $y = \frac{k}{\sqrt{x}}$

$0.08 = \frac{k}{\sqrt{225}}$

$0.08 = \frac{k}{15}$

$15(0.08) = k$

$1.2 = k$

$y = \frac{1.2}{\sqrt{x}}$

$y = \frac{1.2}{\sqrt{625}}$

$y = \frac{1.2}{25} = .048$

Chapter 8 Review

1. $f = \{(1, 3), (2, 5), (4, 9)\}$

Domain $= \{1, 2, 4\}$

3. $f(x) = \frac{3}{x^2 + 4x}$

$f(x) = \frac{3}{x(x+4)}$

$x(x+4) = 0$

$x = 0$ or $x + 4 = 0$

$x = 0$ or $x = -4$

Domain $= \{x \mid x \neq 0 \text{ and } x \neq -4\}$

5. $f(x) = x^2 - 2x - 1$

$f(2) = 2^2 - 2(2) - 1$

$f(2) = 4 - 4 - 1$

$f(2) = -1$

$f(-3) = (-3)^2 - 2(-3) - 1$

$f(-3) = 9 + 6 - 1$

$f(-3) = 14$

$f(a) = a^2 - 2a - 1$

7 - 16. See back of textbook for graphs.

17a. $f(x) = x^2 + 10x - 3$

$f(x) = x^2 + 10x + 25 - 25 - 3$

$f(x) = (x+5)^2 - 28$

vertex $(-5, -28)$

axis of symmetry $x = -5$

b. $f(x) = -2x^2 - 14x + 9$

$f(x) = -2(x^2 + 7x) + 9$

$f(x) = -2\left(x^2 + 7x + \frac{49}{4}\right) + \frac{49}{2} + 9$

$f(x) = -2\left(x + \frac{7}{2}\right)^2 + \frac{67}{2}$

vertex $\left(-\frac{7}{2}, \frac{67}{2}\right)$

axis of symmetry $x = -\frac{7}{2}$

19. $f(x) = x - 4$ and $g(x) = x^2 - 2x + 3$

$(f \circ g)(x) = x^2 - 2x + 3 - 4$

$(f \circ g)(x) = x^2 - 2x - 1$

$(g \circ f)(x) = (x-4)^2 - 2(x-4) + 3$

$(g \circ f)(x) = x^2 - 8x + 16 - 2x + 8 + 3$

$(g \circ f)(x) = x^2 - 10x + 27$

21. $f(x) = 6x - 1$

$y = 6x - 1$

Interchange x and y.

$x = 6y - 1$

$x + 1 = 6y$

$\frac{x+1}{6} = y$

$f^{-1}(x) = \frac{x+1}{6}$

23. $f(x) = -\frac{3}{5}x - \frac{2}{7}$

$y = -\frac{3}{5}x - \frac{2}{7}$

Interchange x and y.

$x = -\frac{3}{5}y - \frac{2}{7}$

$x + \frac{2}{7} = -\frac{3}{5}y$

$-\frac{5}{3}\left(x + \frac{2}{7}\right) = -\frac{5}{3}\left(-\frac{3}{5}y\right)$

$-\frac{5x}{3} - \frac{10}{21} = y$

$f^{-1}(x) = -\frac{5}{3}x - \frac{10}{21}$

$f^{-1}(x) = \frac{-35x - 10}{21}$

25. $y = kx\sqrt{z}$

$60 = k(2)\sqrt{9}$

$60 = 6k$

$10 = k$

$y = 10x\sqrt{z}$

$y = 10(3)\sqrt{16}$

$y = 10(3)(4)$

$y = 120$

27. Let $x =$ one number
$40 - x =$ other number

$f(x) = x(40 - x)$

$f(x) = 40x - x^2$

$f(x) = -x^2 + 40x$

$f(x) = -(x^2 - 40x)$

$f(x) = -(x^2 - 40x + 400) + 400$

$f(x) = -(x - 20)^2 + 400$

The numbers are 20 and 20.

29. Let $x =$ number of additional students

$R(x) = (50 + x)(5 - .05x)$

$R(x) = 250 - 2.5x + 5x - .05x^2$

$R(x) = -.05x^2 + 2.5x + 250$

$R(x) = -.05(x^2 - 50x) + 250$

$R(x) = -.05(x^2 - 50x + 625) + 31.25 + 250$

$R(x) = -.05(x - 25)^2 + 281.25$

There needs to be 25 additional students.

Chapter 8 Test

1. $f(x) = \frac{-3}{2x^2+7x-4}$

   $f(x) = \frac{-3}{(2x-1)(x+4)}$

   $2x-1=0$ or $x+4=0$

   $2x=1$ or $x=-4$

   $x=\frac{1}{2}$ or $x=-4$

   $\text{Domain} = \left\{x \mid x \neq \frac{1}{2} \text{ and } x \neq -4\right\}$

3. $f(x) = -\frac{1}{2}x+\frac{1}{3}$

   $f(-3) = -\frac{1}{2}(-3)+\frac{1}{3}$

   $f(-3) = \frac{3}{2}+\frac{1}{3}$

   $f(-3) = \frac{11}{6}$

5. $f(x) = -2x^2-24x-69$

   $f(x) = -2(x^2+12x)-69$

   $f(x) = -2(x^2+12x+36)+72-69$

   $f(x) = -2(x+6)^2+3$

   vertex $(-6, 3)$

7. $f(x) = -3x+4$ and $g(x) = 7x+2$

   $(f \circ g)(x) = -3(7x+2)+4$

   $(f \circ g)(x) = -21x-6+4$

   $(f \circ g)(x) = -21x-2$

9. $f(x) = \frac{3}{x-2}$ and $g(x) = \frac{2}{x}$

   $(f \circ g)(x) = \frac{3}{\frac{2}{x}-2}$

   $(f \circ g)(x) = \frac{x}{x} \bullet \frac{3}{\left(\frac{2}{x}-2\right)}$

   $(f \circ g)(x) = \frac{3x}{2-2x}$

11. $f(x) = -3x-6$

    $y = -3x-6$

    Interchange x and y.

    $x = -3y-6$

    $x+6 = -3y$

    $y = \frac{x+6}{-3}$

    $y = \frac{-x-6}{3}$

    $f^{-1}(x) = \frac{-x-6}{3}$

13. $y = \frac{k}{x}$

    $\frac{1}{2} = \frac{k}{-8}$

    $-8\left(\frac{1}{2}\right) = k$

    $-4 = k$

15. Let x = one number
    $60-x$ = other number

    $f(x) = x^2+12(60-x)$

    $f(x) = x^2+720-12x$

    $f(x) = x^2-12x+720$

    $f(x) = x^2-12x+36-36+720$

    $f(x) = (x-6)^2+684$

    The numbers are 6 and 54.

17. The graph of $f(x) = (x-6)^3-4$ is the graph of $f(x) = x^3$ translated 6 units to the right and 4 units downward.

19. The graph of $f(x) = -\sqrt{x+5}+7$ is the graph of $f(x) = \sqrt{x}$ reflected across the x-axis and then translated 5 units to the left and 7 units upward.

21 - 25. See back of textbook for graphs.

Chapter 9 Polynomial and Rational Functions

Problem Set 9.1 Synthetic Division

1. $(4x^2-5x-6)\div(x-2)$

$$\begin{array}{r|rrr} 2 & 4 & -5 & -6 \\ & & 8 & 6 \\ \hline & 4 & 3 & 0 \end{array}$$

Q: $4x+3$
R: 0

3. $(2x^2-x-21)\div(x+3)$

$$\begin{array}{r|rrr} -3 & 2 & -1 & -21 \\ & & -6 & 21 \\ \hline & 2 & -7 & 0 \end{array}$$

Q: $2x-7$
R: 0

5. $(3x^2-16x+17)\div(x-4)$

$$\begin{array}{r|rrr} 4 & 3 & -16 & 17 \\ & & 12 & -16 \\ \hline & 3 & -4 & 1 \end{array}$$

Q: $3x-4$
R: 1

7. $(4x^2+19x-32)\div(x+6)$

$$\begin{array}{r|rrr} -6 & 4 & 19 & -32 \\ & & -24 & 30 \\ \hline & 4 & -5 & -2 \end{array}$$

Q: $4x-5$
R: -2

9. $(x^3+2x^2-7x+4)\div(x-1)$

$$\begin{array}{r|rrrr} 1 & 1 & 2 & -7 & 4 \\ & & 1 & 3 & -4 \\ \hline & 1 & 3 & -4 & 0 \end{array}$$

Q: x^2+3x-4
R: 0

11. $(3x^3+8x^2-8)\div(x+2)$

$$\begin{array}{r|rrrr} -2 & 3 & 8 & 0 & -8 \\ & & -6 & -4 & 8 \\ \hline & 3 & 2 & -4 & 0 \end{array}$$

Q: $3x^2+2x-4$
R: 0

13. $(5x^3-9x^2-3x-2)\div(x-2)$

$$\begin{array}{r|rrrr} 2 & 5 & -9 & -3 & -2 \\ & & 10 & 2 & -2 \\ \hline & 5 & 1 & -1 & -4 \end{array}$$

Q: $5x^2+x-1$
R: -4

15. $(x^3+6x^2-8x+1)\div(x+7)$

$$\begin{array}{r|rrrr} -7 & 1 & 6 & -8 & 1 \\ & & -7 & 7 & 7 \\ \hline & 1 & -1 & -1 & 8 \end{array}$$

Q: x^2-x-1
R: 8

17. $(-x^3+7x^2-14x+6)\div(x-3)$

$$\begin{array}{r|rrrr} 3 & -1 & 7 & -14 & 6 \\ & & -3 & 12 & -6 \\ \hline & -1 & 4 & -2 & 0 \end{array}$$

Q: $-x^2+4x-2$
R: 0

19. $(-3x^3+x^2+2x+2)\div(x+1)$

$$\begin{array}{r|rrrr} -1 & -3 & 1 & 2 & 2 \\ & & 3 & -4 & 2 \\ \hline & -3 & 4 & -2 & 4 \end{array}$$

Q: $-3x^2+4x-2$
R: 4

21. $(3x^3-2x-5)\div(x-2)$

$$\begin{array}{r|rrrr} 2 & 3 & 0 & -2 & -5 \\ & & 6 & 12 & 20 \\ \hline & 3 & 6 & 10 & 15 \end{array}$$

Q: $3x^2+6x+10$
R: 15

23. $(2x^4+x^3+3x^2+2x-2)\div(x+1)$

$$\begin{array}{r|rrrrr} -1 & 2 & 1 & 3 & 2 & -2 \\ & & -2 & 1 & -4 & 2 \\ \hline & 2 & -1 & 4 & -2 & 0 \end{array}$$

Q: $2x^3-x^2+4x-2$
R: 0

25. $(x^4+4x^3-7x-1)\div(x-3)$

$$\begin{array}{r|rrrrr} 3 & 1 & 4 & 0 & -7 & -1 \\ & & 3 & 21 & 63 & 168 \\ \hline & 1 & 7 & 21 & 56 & 167 \end{array}$$

Q: $x^3+7x^2+21x+56$
R: 167

27. $(x^4+5x^3-x^2+25)\div(x+5)$

$$\begin{array}{r|rrrrr} -5 & 1 & 5 & -1 & 0 & 25 \\ & & -5 & 0 & 5 & -25 \\ \hline & 1 & 0 & -1 & 5 & 0 \end{array}$$

Q: x^3-x+5
R: 0

29. $(x^4-16)\div(x-2)$

$$\begin{array}{r|rrrrr} 2 & 1 & 0 & 0 & 0 & -16 \\ & & 2 & 4 & 8 & 16 \\ \hline & 1 & 2 & 4 & 8 & 0 \end{array}$$

Q: x^3+2x^2+4x+8
R: 0

31. $(x^5-1)\div(x+1)$

$$\begin{array}{r|rrrrrr} -1 & 1 & 0 & 0 & 0 & 0 & -1 \\ & & -1 & 1 & -1 & 1 & -1 \\ \hline & 1 & -1 & 1 & -1 & 1 & -2 \end{array}$$

Q: $x^4-x^3+x^2-x+1$
R: -2

33. $(x^5+1)\div(x+1)$

$$\begin{array}{r|rrrrrr} -1 & 1 & 0 & 0 & 0 & 0 & 1 \\ & & -1 & 1 & -1 & 1 & -1 \\ \hline & 1 & -1 & 1 & -1 & 1 & 0 \end{array}$$

Q: $x^4-x^3+x^2-x+1$
R: 0

35. $(x^5+3x^4-5x^3-3x^2+3x-4)\div(x+4)$

$$\begin{array}{r|rrrrrr} -4 & 1 & 3 & -5 & -3 & 3 & -4 \\ & & -4 & 4 & 4 & -4 & 4 \\ \hline & 1 & -1 & -1 & 1 & -1 & 0 \end{array}$$

Q: $x^4-x^3-x^2+x-1$
R: 0

37. $(4x^5-6x^4+2x^3+2x^2-5x+2)\div(x-1)$

$$\begin{array}{r|rrrrrr} 1 & 4 & -6 & 2 & 2 & -5 & 2 \\ & & 4 & -2 & 0 & 2 & -3 \\ \hline & 4 & -2 & 0 & 2 & -3 & -1 \end{array}$$

Q: $4x^4-2x^3+2x-3$
R: -1

39. $(9x^3-6x^2+3x-4)\div\left(x-\frac{1}{3}\right)$

$$\begin{array}{r|rrrr} \frac{1}{3} & 9 & -6 & 3 & -4 \\ & & 3 & -1 & \frac{2}{3} \\ \hline & 9 & -3 & 2 & -\frac{10}{3} \end{array}$$

Q: $9x^2-3x+2$
R: $-\frac{10}{3}$

41. $(3x^4-2x^3+5x^2-x-1)\div\left(x+\frac{1}{3}\right)$

$$\begin{array}{r|rrrrr} -\frac{1}{3} & 3 & -2 & 5 & -1 & -1 \\ & & -1 & 1 & -2 & 1 \\ \hline & 3 & -3 & 6 & -3 & 0 \end{array}$$

Q: $3x^3-3x^2+6x-3$
R: 0

Problem Set 9.2 Remainder and Factor Theorems

1. $f(x)=x^2+2x-6$

a. $f(3)=3^2+2(3)-6$
$f(3)=9+6-6$
$f(3)=9$

b.

$$\begin{array}{r|rrr} 3 & 1 & 2 & -6 \\ & & 3 & 15 \\ \hline & 1 & 5 & 9 \end{array}$$

$f(3)=9$

3. $f(x) = x^3 - 2x^2 + 3x - 1$

a. $f(-1) = (-1)^3 - 2(-1)^2 + 3(-1) - 1$

$f(-1) = -1 - 2(1) - 3 - 1$

$f(-1) = -7$

b.
$$\begin{array}{r|rrrr} -1 & 1 & -2 & 3 & -1 \\ & & -1 & 3 & -6 \\ \hline & 1 & -3 & 6 & -7 \end{array}$$

$f(-1) = -7$

5. $f(x) = 2x^4 - x^3 - 3x^2 + 4x - 1$

a. $f(2) = 2(2)^4 - (2)^3 - 3(2)^2 + 4(2) - 1$

$f(2) = 32 - 8 - 12 + 8 - 1$

$f(2) = 19$

b.
$$\begin{array}{r|rrrrr} 2 & 2 & -1 & -3 & 4 & -1 \\ & & 4 & 6 & 6 & 20 \\ \hline & 2 & 3 & 3 & 10 & 19 \end{array}$$

$f(2) = 19$

7. $f(n) = 6n^3 - 35n^2 + 8n - 10$

a. $f(6) = 6(6)^3 - 35(6)^2 + 8(6) - 10$

$f(6) = 1296 - 1260 + 48 - 10$

$f(6) = 74$

b.
$$\begin{array}{r|rrrr} 6 & 6 & -35 & 8 & -10 \\ & & 36 & 6 & 84 \\ \hline & 6 & 1 & 14 & 74 \end{array}$$

$f(6) = 74$

9. $f(n) = 2n^5 - 1$

a. $f(-2) = 2(-2)^5 - 1$

$f(-2) = -64 - 1$

$f(-2) = -65$

b.
$$\begin{array}{r|rrrrrr} -2 & 2 & 0 & 0 & 0 & 0 & -1 \\ & & -4 & 8 & -16 & 32 & -64 \\ \hline & 2 & -4 & 8 & -16 & 32 & -65 \end{array}$$

$f(-2) = -65$

11. $f(x) = 6x^5 - 3x^3 + 2$

$$\begin{array}{r|rrrrrr} -1 & 6 & 0 & -3 & 0 & 0 & 2 \\ & & -6 & 6 & -3 & 3 & -3 \\ \hline & 6 & -6 & 3 & -3 & 3 & -1 \end{array}$$

$f(-1) = -1$

13. $f(x) = 2x^4 - 15x^3 - 9x^2 - 2x - 3$

$$\begin{array}{r|rrrrr} 8 & 2 & -15 & -9 & -2 & -3 \\ & & 16 & 8 & -8 & -80 \\ \hline & 2 & 1 & -1 & -10 & -83 \end{array}$$

$f(8) = -83$

15. $f(n) = 4n^7 + 3$

$f(3) = 4(3)^7 + 3$

$f(3) = 4(2187) + 3$

$f(3) = 8751$

17. $f(n) = 3n^5 + 17n^4 - 4n^3 + 10n^2 - 15n + 13$

$$\begin{array}{r|rrrrrr} -6 & 3 & 17 & -4 & 10 & -15 & 13 \\ & & -18 & 6 & -12 & 12 & 18 \\ \hline & 3 & -1 & 2 & -2 & -3 & 31 \end{array}$$

$f(-6) = 31$

19. $f(x) = -4x^4 - 6x^2 + 7$

$$\begin{array}{r|rrrrr} 4 & -4 & 0 & -6 & 0 & 7 \\ & & -16 & -64 & -280 & -1120 \\ \hline & -4 & -16 & -70 & -280 & -1113 \end{array}$$

$f(4) = -1113$

21.
$$\begin{array}{r|rrr} 2 & 5 & -17 & 14 \\ & & 10 & -14 \\ \hline & 5 & -7 & 0 \end{array}$$

$f(2) = 0$

Yes, it is a factor.

23.
$$\begin{array}{r|rrr} -3 & 6 & 13 & -14 \\ & & -18 & 15 \\ \hline & 6 & -5 & 1 \end{array}$$

$f(-3) = 1$

No, it is not a factor.

25.

$$\begin{array}{r|rrrr} 1 & 4 & -13 & 21 & -12 \\ & & 4 & -9 & 12 \\ \hline & 4 & -9 & 12 & 0 \end{array}$$

$f(1) = 0$

Yes, it is a factor.

27.

$$\begin{array}{r|rrrr} -2 & 1 & 7 & 1 & -18 \\ & & -2 & -10 & 18 \\ \hline & 1 & 5 & -9 & 0 \end{array}$$

$f(-2) = 0$

Yes, it is a factor.

29.

$$\begin{array}{r|rrrr} 3 & 3 & -5 & -17 & 17 \\ & & 9 & 12 & -15 \\ \hline & 3 & 4 & -5 & 2 \end{array}$$

$f(3) = 2$

No, it is not a factor.

31.

$$\begin{array}{r|rrrr} -2 & 1 & 0 & 0 & 8 \\ & & -2 & 4 & -8 \\ \hline & 1 & -2 & 4 & 0 \end{array}$$

$f(-2) = 0$

Yes, it is a factor.

33.

$$\begin{array}{r|rrrrr} 3 & 1 & 0 & 0 & 0 & -81 \\ & & 3 & 9 & 27 & 81 \\ \hline & 1 & 3 & 9 & 27 & 0 \end{array}$$

$f(3) = 0$

Yes, it is a factor.

35.

$$\begin{array}{r|rrrr} 2 & 1 & -6 & -13 & 42 \\ & & 2 & -8 & -42 \\ \hline & 1 & -4 & -21 & 0 \end{array}$$

$(x-2)(x^2-4x-21)$

$(x-2)(x-7)(x+3)$

37.

$$\begin{array}{r|rrrr} -2 & 12 & 29 & 8 & -4 \\ & & -24 & -10 & 4 \\ \hline & 12 & 5 & -2 & 0 \end{array}$$

$(x+2)(12x^2+5x-2)$

$(x+2)(4x-1)(3x+2)$

39.

$$\begin{array}{r|rrrr} -1 & 1 & -2 & -7 & -4 \\ & & -1 & 3 & 4 \\ \hline & 1 & -3 & -4 & 0 \end{array}$$

$(x+1)(x^2-3x-4)$

$(x+1)(x-4)(x+1)$

$(x+1)^2(x-4)$

41.

$$\begin{array}{r|rrrrrr} 6 & 1 & -6 & 0 & 0 & -16 & 96 \\ & & 6 & 0 & 0 & 0 & -96 \\ \hline & 1 & 0 & 0 & 0 & -16 & 0 \end{array}$$

$(x-6)(x^4-16)$

$(x-6)(x^2-4)(x^2+4)$

$(x-6)(x+2)(x-2)(x^2+4)$

43.

$$\begin{array}{r|rrrr} -5 & 9 & 21 & -104 & 80 \\ & & -45 & 120 & -80 \\ \hline & 9 & -24 & 16 & 0 \end{array}$$

$(x+5)(9x^2-24x+16)$

$(x+5)(3x-4)(3x-4)$

$(x+5)(3x-4)^2$

45.

$$\begin{array}{r|rrrr} 1 & k^2 & 3k & 0 & -4 \\ & & k^2 & k^2+3k & k^2+3k \\ \hline & k^2 & k^2+3k & k^2+3k & k^2+3k-4 \end{array}$$

$k^2+3k-4=0$

$(k+4)(k-1)=0$

$k+4=0 \quad \text{or} \quad k-1=0$

$k=-4 \quad \text{or} \quad k=1$

The values are -4 or 1.

47.

$$\begin{array}{r|rrrr} -3 & k & 19 & 1 & -6 \\ & & -3k & 9k-57 & -27k+168 \\ \hline & k & -3k+19 & 9k-56 & -27k+162 \end{array}$$

$-27k+162=0$

$-27k=-162$

$k=6$

The value is 6.

49. $f(c) = 3x^4 + 2x^2 + 5$

$f(c) > 0$ for all values of c.

51. If $f(-1) = 0$, then $(x+1)$ is a factor.

$f(x) = x^n - 1$

$f(-1) = (-1)^n - 1$

$0 = (-1)^n - 1$

$1 = (-1)^n$

This is true for all even positive integral values of n. Therefore, $(x+1)$ is a factor of $x^n - 1$ for all even positive integral values of n.

53a. If $f(y) = 0$, then $(x-y)$ is a factor.

$f(x) = x^n - y^n$

$f(y) = y^n - y^n$

$f(y) = 0$

This is true for all positive integral values of n. Therefore $(x-y)$ is a factor of $x^n - y^n$ for all positive integral values of n.

b. If $f(-y) = 0$, then $(x+y)$ is a factor.

$f(x) = x^n - y^n$

$f(-y) = (-y)^n - y^n$

$0 = (-y)^n - y^n$

$y^n = (-y)^n$

This is true for all even positive integral values on n. Therefore, $(x+y)$ is a factor of $x^n - y^n$ for all even positive integral values of n.

c. If $f(-y) = 0$, then $(x+y)$ is a factor.

$f(x) = x^n + y^n$

$f(-y) = (-y)^n + y^n$

$(-y)^n = -y^n$

This is true for all odd positive integral values of n. Therefore $(x+y)$ is a factor of $x^n + y^n$ for all odd positive integral values of n.

Problem Set 9.3 Polynomial Equations

1. $x^3 - 2x^2 - 11x + 12 = 0$

c: $\pm 1, \pm 2, \pm 3, \pm 4, \pm 6, \pm 12$

d: ± 1

$\frac{c}{d}$: $\pm 1, \pm 2, \pm 3, \pm 4, \pm 6, \pm 12$

$$\begin{array}{r|rrrr} 1 & 1 & -2 & -11 & 12 \\ & & 1 & -1 & -12 \\ \hline & 1 & -1 & -12 & 0 \end{array}$$

$(x-1)(x^2 - x - 12) = 0$

$(x-1)(x-4)(x+3) = 0$

$x - 1 = 0$ or $x - 4 = 0$ or $x + 3 = 0$

$x = 1$ or $x = 4$ or $x = -3$

The solution set is $\{-3, 1, 4\}$.

3. $15x^3 + 14x^2 - 3x - 2 = 0$

c: $\pm 1, \pm 2$

d: $\pm 1, \pm 3, \pm 5, \pm 15$

$\frac{c}{d}$: $\pm 1, \pm 2, \pm \frac{1}{3}, \pm \frac{2}{3}, \pm \frac{1}{5}, \pm \frac{2}{5}, \pm \frac{1}{15}, \pm \frac{2}{15}$

$$\begin{array}{r|rrrr} -1 & 15 & 14 & -3 & -2 \\ & & -15 & 1 & 2 \\ \hline & 15 & -1 & -2 & 0 \end{array}$$

$(x+1)(15x^2 - x - 2) = 0$

$(x+1)(3x+1)(5x-2) = 0$

$x + 1 = 0$ or $3x + 1 = 0$ or $5x - 2 = 0$

$x = -1$ or $3x = -1$ or $5x = 2$

$x = -1$ or $x = -\frac{1}{3}$ or $x = \frac{2}{5}$

The solution set is $\left\{-1, -\frac{1}{3}, \frac{2}{5}\right\}$.

5. $8x^3 - 2x^2 - 41x - 10 = 0$

c: $\pm 1, \pm 2, \pm 5, \pm 10$

d: $\pm 1, \pm 2, \pm 4, \pm 8$

$\frac{c}{d}$: $\pm 1, \pm 2, \pm 5, \pm 10,$
$\pm\frac{1}{2}, \pm\frac{5}{2}, \pm\frac{1}{4}, \pm\frac{5}{4}$
$\pm\frac{1}{8}, \pm\frac{5}{8}$

$$\begin{array}{r|rrrr} -2 & 8 & -2 & -41 & -10 \\ & & -16 & 36 & 10 \\ \hline & 8 & -18 & -5 & 0 \end{array}$$

$(x+2)(8x^2-18x-5)=0$

$(x+2)(2x-5)(4x+1)=0$

$x+2=0$ or $2x-5=0$ or $4x+1=0$

$x=-2$ or $2x=5$ or $4x=-1$

$x=-2$ or $x=\frac{5}{2}$ or $x=-\frac{1}{4}$

The solution set is $\left\{-2, -\frac{1}{4}, \frac{5}{2}\right\}$.

7. $x^3-x^2-8x+12=0$

c: $\pm 1, \pm 2, \pm 3, \pm 4, \pm 6, \pm 12$

d: ± 1

$\frac{c}{d}$: $\pm 1, \pm 2, \pm 3, \pm 4, \pm 6, \pm 12$

$$\begin{array}{r|rrrr} 2 & 1 & -1 & -8 & 12 \\ & & 2 & 2 & -12 \\ \hline & 1 & 1 & -6 & 0 \end{array}$$

$(x-2)(x^2+x-6)=0$

$(x-2)(x+3)(x-2)=0$

$x-2=0$ or $x+3=0$ or $x-2=0$

$x=2$ or $x=-3$ or $x=2$

The solution set is $\{-3, 2\}$.

9. $x^3-4x^2+8=0$

c: $\pm 1, \pm 2, \pm 4, \pm 8$

d: ± 1

$\frac{c}{d}$: $\pm 1, \pm 2, \pm 4, \pm 8$

$$\begin{array}{r|rrrr} 2 & 1 & -4 & 0 & 8 \\ & & 2 & -4 & -8 \\ \hline & 1 & -2 & -4 & 0 \end{array}$$

$(x-2)(x^2-2x-4)=0$

$x-2=0$ or $x^2-2x-4=0$

$x=2$ or $x=\dfrac{-(-2)\pm\sqrt{(-2)^2-4(1)(-4)}}{2(1)}$

$x=\dfrac{2\pm\sqrt{4+16}}{2}$

$x=\dfrac{2\pm\sqrt{20}}{2}$

$x=\dfrac{2\pm 2\sqrt{5}}{2}$

$x=\dfrac{2(1\pm\sqrt{5})}{2}$

$x=1\pm\sqrt{5}$

The solution set is $\{2, 1\pm\sqrt{5}\}$.

11. $x^4+4x^3-x^2-16x-12=0$

c: $\pm 1, \pm 2, \pm 3, \pm 4, \pm 6, \pm 12$

d: ± 1

$\frac{c}{d}$: $\pm 1, \pm 2, \pm 3, \pm 4, \pm 6, \pm 12$

$$\begin{array}{r|rrrrr} -1 & 1 & 4 & -1 & -16 & -12 \\ & & -1 & -3 & 4 & 12 \\ \hline & 1 & 3 & -4 & -12 & 0 \end{array}$$

$(x+1)(x^3+3x^2-4x-12)=0$

c: $\pm 1, \pm 2, \pm 3, \pm 4, \pm 6, \pm 12$

d: ± 1

$\frac{c}{d}$: $\pm 1, \pm 2, \pm 3, \pm 4, \pm 6, \pm 12$

$$\begin{array}{r|rrrr} 2 & 1 & 3 & -4 & -12 \\ & & 2 & 10 & 12 \\ \hline & 1 & 5 & 6 & 0 \end{array}$$

$(x+1)(x-2)(x^2+5x+6)=0$

$(x+1)(x-2)(x+3)(x+2)=0$

$x+1=0$ or $x-2=0$ or $x+3=0$ or $x+2=0$

$x=-1$ or $x=2$ or $x=-3$ or $x=-2$

The solution set is $\{-3, -2, -1, 2\}$.

13. $x^4+x^3-3x^2-17x-30=0$

c: $\pm 1, \pm 2, \pm 3, \pm 5, \pm 6, \pm 10, \pm 15, \pm 30$

d: ± 1

$\frac{c}{d}$: $\pm 1, \pm 2, \pm 3, \pm 5, \pm 6, \pm 10, \pm 15, \pm 30$

$$\begin{array}{r|rrrrr} -2 & 1 & 1 & -3 & -17 & -30 \\ & & -2 & 2 & 2 & 30 \\ \hline & 1 & -1 & -1 & -15 & 0 \end{array}$$

$(x+2)(x^3-x^2-x-15)=0$

c: $\pm 1, \pm 3, \pm 5, \pm 15$

d: ± 1

$\frac{c}{d}$: $\pm 1, \pm 3, \pm 5, \pm 15$

$$\begin{array}{r|rrrr} 3 & 1 & -1 & -1 & -15 \\ & & 3 & 6 & 15 \\ \hline & 1 & 2 & 5 & 0 \end{array}$$

$(x+2)(x-3)(x^2+2x+5)=0$

$x+2=0$ or $x-3=0$

$x=-2$ or $x=3$ or $x^2+2x+5=0$

$$x=\frac{-2\pm\sqrt{2^2-4(1)(5)}}{2(1)}$$

$$x=\frac{-2\pm\sqrt{-16}}{2}$$

$$x=\frac{-2\pm 4i}{2}$$

$$x=-1\pm 2i$$

The solution set is $\{-2, 3, -1\pm 2i\}$.

15. $x^3-x^2+x-1=0$

c: ± 1

d: ± 1

$\frac{c}{d}$: ± 1

$$\begin{array}{r|rrrr} 1 & 1 & -1 & 1 & -1 \\ & & 1 & 0 & 1 \\ \hline & 1 & 0 & 1 & 0 \end{array}$$

$(x-1)(x^2+1)=0$

$x-1=0$ or $x^2+1=0$

$x=1$ or $x^2=-1$

$x=1$ or $x=\pm\sqrt{-1}$

$x=1$ or $x=\pm i$

The solution set is $\{1, \pm i\}$.

17. $2x^4+3x^3-11x^2-9x+15=0$

c: $\pm 1, \pm 3, \pm 5, \pm 15$

d: $\pm 1, \pm 2$

$\frac{c}{d}$: $\pm 1, \pm 3, \pm 5, \pm 15,$
$\pm\frac{1}{2}, \pm\frac{3}{2}, \pm\frac{5}{2}, \pm\frac{15}{2}$

$$\begin{array}{r|rrrrr} 1 & 2 & 3 & -11 & -9 & 15 \\ & & 2 & 5 & -6 & -15 \\ \hline & 2 & 5 & -6 & -15 & 0 \end{array}$$

$(x-1)(2x^3+5x^2-6x-15)=0$

c: $\pm 1, \pm 3, \pm 5, \pm 15$

d: $\pm 1, \pm 2$

$\frac{c}{d}$: $\pm 1, \pm 3, \pm 5, \pm 15,$
$\pm\frac{1}{2}, \pm\frac{3}{2}, \pm\frac{5}{2}, \pm\frac{15}{2}$

$$\begin{array}{r|rrrr} -\frac{5}{2} & 2 & 5 & -6 & -15 \\ & & -5 & 0 & 15 \\ \hline & 2 & 0 & -6 & 0 \end{array}$$

$(x-1)\left(x+\frac{5}{2}\right)(2x^2-6)=0$

$x-1=0$ or $x+\frac{5}{2}=0$ or $2x^2-6=0$

$x=1$ or $x=-\frac{5}{2}$ or $2x^2=6$

$x=1$ or $x=-\frac{5}{2}$ or $x^2=3$

$x=1$ or $x=-\frac{5}{2}$ or $x=\pm\sqrt{3}$

The solution set is $\left\{-\frac{5}{2}, 1, \pm\sqrt{3}\right\}$.

19. $4x^4+12x^3+x^2-12x+4=0$

c: $\pm 1, \pm 2, \pm 4$

d: $\pm 1, \pm 2, \pm 4$

$\frac{c}{d}$: $\pm 1, \pm 2, \pm 4, \pm\frac{1}{2}, \pm\frac{1}{4}$

$$\begin{array}{r|rrrrr} -2 & 4 & 12 & 1 & -12 & 4 \\ & & -8 & -8 & 14 & -4 \\ \hline & 4 & 4 & -7 & 2 & 0 \end{array}$$

$(x+2)(4x^3+4x^2-7x+2)=0$

c: $\pm 1, \pm 2$

d: $\pm 1, \pm 2, \pm 4$

$\frac{c}{d}$: $\pm 1, \pm 2, \pm\frac{1}{2}, \pm\frac{1}{4}$

$$\begin{array}{r|rrrr} \frac{1}{2} & 4 & 4 & -7 & 2 \\ & & 2 & 3 & -2 \\ \hline & 4 & 6 & -4 & 0 \end{array}$$

$(x+2)\left(x-\frac{1}{2}\right)(4x^2+6x-4)=0$

$(x+2)\left(x-\frac{1}{2}\right)(2)(2x^2+3x-2)=0$

$2(x+2)\left(x-\frac{1}{2}\right)(2x-1)(x+2)=0$

$2(x+2)\left(x-\frac{1}{2}\right)(2)\left(x-\frac{1}{2}\right)(x+2)=0$

$4(x+2)^2\left(x-\frac{1}{2}\right)^2=0$

$x+2=0$ or $x-\frac{1}{2}=0$

$x=-2$ or $x=\frac{1}{2}$

The solution set is $\left\{-2, \frac{1}{2}\right\}$.

21. $x^4+3x-2=0$

c: $\pm 1, \pm 2$

d: ± 1

$\frac{c}{d}$: $\pm 1, \pm 2$

$$\begin{array}{r|rrrrr} 1 & 1 & 0 & 0 & 3 & -2 \\ & & 1 & 1 & 1 & 4 \\ \hline & 1 & 1 & 1 & 4 & 2 \end{array}$$

$$\begin{array}{r|rrrrr} -1 & 1 & 0 & 0 & 3 & -2 \\ & & -1 & 1 & -1 & -2 \\ \hline & 1 & -1 & 1 & 2 & -4 \end{array}$$

$$\begin{array}{r|rrrrr} 2 & 1 & 0 & 0 & 3 & -2 \\ & & 2 & 4 & 8 & 22 \\ \hline & 1 & 2 & 4 & 11 & 20 \end{array}$$

$$\begin{array}{r|rrrrr} -2 & 1 & 0 & 0 & 3 & -2 \\ & & -2 & 4 & -8 & 10 \\ \hline & 1 & -2 & 4 & -5 & 8 \end{array}$$

None of the possible rational roots gave a remainder of zero. Therefore, there are no rational roots.

23. $3x^4-4x^3-10x^2+3x-4=0$

c: $\pm 1, \pm 2, \pm 4$

d: $\pm 1, \pm 3$

$\frac{c}{d}$: $\pm 1, \pm 2, \pm 4, \pm\frac{1}{3}, \pm\frac{2}{3}, \pm\frac{4}{3}$

None of the possible rational roots gave a remainder of zero. Therefore, there are no rational roots.

25. $x^5+2x^4-2x^3+5x^2-2x-3=0$

c: $\pm 1, \pm 3$

d: ± 1

$\frac{c}{d}$: $\pm 1, \pm 3$

$$\begin{array}{r|rrrrrr} 1 & 1 & 2 & -2 & 5 & -2 & -3 \\ & & 1 & 3 & 1 & 6 & 4 \\ \hline & 1 & 3 & 1 & 6 & 4 & 1 \end{array}$$

$$\begin{array}{r|rrrrrr} -1 & 1 & 2 & -2 & 5 & -2 & -3 \\ & & -1 & -1 & 3 & -8 & 10 \\ \hline & 1 & 1 & -3 & 8 & -10 & 7 \end{array}$$

$$\begin{array}{r|rrrrrr} 3 & 1 & 2 & -2 & 5 & -2 & -3 \\ & & 3 & 15 & 39 & 132 & 390 \\ \hline & 1 & 5 & 13 & 44 & 130 & 387 \end{array}$$

$$\begin{array}{r|rrrrrr} -3 & 1 & 2 & -2 & 5 & -2 & -3 \\ & & -3 & 3 & -3 & -6 & 24 \\ \hline & 1 & -1 & 1 & 2 & -8 & 21 \end{array}$$

None of the possible rational roots gave a remainder of zero. Therefore, there are no rational roots.

27. $\frac{1}{10}x^3+\frac{1}{5}x^2-\frac{1}{2}x-\frac{3}{5}=0$

$10\left(\frac{1}{10}x^3+\frac{1}{5}x^2-\frac{1}{2}x-\frac{3}{5}\right)=10(0)$

$x^3+2x^2-5x-6=0$

c: $\pm 1, \pm 2, \pm 3, \pm 6$

d: ± 1

$\frac{c}{d}$: $\pm 1, \pm 2, \pm 3, \pm 6$

$$\begin{array}{r|rrrr} -1 & 1 & 2 & -5 & -6 \\ & & -1 & -1 & 6 \\ \hline & 1 & 1 & -6 & 0 \end{array}$$

$(x+1)(x^2+x-6)=0$

$(x+1)(x+3)(x-2)=0$

$x+1=0$ or $x+3=0$ or $x-2=0$

$x=-1$ or $x=-3$ or $x=2$

The solution set is $\{-3, -1, 2\}$.

29. $x^3-\frac{5}{6}x^2-\frac{22}{3}x+\frac{5}{2}=0$

$6\left(x^3-\frac{5}{6}x^2-\frac{22}{3}x+\frac{5}{2}\right)=6(0)$

$6x^3-5x^2-44x+15=0$

c: $\pm 1, \pm 3, \pm 5, \pm 15$

d: $\pm 1, \pm 2, \pm 3, \pm 6$

$\frac{c}{d}$: $\pm 1, \pm 3, \pm 5, \pm 15$

$\pm\frac{1}{2}, \pm\frac{3}{2}, \pm\frac{5}{2}, \pm\frac{15}{2}$

$\pm\frac{1}{3}, \pm\frac{5}{3}, \pm\frac{1}{6}, \pm\frac{5}{6}$

$$\begin{array}{r|rrrr} 3 & 6 & -5 & -44 & 15 \\ & & 18 & 39 & -15 \\ \hline & 6 & 13 & -5 & 0 \end{array}$$

$(x-3)(6x^2+13x-5)=0$

$(x-3)(2x+5)(3x-1)=0$

$x-3=0$ or $2x+5=0$ or $3x-1=0$

$x=3$ or $2x=-5$ or $3x=1$

$x=3$ or $x=-\frac{5}{2}$ or $x=\frac{1}{3}$

The solution set is $\left\{-\frac{5}{2}, \frac{1}{3}, 3\right\}$.

31. $6x^2+7x-20=0$

There is one variation in sign,
so there is one positive real solution.

Replacing x with $-x$:

$6(-x)^2+7(-x)-20$

$6x^2-7x-20$

There is one variation in sign,
so there is one negative real solution.

33. $2x^3+x-3=0$

There is one variation in sign,
so there is one positive real solution.

Replacing x with $-x$:

$2(-x)^3+(-x)-3$

$-2x^3-x-3$

No variations in sign,
so there are no negative real solutions.

So, there is one positive real solution
and two nonreal complex solutions.

35. $3x^3-2x^2+6x+5=0$

There are two variations in sign, so
there are two or no positive real solutions.

Replacing x with $-x$:

$3(-x)^3-2(-x)^2+6(-x)+5$

$-3x^3-2x^2-6x+5$

There is one variation in sign,
so there is one negative real solution.

So, there are two cases.

I: 1 negative and 2 positive
real solutions

or

II: 1 negative real solution and
2 nonreal complex solutions.

37. $x^5-3x^4+5x^3-x^2+2x-1=0$

There are 5 variations in sign, so there
could be 5, 3, or 1 positive real solutions.

Replacing x with $-x$:

$(-x)^5-3(-x)^4+5(-x)^3-(-x)^2+2(-x)-1$

$-x^5-3x^4-5x^3-x^2-2x-1$

There are no variations in sign,
so there are no negative real solutions.

There are three cases.

I: 5 positive real solutions

or

II: 3 positive real solutions and
2 nonreal complex solutions.

or

III: 1 positive real solution and
4 nonreal complex solutions.

39. $x^5 + 32 = 0$

There are no variations in sign, so there are no positive real solutions.

Replacing x with $-x$:

$(-x)^5 + 32$

$-x^5 + 32$

There is one variation in sign, so there is one negative real solution.

So there is one negative real solution and four complex solutions.

Problem Set 9.4 Graphing Polynomial Functions

1 - 21. See back of textbook for graphs.

23. $f(x) = -x^3 - x^2 + 6x$

$f(x) = -x(x^2 + x - 6)$

$f(x) = -x(x+3)(x-2)$

25. $f(x) = x^4 - 5x^3 + 6x^2$

$f(x) = x^2(x^2 - 5x + 6)$

$f(x) = x^2(x-3)(x-2)$

27. $f(x) = x^3 + 2x^2 - x - 2$

$f(x) = x^2(x+2) - 1(x+2)$

$f(x) = (x+2)(x^2 - 1)$

$f(x) = (x+2)(x-1)(x+1)$

29. $f(x) = x^3 - 8x^2 + 19x - 12$

$$\begin{array}{r|rrrr} 1 & 1 & -8 & 19 & -12 \\ & & 1 & -7 & 12 \\ \hline & 1 & -7 & 12 & 0 \end{array}$$

$f(x) = (x-1)(x^2 - 7x + 12)$

$f(x) = (x-1)(x-3)(x-4)$

31. $f(x) = 2x^3 - 3x^2 - 3x + 2$

$$\begin{array}{r|rrrr} -1 & 2 & -3 & -3 & 2 \\ & & -2 & 5 & -2 \\ \hline & 2 & -5 & 2 & 0 \end{array}$$

$f(x) = (x+1)(2x^2 - 5x + 2)$

$f(x) = (x+1)(2x-1)(x-2)$

33. $f(x) = x^4 - 5x^2 + 4$

$f(x) = (x^2 - 4)(x^2 - 1)$

$f(x) = (x+2)(x-2)(x+1)(x-1)$

35. $f(x) = (x+3)(x-6)(8-x)$

a. y-intercepts, let $x = 0$

$f(0) = (0+3)(0-6)(8-0)$

$f(0) = (3)(-6)(8) = -144$

b. x-intercepts, let $y = 0$

$0 = (x+3)(x-6)(8-x)$

$x + 3 = 0$ or $x - 6 = 0$ or $8 - x = 0$

$x = -3$ or $x = 6$ or $x = 8$

c.

Interval	Test Value	Sign of f(x)	Location of graph
$x < -3$	$f(-4) = 120$	pos	above x-axis
$-3 < x < 6$	$f(0) = -144$	neg	below x-axis
$6 < x < 8$	$f(7) = 10$	pos	above x-axis
$x > 8$	$f(9) = -36$	neg	below x-axis

$f(x) > 0$ for $\{x \mid x < -3 \text{ or } 6 < x < 8\}$

$f(x) < 0$ for $\{x \mid -3 < x < 6 \text{ or } x > 8\}$

37. $f(x) = (x+3)^4(x-1)^3$

a. y-intercept, let $x = 0$

$f(0) = (0+3)^4(0-1)^3$

$f(0) = (81)(-1) = -81$

b. x-intercept, let $y = 0$

$0 = (x+3)^4(x-1)^3$

$x + 3 = 0$ or $x - 1 = 0$

$x = -3$ or $x = 1$

c.

Interval	Test Value	Sign of f(x)	Location of graph
$x < -3$	$f(-4)=-125$	neg	below x-axis
$-3 < x < 1$	$f(0) = -81$	neg	below x-axis
$x > 1$	$f(2) = 625$	pos	above x-axis

$f(x) > 0$ for $\{x \mid x > 1\}$

$f(x) < 0$ for $\{x \mid x < -3 \text{ or } -3 < x < 1\}$

39. $f(x) = x(x-6)^2(x+4)$

a. y-intercept, let $x = 0$

$f(0) = 0(0-6)^2(0+4)^2$

$f(0) = 0$

b. x-intercept, let $y = 0$

$0 = x(x-6)^2(x+4)$

$x = 0$ or $x - 6 = 0$ or $x + 4 = 0$

$x = 0$ or $x = 6$ or $x = -4$

c.

Interval	Test Value	Sign of f(x)	Location of graph
$x < -4$	$f(-5) = 605$	pos	above x-axis
$-4 < x < 0$	$f(-3)=-243$	neg	below x-axis
$0 < x < 6$	$f(1) = 125$	pos	above x-axis
$x > 6$	$f(7) = 77$	pos	above x-axis

$f(x) > 0$ for $\{x \mid x < -4 \text{ or } 0 < x < 6 \text{ or } x > 6\}$

$f(x) < 0$ for $\{x \mid -4 < x < 0\}$

41. $f(x) = x^2(2-x)(x+3)$

a. y-intercept, let $x = 0$

$f(0) = 0^2(2-0)(0+3)$

$f(0) = 0$

b. x-intercepts, let $x = 0$

$0 = x^2(2-x)(x+3)$

$x^2 = 0$ or $2 - x = 0$ or $x + 3 = 0$

$x = 0$ or $x = 2$ or $x = -3$

c.

Interval	Test Value	Sign of f(x)	Location of graph
$x < -3$	$f(-4) = -96$	neg	below x-axis
$-3 < x < 0$	$f(-1) = 6$	pos	above x-axis
$0 < x < 2$	$f(1) = 4$	pos	above x-axis
$x > 2$	$f(3) = -54$	neg	below x-axis

$f(x) > 0$ for $\{x \mid -3 < x < 0 \text{ or } 0 < x < 2\}$

$f(x) < 0$ for $\{x \mid x < -3 \text{ or } x > 2\}$

Problem Set 9.5 Graphing Rational Functions

1 - 21. See back of textbook for graphs.

Chapter 9 Review

1. $(3x^3 - 4x^2 + 6x - 2) \div (x - 1)$

$$\begin{array}{r|rrrr} 1 & 3 & -4 & 6 & -2 \\ & & 3 & -1 & 5 \\ \hline & 3 & -1 & 5 & 3 \end{array}$$

Q: $3x^2 - x + 5$
R: 3

3. $(-2x^4 + x^3 - 2x^2 - x - 1) \div (x + 4)$

$$\begin{array}{r|rrrrr} -4 & -2 & 1 & -2 & -1 & -1 \\ & & 8 & -36 & 152 & -604 \\ \hline & -2 & 9 & -38 & 151 & -605 \end{array}$$

Q: $-2x^3 + 9x^2 - 38x + 151$
R: -605

5. $f(x) = 4x^5 - 3x^3 + x^2 - 1$

$f(1) = 4(1)^5 - 3(1)^3 + 1^2 - 1$

$f(1) = 4 - 3 + 1 - 1 = 1$

7. $f(x) = -x^4 + 9x^2 - x - 2$

$$\begin{array}{r|rrrrr} -2 & -1 & 0 & 9 & -1 & -2 \\ & & 2 & -4 & -10 & 22 \\ \hline & -1 & 2 & 5 & -11 & 20 \end{array}$$

$f(-2) = 20$

9.

$$\begin{array}{r|rrrr} -2 & 2 & 1 & -7 & -2 \\ & & -4 & 6 & 2 \\ \hline & 2 & -3 & -1 & 0 \end{array}$$

Since $f(-2)=0$, $(x+2)$ is a factor of $(2x^3+x^2-7x-2)$.

11.

$$\begin{array}{r|rrrrrr} 4 & 1 & 0 & 0 & 0 & 0 & -1024 \\ & & 4 & 16 & 64 & 256 & 1024 \\ \hline & 1 & 4 & 16 & 64 & 256 & 0 \end{array}$$

Since $f(4)=0$, $(x-4)$ is a factor of (x^5-1024).

13. $x^3-3x^2-13x+15=0$

c: $\pm1, \pm3, \pm5, \pm15$

d: ±1

$\frac{c}{d}$: $\pm1, \pm3, \pm5, \pm15$

$$\begin{array}{r|rrrr} 1 & 1 & -3 & -13 & 15 \\ & & 1 & -2 & -15 \\ \hline & 1 & -2 & -15 & 0 \end{array}$$

$(x-1)(x^2-2x-15)=0$

$(x-1)(x-5)(x+3)=0$

$x-1=0$ or $x-5=0$ or $x+3=0$

$x=1$ or $x=5$ or $x=-3$

The solution set is $\{-3, 1, 5\}$.

15. $x^4-5x^3+34x^2-82x+52=0$

c: $\pm1, \pm2, \pm4, \pm13, \pm26, \pm52$

d: ±1

$\frac{c}{d}$: $\pm1, \pm2, \pm4, \pm13, \pm26, \pm52$

$$\begin{array}{r|rrrrr} 1 & 1 & -5 & 34 & -82 & 52 \\ & & 1 & -4 & 30 & -52 \\ \hline & 1 & -4 & 30 & -52 & 0 \end{array}$$

$(x-1)(x^3-4x^2+30x-52)=0$

c: $\pm1, \pm2, \pm4, \pm13, \pm26, \pm52$

d: ±1

$\frac{c}{d}$: $\pm1, \pm2, \pm4, \pm13, \pm26, \pm52$

$$\begin{array}{r|rrrr} 2 & 1 & -4 & 30 & -52 \\ & & 2 & -4 & 52 \\ \hline & 1 & -2 & 26 & 0 \end{array}$$

$(x-1)(x-2)(x^2-2x+26)=0$

$x-1=0$ or $x-2=0$ or $x^2-2x+26=0$

$x=1$ or $x=2$ or $x=\frac{-(-2)\pm\sqrt{(-2)^2-4(1)(26)}}{2(1)}$

$x=\frac{2\pm\sqrt{4-104}}{2}$

$x=\frac{2\pm\sqrt{-100}}{2}$

$x=\frac{2\pm10i}{2}$

$x=1\pm5i$

The solution set is $\{1, 2, 1\pm5i\}$.

17. $4x^4-3x^3+2x^2+x+4=0$

There are two variations in sign, so there are two or no positive real solutions.

Replacing x with $-x$:

$4(-x)^4-3(-x)^3+2(-x)^2-x+4$

$4x^4+3x^3+2x^2-x+4$

There are two variations in sign, so there are two of no negative real solutions.

So, there are 4 cases.

I: 2 positive and 2 negative real solutions

or

II: 2 positive real solutions and 2 nonreal complex solutions.

or

III: 2 negative real solutions and 2 nonreal complex solutions

or

IV: 4 nonreal complex solutions.

19 - 21. See back of textbook for graphs.

23. $f(x) = \frac{2x}{x-3}$

$f(x) = \frac{\frac{2x}{x}}{\frac{x}{x} - \frac{3}{x}}$

$f(x) = \frac{2}{1 - \frac{3}{x}}$

Horizontal asymptote of $y = \frac{2}{1} = 2$.

Vertical asymptote of $x = 3$.

Chapter 9 Test

1. $(3x^3 + 5x^2 - 14x - 6) \div (x+3)$

$$\begin{array}{r|rrrr} -3 & 3 & 5 & -14 & -6 \\ & & -9 & 12 & 6 \\ \hline & 3 & -4 & -2 & 0 \end{array}$$

Q: $3x^2 - 4x - 2$
R: 0

3. $f(x) = x^5 - 8x^4 + 9x^3 - 13x^2 - 9x - 10$

$$\begin{array}{r|rrrrrr} 7 & 1 & -8 & 9 & -13 & -9 & -10 \\ & & 7 & -7 & 14 & 7 & -14 \\ \hline & 1 & -1 & 2 & 1 & -2 & -24 \end{array}$$

$f(7) = -24$

5. $f(x) = x^5 - 35x^3 - 32x + 15$

$$\begin{array}{r|rrrrrr} 6 & 1 & 0 & -35 & 0 & -32 & 15 \\ & & 6 & 36 & 6 & 36 & 24 \\ \hline & 1 & 6 & 1 & 6 & 4 & 39 \end{array}$$

$f(6) = 39$

7.
$$\begin{array}{r|rrrr} -2 & 5 & 9 & -9 & -17 \\ & & -10 & 2 & 14 \\ \hline & 5 & -1 & -7 & -3 \end{array}$$

Since $f(-2) = -3$, $(x+2)$ is not a factor of $(5x^3 + 9x^2 - 9x - 17)$.

9.
$$\begin{array}{r|rrrrr} 6 & 1 & 0 & -2 & 3 & -12 \\ & & 6 & 36 & 204 & 1242 \\ \hline & 1 & 6 & 34 & 207 & 1230 \end{array}$$

Since $f(6) = 1230$, $(x-6)$ is not a factor of $(x^4 - 2x^2 + 3x - 12)$.

11. $2x^3 + 5x^2 - 13x - 4 = 0$

c: $\pm 1, \pm 2, \pm 4$

d: $\pm 1, \pm 2$

$\frac{c}{d}$: $\pm 1, \pm 2, \pm 4, \pm \frac{1}{2}$

$$\begin{array}{r|rrrr} -4 & 2 & 5 & -13 & -4 \\ & & -8 & 12 & 4 \\ \hline & 2 & -3 & -1 & 0 \end{array}$$

$(x+4)(2x^2 - 3x - 1) = 0$

$x + 4 = 0$ or $2x^2 - 3x - 1 = 0$

$x = -4$ or $x = \frac{-(-3) \pm \sqrt{(-3)^2 - 4(2)(-1)}}{2(2)}$

$x = \frac{3 \pm \sqrt{17}}{4}$

The solution set is $\left\{-4, \frac{3 \pm \sqrt{17}}{4}\right\}$.

13. $2x^3 + 3x^2 - 17x + 12 = 0$

c: $\pm 1, \pm 2, \pm 3, \pm 4, \pm 6, \pm 12$

d: $\pm 1, \pm 2$

$\frac{c}{d}$: $\pm 1, \pm 2, \pm 3, \pm 4, \pm 6, \pm 12,$

$\pm \frac{1}{2}, \pm \frac{3}{2}$

$$\begin{array}{r|rrrr} 1 & 2 & 3 & -17 & 12 \\ & & 2 & 5 & -12 \\ \hline & 2 & 5 & -12 & 0 \end{array}$$

$(x-1)(2x^2 + 5x - 12) = 0$

$(x-1)(2x-3)(x+4) = 0$

$x - 1 = 0$ or $2x - 3 = 0$ or $x + 4 = 0$

$x = 1$ or $2x = 3$ or $x = -4$

$x = 1$ or $x = \frac{3}{2}$ or $x = -4$

The solution set is $\left\{-4, 1, \frac{3}{2}\right\}$.

15. $5x^4 + 3x^3 - x^2 - 9 = 0$

There is one variation in sign, so there is one positive real solution.

Replacing x with $-x$:

$5(-x)^4+3(-x)^3-(-x)^2-9$

$5x^4-3x^3-x^2-9$

There is one variation in sign, so there is one negative real solution.

Therefore, there is 1 positive, 1 negative, and 2 nonreal complex solutions.

17. $f(x)=\frac{5x}{x+3}$

Vertical asymptote

$x=-3$

19. $f(x)=\frac{x^2}{x^2+2}$

Replace x with $-x$:

$f(-x)=\frac{(-x)^2}{(-x)^2+2}$

$f(-x)=\frac{x^2}{x^2+2}$

$f(x)=f(-x)$

It is symmetric to the y-axis.

Replace y with $-y$:

$-y=\frac{x^2}{x^2+2}$

$y=\frac{-x^2}{x^2+2}$

It is not symmetric to the x-axis.

Replace x with $-x$ and y with $-y$:

$-y=\frac{(-x)^2}{(-x)^2+2}$

$-y=\frac{x^2}{x^2+2}$

$y=\frac{-x^2}{x^2+2}$

It is not symmetric with respect to the origin.

21 - 25. See back of textbook for graphs.

Chapter 10 Exponential and Logarithmic Functions

Problem Set 10.1 Exponents and Exponential Functions

1. $2^x = 64$

   $2^x = 2^6$

   $x = 6$

   The solution set is {6}.

3. $3^{2x} = 27$

   $3^{2x} = 3^3$

   $2x = 3$

   $x = \frac{3}{2}$

   The solution set is $\left\{\frac{3}{2}\right\}$.

5. $\left(\frac{1}{2}\right)^x = \frac{1}{128}$

   $\left(\frac{1}{2}\right)^x = \left(\frac{1}{2}\right)^7$

   $x = 7$

   The solution set is {7}.

7. $3^{-x} = \frac{1}{243}$

   $3^{-x} = \frac{1}{3^5}$

   $3^{-x} = 3^{-5}$

   $-x = -5$

   $x = 5$

   The solution set is {5}.

9. $6^{3x-1} = 36$

   $6^{3x-1} = 6^2$

   $3x - 1 = 2$

   $3x = 3$

   $x = 1$

   The solution set is {1}.

11. $\left(\frac{3}{4}\right)^n = \frac{64}{27}$

    $\left(\frac{3}{4}\right)^n = \left(\frac{4}{3}\right)^3$

    $\left(\frac{3}{4}\right)^n = \left(\frac{3}{4}\right)^{-3}$

    $n = -3$

    The solution set is {−3}.

13. $16^x = 64$

    $(2^4)^x = 2^6$

    $2^{4x} = 2^6$

    $4x = 6$

    $x = \frac{6}{4} = \frac{3}{2}$

    The solution set is $\left\{\frac{3}{2}\right\}$.

15. $27^{4x} = 9^{x+1}$

    $(3^3)^{4x} = (3^2)^{x+1}$

    $3^{12x} = 3^{2x+2}$

    $12x = 2x + 2$

    $10x = 2$

    $x = \frac{2}{10} = \frac{1}{5}$

    The solution set is $\left\{\frac{1}{5}\right\}$.

17. $9^{4x-2} = \frac{1}{81}$

    $9^{4x-2} = \frac{1}{9^2}$

    $9^{4x-2} = 9^{-2}$

    $4x - 2 = -2$

    $4x = 0$

    $x = 0$

    The solution set is {0}.

19. $10^x = .1$

    $10^x = 10^{-1}$

    $x = -1$

    The solution set is {−1}.

21. $(2^{x+1})(2^x) = 64$

    $2^{2x+1} = 2^6$

$2x+1=6$

$2x=5$

$x=\frac{5}{2}$

The solution set is $\left\{\frac{5}{2}\right\}$.

23. $(27)(3^x)=9^x$

$(3^3)(3^x)=(3^2)^x$

$3^{x+3}=3^{2x}$

$x+3=2x$

$3=x$

The solution set is $\{3\}$.

25. $(4^x)(16^{3x-1})=8$

$(2^2)^x(2^4)^{3x-1}=2^3$

$(2^{2x})(2^{12x-4})=2^3$

$2^{14x-4}=2^3$

$14x-4=3$

$14x=7$

$x=\frac{7}{14}=\frac{1}{2}$

The solution set is $\left\{\frac{1}{2}\right\}$.

27 - 45. See back of textbook for graphs.

Problem Set 10.2 Applications of Exponential Functions

1. $P=P_0(1.04)^t$

a. $P=.55(1.04)^3$

$P=.55(1.125)$

$P=\$.62$

b. $P=3.43(1.04)^5$

$P=3.43(1.217)$

$P=\$4.17$

c. $P=1.76(1.04)^4$

$P=1.76(1.170)$

$P=\$2.06$

d. $P=.44(1.04)^{10}$

$P=.44(1.480)$

$P=\$.65$

e. $P=9000(1.04)^5$

$P=9000(1.217)$

$P=\$10,950$

f. $P=50,000(1.04)^8$

$P=50,000(1.369)$

$P=\$68,428$

g. $P=500(1.04)^7$

$P=500(1.316)$

$P=\$658$

3. $A=P(1+\frac{r}{n})^{nt}$

$A=200\left(1+\frac{.06}{1}\right)^{1(6)}$

$A=200(1.06)^6$

$A=200(1.419)$

$A=\$283.70$

5. $A=P(1+\frac{r}{n})^{nt}$

$A=500\left(1+\frac{.08}{2}\right)^{2(7)}$

$A=500(1.04)^{14}$

$A=500(1.732)$

$A=\$865.84$

7. $A=P(1+\frac{r}{n})^{nt}$

$A=800\left(1+\frac{.09}{4}\right)^{4(9)}$

$A=800(1.0225)^{36}$

$A=800(2.228)$

$A=\$1782.25$

9. $A=P(1+\frac{r}{n})^{nt}$

$A=1500\left(1+\frac{.12}{12}\right)^{12(5)}$

$A=1500(1.01)^{60}$

$A = 1500(1.817)$

$A = \$2725.05$

11. $A = P\left(1+\frac{r}{t}\right)^{nt}$

$A = 5000\left(1+\frac{.085}{1}\right)^{1(15)}$

$A = 5000(1.085)^{15}$

$A = 5000(3.3997)$

$A = \$16,998.71$

13. $A = P\left(1+\frac{r}{n}\right)^{nt}$

$A = 8000\left(1+\frac{.105}{4}\right)^{4(10)}$

$A = 8000(1.02625)^{40}$

$A = 8000(2.819)$

$A = \$22,553.65$

15. $A = Pe^{rt}$

$A = 400e^{.07(5)}$

$A = 400e^{.35}$

$A = 400(1.419)$

$A = \$567.63$

17. $A = Pe^{rt}$

$A = 750e^{.08(8)}$

$A = 750e^{.64}$

$A = 750(1.896)$

$A = \$1422.36$

19. $A = Pe^{rt}$

$A = 2000e^{.10(15)}$

$A = 2000e^{1.5}$

$A = 2000(4.482)$

$A = \$8963.38$

21. $A = Pe^{rt}$

$A = 7500e^{.085(10)}$

$A = 7500e^{.85}$

$A = 7500(2.3396)$

$A = \$17,547.35$

23. $A = Pe^{rt}$

$A = 15000e^{.0775(10)}$

$A = 15000e^{.775}$

$A = 15000(2.171)$

$A = \$32558.88$

25. $A = P\left(1+\frac{r}{n}\right)^{nt}$

$2700 = 1500\left(1+\frac{r}{4}\right)^{4(10)}$

$1.8 = \left(1+\frac{r}{4}\right)^{40}$

$(1.8)^{\frac{1}{40}} = \left[\left(1+\frac{r}{4}\right)^{40}\right]^{\frac{1}{40}}$

$1.0148 = 1+\frac{r}{4}$

$.0148 = \frac{r}{4}$

$.0592 = r$

$r = 5.9\%$

27. $P(1+r) = Pe^{r}$

$1+r = e^{r}$

$1+r = e^{.0775}$

$1+r = 1.0806$

$r = .0806$

$r = 8.06\%$

29. $A = P\left(1+\frac{r}{n}\right)^{nt}$

$A = P\left(1+\frac{.0825}{4}\right)^{4}$

$A = P(1.020625)^{4}$

$A = P(1.08509)$

$A = P\left(1+\frac{.083}{2}\right)^{2}$

$A = P(1.0415)^{2}$

$A = P(1.08472)$

8.25% compounded quarterly will yield more.

31. $Q = Q_0\left(\frac{1}{2}\right)^{\frac{t}{n}}$

$Q = 400\left(\frac{1}{2}\right)^{\frac{87}{29}}$

$Q = 400\left(\frac{1}{2}\right)^{3}$

$Q = 400\left(\frac{1}{8}\right)$

$Q = 50$

There will be 50 grams after 87 years.

$Q = 400\left(\frac{1}{2}\right)^{\frac{100}{29}}$

$Q = 400\left(\frac{1}{2}\right)^{3.4483}$

$Q = 37$

There will be 37 grams after 100 years.

33. $Q(t) = 1000e^{.4t}$

$Q(2) = 1000e^{.4(2)}$

$Q(2) = 1000e^{.8}$

$Q(2) = 1000(2.2255)$

$Q(2) = 2226$

$Q(3) = 1000e^{.4(3)}$

$Q(3) = 1000e^{1.2}$

$Q(3) = 1000(3.3201)$

$Q(3) = 3320$

$Q(5) = 1000e^{.4(5)}$

$Q(5) = 1000e^{2}$

$Q(5) = 1000(7.3891)$

$Q(5) = 7389$

35. $Q = Q_0e^{.3t}$

$6640 = Q_0e^{.3(4)}$

$6640 = Q_0e^{1.2}$

$6640 = Q_0(3.3201)$

$2000 = Q_0$

There were initially 2000 bacteria.

37. $P(a) = 14.7e^{-.21a}$

a. $P(3.85) = 14.7e^{-.21(3.85)}$

$P(3.85) = 14.7e^{-.8085}$

$P(3.85) = 14.7(.4455)$

$P(3.85) = 6.5$ lbs. per sq. in.

b. $P(1) = 14.7e^{-.21(1)}$

$P(1) = 14.7e^{-.21}$

$P(1) = 14.7(.8106)$

$P(1) = 11.9$ lbs. per sq. in.

c. $1985 \text{ feet} \times \frac{1 \text{ mile}}{5280 \text{ feet}} = .376$ miles

$P(.376) = 14.7e^{-.21(.376)}$

$P(.376) = 14.7e^{-.07896}$

$P(.376) = 14.7(.9241)$

$P(.376) = 13.6$ lbs. per sq. in.

d. $1090 \text{ feet} \times \frac{1 \text{ mile}}{5280 \text{ feet}} = .2064$

$P(.2064) = 14.7e^{-.21(.2064)}$

$P(.2064) = 14.7e^{-.0433}$

$P(.2064) = 14.7(.9576)$

$P(.2064) = 14.1$ lbs. per sq. in.

39 - 43. See back of textbook for graphs.

Problem Set 10.3 Logarithms

1. $2^7 = 128$

$\log_2 128 = 7$

3. $5^3 = 125$

$\log_5 125 = 3$

5. $10^3 = 1000$

$\log_{10} 1000 = 3$

7. $2^{-2} = \frac{1}{4}$

$\log_2 \frac{1}{4} = -2$

9. $10^{-1} = .1$
$\log_{10} .1 = -1$

11. $\log_3 81 = 4$
$3^4 = 81$

13. $\log_4 64 = 3$
$4^3 = 64$

15. $\log_{10} 10000 = 4$
$10^4 = 10000$

17. $\log_2\left(\frac{1}{16}\right) = -4$
$2^{-4} = \frac{1}{16}$

19. $\log_{10} .001 = -3$
$10^{-3} = .001$

21. $\log_2 16 = x$
$2^x = 16$
$2^x = 2^4$
$x = 4$

23. $\log_3 81 = x$
$3^x = 81$
$3^x = 3^4$
$x = 4$

25. $\log_6 216 = x$
$6^x = 216$
$6^x = 6^3$
$x = 3$

27. $\log_7 \sqrt{7} = x$
$7^x = \sqrt{7}$
$7^x = 7^{\frac{1}{2}}$
$x = \frac{1}{2}$

29. $\log_{10} 1 = x$
$10^x = 1$
$10^x = 10^0$
$x = 0$

31. $\log_{10} .1 = x$
$10^x = .1$
$10^x = 10^{-1}$
$x = -1$

33. $10^{\log_{10} 5} = x$
$5 = x$

35. $\log_2\left(\frac{1}{32}\right) = x$
$2^x = \frac{1}{32}$
$2^x = \frac{1}{2^5}$
$2^x = 2^{-5}$
$x = -5$

37. $\log_5 (\log_2 32) = x$
$\log_5 5 = x$
$5^x = 5$
$5^x = 5^1$
$x = 1$

39. $\log_{10} (\log_7 7) = x$
$\log_{10} 1 = x$
$10^x = 1$
$10^x = 10^0$
$x = 0$

41. $\log_7 x = 2$
$7^2 = x$
$49 = x$
The solution set is $\{49\}$.

43. $\log_8 x = \frac{4}{3}$

$8^{\frac{4}{3}} = x$

$(8^{\frac{1}{3}})^4 = x$

$(2)^4 = x$

$16 = x$

The solution set is $\{16\}$.

45. $\log_9 x = \frac{3}{2}$

$9^{\frac{3}{2}} = x$

$(9^{\frac{1}{2}})^3 = x$

$3^3 = x$

$27 = x$

The solution set is $\{27\}$.

47. $\log_4 x = -\frac{3}{2}$

$4^{-\frac{3}{2}} = x$

$\frac{1}{4^{\frac{3}{2}}} = x$

$\frac{1}{(4^{\frac{1}{2}})^3} = x$

$\frac{1}{2^3} = x$

$\frac{1}{8} = x$

The solution set is $\left\{\frac{1}{8}\right\}$.

49. $\log_x 2 = \frac{1}{2}$

$x^{\frac{1}{2}} = 2$

$(x^{\frac{1}{2}})^2 = 2^2$

$x = 4$

The solution set is $\{4\}$.

51. $\log_2 35$

$\log_2 (5 \bullet 7)$

$\log_2 5 + \log_2 7$

$2.3219 + 2.8074$

5.1293

53. $\log_2 125$

$\log_2 5^3$

$3 \log_2 5$

$3(2.3219)$

6.9657

55. $\log_2 \sqrt{7}$

$\log_2 7^{\frac{1}{2}}$

$\frac{1}{2} \log_2 7$

$\frac{1}{2}(2.8074)$

1.4037

57. $\log_2 175$

$\log_2 (5^2 \bullet 7)$

$\log_2 5^2 + \log_2 7$

$2 \log_2 5 + \log_2 7$

$2(2.3219) + 2.8074$

$4.6438 + 2.8074$

7.4512

59. $\log_2 80$

$\log_2 (16 \bullet 5)$

$\log_2 16 + \log_2 5$

$\log_2 2^4 + 2.3219$

$4 + 2.3219$

6.3219

61. $\log_8 \left(\frac{5}{11}\right)$

$\log_8 5 - \log_8 11$

$.7740 - 1.1531$

$-.3791$

63. $\log_8 \sqrt{11}$

$\log_8 11^{\frac{1}{2}}$

$\frac{1}{2}\log_8 11$

$\frac{1}{2}(1.1531)$

$.5766$

65. $\log_8 88$

$\log_8 (8 \bullet 11)$

$\log_8 8 + \log_8 11$

$1 + 1.1531$

2.1531

67. $\log_8 \left(\frac{25}{11}\right)$

$\log_8 25 - \log_8 11$

$\log_8 5^2 - 1.1531$

$2 \log_8 5 - 1.1531$

$2(.7740) - 1.1531$

$1.5480 - 1.1531$

$.3949$

69. $\log_b xyz$

$\log_b x + \log_b y + \log_b z$

71. $\log_b \left(\frac{y}{z}\right)$

$\log_b y - \log_b z$

73. $\log_b y^3 z^4$

$\log_b y^3 + \log_b z^4$

$3 \log_b y + 4 \log_b z$

75. $\log_b \left(\frac{x^{\frac{1}{2}} y^{\frac{1}{3}}}{z^4}\right)$

$\log_b x^{\frac{1}{2}} + \log_b y^{\frac{1}{3}} - \log_b z^4$

$\frac{1}{2}\log_b x + \frac{1}{3}\log_b y - 4 \log_b z$

77. $\log_b \sqrt[3]{x^2 z}$

$\log_b x^{\frac{2}{3}} z^{\frac{1}{3}}$

$\log_b x^{\frac{2}{3}} + \log_b z^{\frac{1}{3}}$

$\frac{2}{3}\log_b x + \frac{1}{3}\log_b z$

79. $\log_b \left(x\sqrt{\frac{x}{y}}\right)$

$\log_b \frac{x \bullet x^{\frac{1}{2}}}{y^{\frac{1}{2}}}$

$\log_b \frac{x^{\frac{3}{2}}}{y^{\frac{1}{2}}}$

$\log_b x^{\frac{3}{2}} - \log_b y^{\frac{1}{2}}$

$\frac{3}{2}\log_b x - \frac{1}{2}\log_b y$

81. $2 \log_b x - 4 \log_b y$

$\log_b x^2 - \log_b y^4$

$\log_b \left(\frac{x^2}{y^4}\right)$

83. $\log_b x - (\log_b y - \log_b z)$

$\log_b x - \log_b y + \log_b z$

$\log_b x + \log_b z - \log_b y$

$\log_b (xz) - \log_b y$

$\log_b \left(\frac{xz}{y}\right)$

85. $2 \log_b x + 4 \log_b y - 3 \log_b z$

$\log_b x^2 + \log_b y^4 - \log_b z^3$

$\log_b (x^2 y^4) - \log_b z^3$

$\log_b \left(\frac{x^2 y^4}{z^3}\right)$

87. $\frac{1}{2}\log_b x - \log_b x + 4 \log_b y$

$\log_b \sqrt{x} - \log_b x + \log_b y^4$

$\log_b \sqrt{x} + \log_b y^4 - \log_b x$

$\log_b (y^4\sqrt{x}) - \log_b x$

$\log_b \left(\frac{y^4\sqrt{x}}{x}\right)$

89. $\log_3 x + \log_3 4 = 2$

$\log_3 4x = 2$

$3^2 = 4x$

$9 = 4x$

$\frac{9}{4} = x$

The solution set is $\left\{\frac{9}{4}\right\}$.

91. $\log_{10} x + \log_{10} (x-21) = 2$

$\log_{10} x(x-21) = 2$

$10^2 = x(x-21)$

$100 = x^2 - 21x$

$x^2 - 21x - 100 = 0$

$(x-25)(x+4) = 0$

$x - 25 = 0$ or $x + 4 = 0$

$x = 25$ or $x = -4$

Discard the root $x = -4$.
The solution set is $\{25\}$.

93. $\log_2 x + \log_2 (x-3) = 2$

$\log_2 x(x-3) = 2$

$2^2 = x(x-3)$

$4 = x^2 - 3x$

$x^2 - 3x - 4 = 0$

$(x-4)(x+1) = 0$

$x - 4 = 0$ or $x + 1 = 0$

$x = 4$ or $x = -1$

Discard the root $x = -1$.
The solution set is $\{4\}$.

95. $\log_{10} (2x-1) - \log_{10} (x-2) = 1$

$\log_{10}\left(\frac{2x-1}{x-2}\right) = 1$

$10^1 = \frac{2x-1}{x-2};\ x \neq 2$

$(x-2)(10) = (x-2)\left(\frac{2x-1}{x-2}\right)$

$10x - 20 = 2x - 1$

$8x = 19$

$x = \frac{19}{8}$

The solution set is $\left\{\frac{19}{8}\right\}$.

97. $\log_5 (3x-2) = 1 + \log_5 (x-4)$

$\log_5 (3x-2) - \log_5 (x-4) = 1$

$\log_5 \left(\frac{3x-2}{x-4}\right) = 1$

$5^1 = \frac{3x-2}{x-4};\ x \neq 4$

$5(x-4) = 3x - 2$

$5x - 20 = 3x - 2$

$2x = 18$

$x = 9$

The solution set is $\{9\}$.

99. $\log_8 (x+7) + \log_8 x = 1$

$\log_8 (x+7)(x) = 1$

$8^1 = (x+7)(x)$

$8 = x^2 + 7x$

$x^2 + 7x - 8 = 0$

$(x+8)(x-1) = 0$

$x + 8 = 0$ or $x - 1 = 0$

$x = -8$ or $x = 1$

Discard the root $x = -8$.
The solution set is $\{1\}$.

101. $\log_b \left(\frac{r}{s}\right) = \log_b r - \log_b s$

Let $m = \log_b r$ and $n = \log_b s$.

$m = \log_b r$ becomes $b^m = r$.

$n = \log_b s$ becomes $b^n = s$.

Then the quotient $\frac{r}{s}$ becomes

$\frac{r}{s} = \frac{b^m}{b^n} = b^{m-n}$.

Changing $\frac{r}{s} = b^{m-n}$ to logarithmic form

$\log_b\left(\frac{r}{s}\right) = m - n$

$\log_b\left(\frac{r}{s}\right) = \log_b r - \log_b s$

Problem Set 10.4 Logarithmic Functions

1. $\log 7.24$
 .8597

3. $\log 52.23$
 1.7179

5. $\log 3214.1$
 3.5071

7. $\log .729$
 −.1373

9. $\log 0.00034$
 −3.4685

11. $\log x = 2.6143$
 $x = 411.43$

13. $\log x = 4.9547$
 $x = 90095$

15. $\log x = 1.9006$
 $x = 79.543$

17. $\log x = -1.3148$
 $x = 0.048440$

19. $\log x = -2.1928$
 $x = 0.0064150$

21. $\ln 5$
 1.6094

23. $\ln 32.6$
 3.4843

25. $\ln 430$
 6.0638

27. $\ln .46$
 −.7765

29. $\ln .0314$
 −3.4609

31. $\ln x = .4721$
 1.6034

33. $\ln x = 1.1425$
 3.1346

35. $\ln x = 4.6873$
 108.56

37. $\ln x = -.7284$
 .48268

39. $\ln x = -3.3244$
 .035994

41a.

x	0.1	0.5	1	2	4	8	10
log x	−1.0	−.3	0	.3	.6	.9	1.0

b.

x	−1	−.3	0	.3	.6	.9	1
10^x	.1	.5	1.0	2.0	4.0	8.0	10.0

43 - 53. See back of textbook for graphs.

55. $\frac{\ln 2}{\ln 7} = \frac{.6931}{1.946} = .36$

57. $\frac{\ln 5}{2\ln 3} = \frac{1.6094}{2(1.0986)} = \frac{1.6094}{2.1972} = .73$

59. $\frac{\ln 2}{.03} = \frac{.6931}{.03} = 23.10$

61. $\frac{\log 5}{3\log 1.07} = \frac{.6990}{3(.0294)} = \frac{.6990}{.0882} = 7.93$

Problem Set 10.5 Exponential Equations, Logarithmic Equations, and Problem Solving

1. $3^x = 13$
$\log 3^x = \log 13$
$x \log 3 = \log 13$
$x(.4771) = 1.1139$
$x = 2.33$
The solution set is {2.33}.

3. $4^n = 35$
$\log 4^n = \log 35$
$n \log 4 = \log 35$
$n(.6021) = 1.5441$
$n = 2.56$

5. $2^x + 7 = 50$
$2^x = 43$
$\log 2^x = \log 43$
$x \log 2 = \log 43$
$x(.3010) = 1.6335$
$x = 5.43$

7. $3^{x-2} = 11$
$\log 3^{x-2} = \log 11$
$(x-2)(\log 3) = \log 11$
$(x-2)(.4771) = 1.0414$
$.4771x - .9542 = 1.0414$
$.4771x = 1.9956$
$x = 4.18$

9. $5^{3t+1} = 9$
$\log 5^{3t+1} = \log 9$
$(3t+1)(\log 5) = \log 9$
$(3t+1)(.6990) = .9542$
$2.097t + .6990 = .9542$
$2.097t = .2552$
$t = .12$

11. $e^x = 27$
$\ln e^x = \ln 27$
$x(\ln e) = \ln 27$
$x(1) = \ln 27$
$x = \ln 27$
$x = 3.30$

13. $e^{x-2} = 13.1$
$\ln e^{x-2} = \ln 13.1$
$x - 2 = 2.57$
$x = 4.57$

15. $3e^x - 1 = 17$
$3e^x = 18$
$e^x = 6$
$\ln e^x = \ln 6$
$x = 1.79$

17. $5^{2x+1} = 7^{x+3}$
$\log 5^{2x+1} = \log 7^{x+3}$
$(2x+1)(\log 5) = (x+3)(\log 7)$
$(2x+1)(.6990) = (x+3)(.8451)$
$1.398x + .6990 = .8451x + 2.5353$
$.5529x = 1.8363$
$x = 3.32$

19. $3^{2x+1} = 2^{3x+2}$
$\log 3^{2x+1} = \log 2^{3x+2}$
$(2x+1)(\log 3) = (3x+2)(\log 2)$
$(2x+1)(.4771) = (3x+2)(.3010)$
$.9542x + .4771 = .9030x + .6020$
$.0512x = .1249$
$x = 2.44$

21. $\log x + \log(x+21) = 2$
$\log x(x+21) = 2$
$10^2 = x(x+21)$

$100 = x^2 + 21x$

$0 = x^2 + 21x - 100$

$0 = (x+25)(x-4)$

$x + 25 = 0$ or $x - 4 = 0$

$x = -25$ or $x = 4$

Discard the root $x = -25$.

The solution set is $\{4\}$.

23. $\log(3x-1) = 1 + \log(5x-2)$

$\log(3x-1) - \log(5x-2) = 1$

$\log\frac{3x-1}{5x-2} = 1$

$10^1 = \frac{3x-1}{5x-2}$

$10(5x-2) = 3x - 1$

$50x - 20 = 3x - 1$

$47x = 19$

$x = \frac{19}{47}$

The solution set is $\left\{\frac{19}{47}\right\}$.

25. $\log(x+1) = \log 3 - \log(2x-1)$

$\log(x+1) + \log(2x-1) = \log 3$

$\log(x+1)(2x-1) = \log 3$

$(x+1)(2x-1) = 3$

$2x^2 + x - 1 = 3$

$2x^2 + x - 4 = 0$

$$x = \frac{-1 \pm \sqrt{1^2 - 4(2)(-4)}}{2(2)}$$

$$x = \frac{-1 \pm \sqrt{33}}{4}$$

Discard the root $x = \frac{-1-\sqrt{33}}{4}$.

The solution set is $\left\{\frac{-1+\sqrt{33}}{4}\right\}$.

27. $\log(x+2) - \log(2x+1) = \log x$

$\log(x+2) = \log x + \log(2x+1)$

$\log(x+2) = \log x(2x+1)$

$x + 2 = x(2x+1)$

$x + 2 = 2x^2 + x$

$2 = 2x^2$

$x^2 = 1$

$x = \pm\sqrt{1} = \pm 1$

Discard the root $x = -1$.

The solution set is $\{1\}$.

29. $\ln(2t+5) = \ln 3 + \ln(t-1)$

$\ln(2t+5) = \ln 3(t-1)$

$2t + 5 = 3(t-1)$

$2t + 5 = 3t - 3$

$8 = t$

The solution set is $\{8\}$.

31. $\log\sqrt{x} = \sqrt{\log x}$

$\log x^{\frac{1}{2}} = (\log x)^{\frac{1}{2}}$

$\frac{1}{2}(\log x) = (\log x)^{\frac{1}{2}}$

$2\left[\frac{1}{2}(\log x)\right] = 2(\log x)^{\frac{1}{2}}$

$\log x = 2(\log x)^{\frac{1}{2}}$

$\log x - 2(\log x)^{\frac{1}{2}} = 0$

let $u = (\log x)^{\frac{1}{2}}$

$u^2 - 2u = 0$

$u(u-2) = 0$

$u = 0$ or $u - 2 = 0$

$u = 0$ or $u = 2$

$(\log x)^{\frac{1}{2}} = 0$ or $(\log x)^{\frac{1}{2}} = 2$

$[(\log x)^{\frac{1}{2}}]^2 = 0^2$ or $[(\log x)^{\frac{1}{2}}]^2 = 2^2$

$\log x = 0$ or $\log x = 4$

$10^0 = x$ or $10^4 = x$

$1 = x$ or $10{,}000 = x$

The solution set is $\{1, 10000\}$.

33. $\log_2 40 = \frac{\log 40}{\log 2} = \frac{1.6020}{.3010} = 5.322$

35. $\log_3 16 = \frac{\log 16}{\log 3} = \frac{1.2041}{.4771} = 2.524$

37. $\log_4 1.6 = \frac{\log 1.6}{\log 4} = \frac{.2041}{.6021} = .339$

39. $\log_5 .26 = \frac{\log .26}{\log 5} = \frac{-.5850}{.6990} = -.837$

41. $\log_7 500 = \frac{\log 500}{\log 7} = \frac{2.6990}{.8451} = 3.194$

43. $A = P(1 + \frac{r}{n})^{nt}$

$1000 = 750\left(1 + \frac{.12}{4}\right)^{4t}$

$1.3333 = (1.03)^{4t}$

$\log 1.3333 = \log (1.03)^{4t}$

$.1249 = 4t \log (1.03)$

$.1249 = 4t(.0128)$

$.1249 = .0512t$

$2.4 = t$

It will take 2.4 years.

45. $A = Pe^{rt}$

$4000 = 2000e^{.13t}$

$2 = e^{.13t}$

$\ln 2 = \ln e^{.13t}$

$.6931 = .13t$

$5.3 = t$

It will take 5.3 years.

47. $A = Pe^{rt}$

$900 = 500e^{r(10)}$

$1.8 = e^{10r}$

$\ln 1.8 = \ln e^{10r}$

$.5878 = 10r$

$.05878 = r$

$5.9\% = r$

The interest rate would need to be 5.9%.

49. $Q = Q_0e^{.34t}$

$4000 = 400e^{.34t}$

$10 = e^{.34t}$

$\ln 10 = \ln e^{.34t}$

$2.3026 = .34t$

$6.8 = t$

It will take 6.8 hours.

51. $P(a) = 14.7e^{-.21a}$

$11.53 = 14.7e^{-.21a}$

$.7844 = e^{-.21a}$

$\ln .7844 = \ln e^{-.21a}$

$-.2428 = -.21a$

$1.156 = a$

$1.156 \text{ miles} \times \frac{5280 \text{ feet}}{1 \text{ mile}} = 6104 \text{ feet}$

Cheyenne is approximately 6100 feet above sea level.

53. $Q(t) = Q_0e^{.4t}$

$2000 = 500e^{.4t}$

$4 = e^{.4t}$

$\ln 4 = \ln e^{.4t}$

$1.3863 = .4t$

$3.5 = t$

It will take 3.5 hours.

55. $R = \log \frac{I}{I_0}$

$R = \log \frac{5{,}000{,}000I_0}{I_0}$

$R = \log 5{,}000{,}000$

$R = 6.7$

57. For $R = 7.3$

$I = (10^{7.3})I_0$

For $R = 6.4$

$I = (10^{6.4})I_0$

$\frac{(10^{7.3})I_0}{(10^{6.4})I_0} = 10^{7.3-6.4} = 10^{.9}$

$10^{.9} = 7.9$

It is approximately
8 times more intense.

Chapter 10 Review

1. $8^{\frac{5}{3}}$

$(8^{\frac{1}{3}})^5 = 2^5 = 32$

3. $(-27)^{\frac{4}{3}}$

$[(-27)^{\frac{1}{3}}]^4 = (-3)^4 = 81$

5. $\log_7\left(\frac{1}{49}\right) = x$

$7^x = \frac{1}{49}$

$7^x = \frac{1}{7^2}$

$7^x = 7^{-2}$

$x = -2$

7. $\log_2\left(\frac{\sqrt[4]{32}}{2}\right) = x$

$2^x = \frac{\sqrt[4]{32}}{2}$

$2^x = \frac{\sqrt[4]{2^5}}{2}$

$2^x = \frac{2^{\frac{5}{4}}}{2^1}$

$2^x = 2^{\frac{5}{4}-1}$

$2^x = 2^{\frac{1}{4}}$

$x = \frac{1}{4}$

9. $\ln e = x$

$e^x = e$

$x = 1$

11. $\log_{10} 2 + \log_{10} x = 1$

$\log_{10} 2x = 1$

$10^1 = 2x$

$5 = x$

The solution set is $\{5\}$.

13. $4^x = 128$

$(2^2)^x = 2^7$

$2^{2x} = 2^7$

$2x = 7$

$x = \frac{7}{2}$

The solution set is $\left\{\frac{7}{2}\right\}$.

15. $\log_2 x = 3$

$2^3 = x$

$8 = x$

The solution set is $\{8\}$.

17. $2e^x = 14$

$e^x = 7$

$\ln e^x = \ln 7$

$x = 1.95$

The solution set is $\{1.95\}$.

19. $\ln(x+4) - \ln(x+2) = \ln x$

$\ln(x+4) = \ln x + \ln(x+2)$

$\ln(x+4) = \ln x(x+2)$

$x+4 = x(x+2)$

$x+4 = x^2+2x$

$0 = x^2+x-4$

$x = \frac{-1 \pm \sqrt{1^2-4(1)(-4)}}{2(1)}$

$x = \frac{-1 \pm \sqrt{17}}{2}$

$x = \frac{-1+\sqrt{17}}{2}$ or $x = \frac{-1-\sqrt{17}}{2}$

$x = 1.56$ or $x = -2.56$

Discard the root $x = -2.56$.

The solution set is $\{1.56\}$.

21. $\log(\log x) = 2$

$10^2 = \log x$

$100 = \log x$

$10^{100} = x$

The solution set is $\{10^{100}\}$.

23. $\ln(2t-1) = \ln 4 + \ln(t-3)$

$\ln(2t-1) = \ln 4(t-3)$

$2t - 1 = 4(t-3)$

$2t - 1 = 4t - 12$

$11 = 2t$

$\frac{11}{2} = t$

The solution set is $\left\{\frac{11}{2}\right\}$.

25. $\log\left(\frac{7}{3}\right)$

$\log 7 - \log 3$

$0.8451 - 0.4771$

$.3680$

27. $\log 27$

$\log 3^3$

$3 \log 3$

$3(0.4771)$

1.4313

29a. $\log_b\left(\frac{x}{y^2}\right)$

$\log_b x - \log_b y^2$

$\log_b x - 2\log_b y$

b. $\log_b \sqrt[4]{xy^2}$

$\log_b (x^{\frac{1}{4}} y^{\frac{2}{4}})$

$\log_b x^{\frac{1}{4}} + \log_b y^{\frac{2}{4}}$

$\log_b x^{\frac{1}{4}} + \log_b y^{\frac{1}{2}}$

$\frac{1}{4}\log_b x + \frac{1}{2}\log_b y$

c. $\log_b\left(\frac{\sqrt{x}}{y^3}\right)$

$\log_b \sqrt{x} - \log_b y^3$

$\log_b x^{\frac{1}{2}} - \log_b y^3$

$\frac{1}{2}\log_b x - 3\log_b y$

31. $\log_2 3 = \frac{\log 3}{\log 2} = \frac{.47712}{.30103} = 1.58$

33. $\log_4 191 = \frac{\log 191}{\log 4} = \frac{2.28103}{.60206} = 3.79$

35 - 41. See back of textbook for graphs.

43. $A = P(1 + \frac{r}{n})^{nt}$

$A = 750\left(1 + \frac{.11}{4}\right)^{4(10)}$

$A = 750(1 + .0275)^{40}$

$A = 750(1.0275)^{40}$

$A = 750(2.959874)$

$A = \$2219.91$

45. $A = P(1 + \frac{r}{n})^{nt}$

$A = 2500\left(1 + \frac{.095}{2}\right)^{2(20)}$

$A = 2500(1 + .0475)^{40}$

$A = 2500(1.0475)^{40}$

$A = 2500(6.399724)$

$A = \$15.999.31$

47. $A = P(1 + \frac{r}{n})^{nt}$

$3500 = 1000\left(1 + \frac{.105}{4}\right)^{4t}$

$3.5 = (1 + .02625)^{4t}$

$3.5 = (1.02625)^{4t}$

$\log 3.5 = \log (1.02625)^{4t}$

$.544068 = 4t \log(1.02625)$

$.544068 = 4t(.011253)$

$.544068 = .045012t$

$12.1 = t$

It will take 12.1 years.

49. $P(t) = P_0 e^{.02t}$

$P(10) = 50000e^{.02(10)}$

$P(10) = 50000e^{.2}$

$P(10) = 50000(1.2214)$

$P(10) = 61{,}070$

$P(15) = 50000e^{.02(15)}$

$P(15) = 50000e^{.3}$

$P(15) = 50000(1.34986)$

$P(15) = 67{,}493$

$P(20) = 50000e^{.02(20)}$

$P(20) = 50000e^{.4}$

$P(20) = 50000(1.49182)$

$P(20) = 74{,}591$

51. $Q = Q_0\left(\frac{1}{2}\right)^{\frac{t}{n}}$

$Q = 750\left(\frac{1}{2}\right)^{\frac{100}{40}}$

$Q = 750(.5)^{2.5}$

$Q = 750(.1767767)$

$Q = 133$

There will be 133 grams left after 100 days.

Chapter 10 Test

1. $\log_3 \sqrt{3} = x$

$3^x = \sqrt{3}$

$3^x = 3^{\frac{1}{2}}$

$x = \frac{1}{2}$

3. $-2 + \ln e^3$

$-2 + 3$

1

5. $4^x = \frac{1}{64}$

$4^x = \frac{1}{4^3}$

$4^x = 4^{-3}$

$x = -3$

The solution set is $\{-3\}$.

7. $2^{3x-1} = 128$

$2^{3x-1} = 2^7$

$3x - 1 = 7$

$3x = 8$

$x = \frac{8}{3}$

The solution set is $\left\{\frac{8}{3}\right\}$.

9. $\log x + \log (x+48) = 2$

$\log x(x+48) = 2$

$10^2 = x(x+48)$

$100 = x^2 + 48x$

$0 = x^2 + 48x - 100$

$0 = (x+50)(x-2)$

$x + 50 = 0$ or $x - 2 = 0$

$x = -50$ or $x = 2$

Discard the root $x = -50$.
The solution set is $\{2\}$.

11. $\log_3 100$

$\log_3 (5^2 \bullet 4)$

$\log_3 5^2 + \log_3 4$

$2 \log_3 5 + \log_3 4$

$2(1.4650) + 1.2619$

$2.9300 + 1.2619$

4.1919

13. $\log_3 \sqrt{5}$

$\log_3 5^{\frac{1}{2}}$

$\frac{1}{2}\log_3 5$

$\frac{1}{2}(1.4650)$

$.7325$

15. $e^x = 176$

$\ln e^x = \ln 176$

$x = 5.17$

The solution set is $\{5.17\}$.

17. $\log_5 632 = \frac{\log 632}{\log 5} = \frac{2.800717}{.698970} = 4.0069$

19. $A = P\left(1 + \frac{r}{n}\right)^{nt}$

$A = 3500\left(1 + \frac{.075}{4}\right)^{4(8)}$

$A = 3500(1.01875)^{32}$

$A = 3500(1.812024)$

$A = \$6342.08$

21. $Q(t) = Q_0 e^{.23t}$

$2400 = 400e^{.23t}$

$6 = e^{.23t}$

$\ln 6 = \ln e^{.23t}$

$1.79176 = .23t$

$7.8 = t$

It will take 7.8 hours.

23 - 25. See back of textbook for graphs.

Chapter 11 Systems of Equations and Inequalities

Problem Set 11.1 Systems of Two Linear Equations in Two Variables

1. $\begin{pmatrix} x-y=1 \\ 2x+y=8 \end{pmatrix}$

$\{(3, 2)\}$

Checking 1st equation

$3-2 \stackrel{?}{=} 1$

$1=1$

Checking 2nd equation

$2(3)+2 \stackrel{?}{=} 8$

$8=8$

3. $\begin{pmatrix} 4x+3y=-5 \\ 2x-3y=-7 \end{pmatrix}$

$\{(-2, 1)\}$

Checking 1st equation

$4(-2)+3(1) \stackrel{?}{=} -5$

$-8+3 \stackrel{?}{=} -5$

$-5=-5$

Checking 2nd equation

$2(-2)-3(1) \stackrel{?}{=} -7$

$-4-3 \stackrel{?}{=} -7$

$-7=-7$

5. $\begin{pmatrix} \frac{1}{2}x+\frac{1}{4}y=9 \\ 4x+2y=72 \end{pmatrix}$

Dependent

7. $\begin{pmatrix} \frac{1}{2}x-\frac{1}{3}y=3 \\ x+4y=-8 \end{pmatrix}$

$\{(4, -3)\}$

Checking 1st equation

$\frac{1}{2}(4)-\frac{1}{3}(-3) \stackrel{?}{=} 3$

$2+1 \stackrel{?}{=} 3$

$3=3$

Checking 2nd equation

$4+4(-3) \stackrel{?}{=} -8$

$4-12 \stackrel{?}{=} -8$

$-8=-8$

9. $\begin{pmatrix} x-\frac{1}{2}y=-4 \\ 8x-4y=-1 \end{pmatrix}$

Inconsistent

11. $\begin{pmatrix} x+y=20 \\ x=y-4 \end{pmatrix}$

$x+y=20$

$(y-4)+y=20$

$2y-4=20$

$2y=24$

$y=12$

$x=y-4$

$x=12-4$

$x=8$

The solution set is $\{(8, 12)\}$.

13. $\begin{pmatrix} y=-3x-18 \\ 5x-2y=-8 \end{pmatrix}$

$5x-2y=-8$

$5x-2(-3x-18)=-8$

$5x+6x+36=-8$

$11x+36=-8$

$11x=-44$

$x=-4$

$y=-3x-18$

$y=-3(-4)-18$

$y=12-18$

$y=-6$

The solution set is $\{(-4, -6)\}$.

15. $\begin{pmatrix} x=-3y \\ 7x-2y=-69 \end{pmatrix}$

$7x-2y=-69$

$7(-3y)-2y=-69$

$-21y - 2y = -69$

$-23y = -69$

$y = 3$

$x = -3y$

$x = -3(3)$

$x = -9$

The solution set is $\{(-9, 3)\}$.

17. $\begin{pmatrix} 2x + 3y = 11 \\ 3x - 2y = -3 \end{pmatrix}$

Solve equation 2 for x.

$3x - 2y = -3$

$3x = 2y - 3$

$x = \frac{2}{3}y - 1$

Now substitute in equation 1.

$2x + 3y = 11$

$2\left(\frac{2}{3}y - 1\right) + 3y = 11$

$\frac{4}{3}y - 2 + 3y = 11$

$\frac{13}{3}y - 2 = 11$

$\frac{13}{3}y = 13$

$\frac{3}{13}\left(\frac{13}{3}y\right) = \frac{3}{13}(13)$

$y = 3$

$x = \frac{2}{3}y - 1$

$x = \frac{2}{3}(3) - 1$

$x = 2 - 1$

$x = 1$

The solution set is $\{(1, 3)\}$.

19. $\begin{pmatrix} 3x - 4y = 9 \\ x = 4y - 1 \end{pmatrix}$

$3x - 4y = 9$

$3(4y - 1) - 4y = 9$

$12y - 3 - 4y = 9$

$8y - 3 = 9$

$8y = 12$

$y = \frac{12}{8} = \frac{3}{2}$

$x = 4y - 1$

$x = 4\left(\frac{3}{2}\right) - 1$

$x = 6 - 1$

$x = 5$

The solution set is $\left\{\left(5, \frac{3}{2}\right)\right\}$.

21. $\begin{pmatrix} y = \frac{2}{5}x - 1 \\ 3x + 5y = 4 \end{pmatrix}$

$3x + 5y = 4$

$3x + 5\left(\frac{2}{5}x - 1\right) = 4$

$3x + 2x - 5 = 4$

$5x - 5 = 4$

$5x = 9$

$x = \frac{9}{5}$

$y = \frac{2}{5}x - 1$

$y = \frac{2}{5}\left(\frac{9}{5}\right) - 1$

$y = \frac{18}{25} - 1$

$y = -\frac{7}{25}$

The solution set is $\left\{\left(\frac{9}{5}, -\frac{7}{25}\right)\right\}$.

23. $\begin{pmatrix} 7x - 3y = -2 \\ x = \frac{3}{4}y + 1 \end{pmatrix}$

$7x - 3y = -2$

$7\left(\frac{3}{4}y + 1\right) - 3y = -2$

$\frac{21}{4}y + 7 - 3y = -2$

$\frac{9}{4}y + 7 = -2$

$\frac{9}{4}y = -9$

$\frac{4}{9}\left(\frac{9}{4}y\right) = \frac{4}{9}(-9)$

$y = -4$

$x = \frac{3}{4}y + 1$

$x = \frac{3}{4}(-4) + 1$

$x = -3 + 1 = -2$

The solution set is $\{(-2, -4)\}$.

25. $\begin{pmatrix} 2x + y = 12 \\ 3x - y = 13 \end{pmatrix}$

Solve equation 1 for y.

$2x + y = 12$

$y = -2x + 12$

Substitute into equation 2.

$3x - y = 13$

$3x - (-2x + 12) = 13$

$3x + 2x - 12 = 13$

$5x - 12 = 13$

$5x = 25$

$x = 5$

$y = -2x + 12$

$y = -2(5) + 12$

$y = -10 + 12$

$y = 2$

The solution set is $\{(5, 2)\}$.

27. $\begin{pmatrix} 4x + 3y = -40 \\ 5x - y = -12 \end{pmatrix}$

Solve equation 2 for y.

$5x - y = -12$

$-y = -5x - 12$

$y = 5x + 12$

Substitute into equation 1.

$4x + 3y = -40$

$4x + 3(5x + 12) = -40$

$4x + 15x + 36 = -40$

$19x + 36 = -40$

$19x = -76$

$x = -4$

$y = 5x + 12$

$y = 5(-4) + 12$

$y = -20 + 12$

$y = -8$

The solution set is $\{(-4, -8)\}$.

29. $\begin{pmatrix} .06x + .07y = 86 \\ y = x + 300 \end{pmatrix}$

$.06x + .07y = 86$

$.06x + .07(x + 300) = 86$

$.06x + .07x + 21 = 86$

$.13x + 21 = 86$

$.13x = 65$

$x = \frac{65}{.13} = 500$

$y = x + 300$

$y = 500 + 300$

$y = 800$

The solution set is $\{(500, 800)\}$.

31. $\begin{pmatrix} 3x + 5y = 22 \\ 4x - 7y = -39 \end{pmatrix}$

Solve equation 2 for x.

$4x - 7y = -39$

$4x = 7y - 39$

$x = \frac{7}{4}y - \frac{39}{4}$

Substitute into equation 1.

$3x + 5y = 22$

$3\left(\frac{7}{4}y - \frac{39}{4}\right) + 5y = 22$

$\frac{21}{4}y - \frac{117}{4} + 5y = 22$

$\frac{41}{4}y - \frac{117}{4} = 22$

$\frac{41}{4}y = 22 + \frac{117}{4}$

$\frac{41}{4}y = \frac{205}{4}$

$\frac{4}{41}\left(\frac{41}{4}y\right) = \frac{4}{41}\left(\frac{205}{4}\right)$

$y = 5$

$x = \frac{7}{4}y - \frac{39}{4}$

$x = \frac{7}{4}(5) - \frac{39}{4}$

$x = \frac{35}{4} - \frac{39}{4}$

$x = -\frac{4}{4} = -1$

The solution set is $\{(-1, 5)\}$.

33. $\begin{pmatrix} 4x - 5y = 3 \\ 8x + 15y = -24 \end{pmatrix}$

Solve equation 1 for x.

$4x - 5y = 3$

$4x = 5y + 3$

$x = \frac{5}{4}y + \frac{3}{4}$

Substitute into equation 2.

$8x + 15y = -24$

$8\left(\frac{5}{4}y + \frac{3}{4}\right) + 15y = -24$

$10y + 6 + 15y = -24$

$25y + 6 = -24$

$25y = -30$

$y = -\frac{30}{25} = -\frac{6}{5}$

$x = \frac{5}{4}y + \frac{3}{4}$

$x = \frac{5}{4}\left(-\frac{6}{5}\right) + \frac{3}{4}$

$x = -\frac{6}{4} + \frac{3}{4}$

$x = -\frac{3}{4}$

The solution set is $\left\{\left(-\frac{3}{4}, -\frac{6}{5}\right)\right\}$.

35. $\begin{pmatrix} x + y = 750 \\ .06x + .07y = 50 \end{pmatrix}$

$x + y = 750$

$x = 750 - y$

$.06x + .07y = 50$

$.06(750 - y) + .07y = 50$

$45 - .06y + .07y = 50$

$45 + .01y = 50$

$.01y = 5$

$y = \frac{5}{.01} = 500$

$x = 750 - y$

$x = 750 - 500$

$x = 250$

The solution set is $\{(250, 500)\}$.

37. Let $x =$ money at 7%.
$y =$ money at 8%.

$\begin{pmatrix} x + 6000 = y \\ .07x + .08y = 780 \end{pmatrix}$

$.07x + .08y = 780$

$.07x + .08(x + 6000) = 780$

$100[.07x + .08(x + 6000)] = 100(780)$

$7x + 8(x + 6000) = 78000$

$7x + 8x + 48000 = 78000$

$15x = 30000$

$x = 2000$

$y = x + 6000$

$y = 2000 + 6000$

$y = 8000$

She invested \$2000 at 7% and \$8000 at 8%.

39. Let $x =$ tens place digit
$y =$ ones place digit

$\begin{pmatrix} x + y = 11 \\ x = 1 + 4y \end{pmatrix}$

$x + y = 11$

$(1 + 4y) + y = 11$

$5y + 1 = 11$

$5y = 10$

$y = 2$

$x = 1 + 4y$

$x = 1 + 4(2)$

$x = 9$

The number is 92.

41. Let $x =$ one number
$y =$ other number

$\begin{pmatrix} x + y = 131 \\ x = 3y - 5 \end{pmatrix}$

$x + y = 131$

$(3y - 5) + y = 131$

$4y - 5 = 131$

$4y = 136$

$y = 34$

$x = 3y - 5$

$x = 3(34) - 5$

$x = 102 - 5$

$x = 97$

The numbers are 34 and 97.

43. Let $x =$ number of females
$y =$ number of males

$\begin{pmatrix} x + y = 50 \\ x = 5y + 2 \end{pmatrix}$

$x + y = 50$

$(5y + 2) + y = 50$

$6y + 2 = 50$

$6y = 48$

$y = 8$

$x = 5y + 2$

$x = 5(8) + 2$

$x = 42$

There are 42 females.

45. Let $x =$ width
$y =$ length

$\begin{pmatrix} 2x + 2y = 94 \\ y = x + 7 \end{pmatrix}$

$2x + 2y = 94$

$2x + 2(x + 7) = 94$

$2x + 2x + 14 = 94$

$4x = 80$

$x = 20$

$y = x + 7$

$y = 20 + 7$

$y = 27$

The length is 27 inches and the width is 20 inches.

47. Let $x =$ number of \$5 bills
$y =$ number of \$10 bills

$\begin{pmatrix} x + y = 100 \\ 5x + 10y = 700 \end{pmatrix}$

Solve equation 1 for x.

$x + y = 100$

$x = 100 - y$

Substitute into equation 2.

$5x + 10y = 700$

$5(100 - y) + 10y = 700$

$500 - 5y + 10y = 700$

$5y = 200$

$y = 40$

$x = 100 - y$

$x = 100 - 40$

$x = 60$

The are 60 five-dollar bills and 40 ten-dollar bills.

49. Let $x =$ number of student tickets
$y =$ number of non-student tickets

$\begin{pmatrix} x + y = 3000 \\ 3x + 5y = 10000 \end{pmatrix}$

Solve equation 1 for x.

$x + y = 3000$

$x = 3000 - y$

Substitute into equation 2.

$3x + 5y = 10000$

$3(3000 - y) + 5y = 10000$

$9000 - 3y + 5y = 10000$

$2y = 1000$

$y = 500$

$x = 3000 - y$

$x = 3000 - 500$

$x = 2500$

There were 2500 student tickets sold and 500 non-student tickets sold.

Problem Set 11.2 Elimination-by-Addition Method

1. $\begin{pmatrix} 2x + 3y = -1 \\ 5x - 3y = 29 \end{pmatrix}$

$$\begin{array}{r} 2x + 3y = -1 \\ 5x - 3y = 29 \\ \hline 7x = 28 \end{array}$$

$x = 4$

$2x + 3y = -1$

$2(4) + 3y = -1$

$8 + 3y = -1$

$3y = -9$

$y = -3$

The solution set is $\{(4, -3)\}$.

3. $\begin{pmatrix} 6x - 7y = 15 \\ 6x + 5y = -21 \end{pmatrix}$

Multiply equation 2 by -1, then add to equation 1.

$$\begin{array}{r} 6x - 7y = 15 \\ -6x - 5y = 21 \\ \hline -12y = 36 \end{array}$$

$y = -3$

$6x - 7y = 15$

$6x - 7(-3) = 15$

$6x + 21 = 15$

$6x = -6$

$x = -1$

The solution set is $\{(-1, -3)\}$.

5. $\begin{pmatrix} x - 2y = -12 \\ 2x + 9y = 2 \end{pmatrix}$

Multiply equation 1 by -2 and add to equation 2.

$$\begin{array}{r} -2x + 4y = 24 \\ 2x + 9y = 2 \\ \hline 13y = 26 \end{array}$$

$y = 2$

$x - 2y = -12$

$x - 2(2) = -12$

$x - 4 = -12$

$x = -8$

The solution set is $\{(-8, 2)\}$.

7. $\begin{pmatrix} 4x + 7y = -16 \\ 6x - y = -24 \end{pmatrix}$

Multiply equation 2 by 7 and add to equation 1.

$$\begin{array}{r} 4x + 7y = -16 \\ 42x - 7y = -168 \\ \hline 46x \quad = -184 \end{array}$$

$x = -4$

$4x + 7y = -16$

$4(-4) + 7y = -16$

$-16 + 7y = -16$

$7y = 0$

$y = 0$

The solution set is $\{(-4, 0)\}$.

9. $\begin{pmatrix} 3x - 2y = 5 \\ 2x + 5y = -3 \end{pmatrix}$

Multiply equation 1 by 5 and multiply equation 2 by 2. Then add the equations.

$$\begin{array}{r} 15x - 10y = 25 \\ 4x + 10y = -6 \\ \hline 19x \quad = 19 \end{array}$$

$x = 1$

$3x - 2y = 5$

$3(1) - 2y = 5$

$3 - 2y = 5$

$-2y = 2$

$y = -1$

The solution set is $\{(1, -1)\}$.

11. $\begin{pmatrix} 7x-2y=4 \\ 7x-2y=9 \end{pmatrix}$

Multiply equation 1 by -1, then add to equation 2.

$$\begin{array}{r} -7x+2y=-4 \\ \underline{7x-2y=9} \\ 0=5 \end{array}$$

Since $0 \neq 5$, the system is inconsistent.

The solution set is $\emptyset$.

13. $\begin{pmatrix} 5x+4y=1 \\ 3x-2y=-1 \end{pmatrix}$

Multiply equation 2 by 2, then add to equation 1.

$$\begin{array}{r} 5x+4y=1 \\ \underline{6x-4y=-2} \\ 11x=-1 \end{array}$$

$x=-\frac{1}{11}$

$5x+4y=1$

$5\left(-\frac{1}{11}\right)+4y=1$

$-\frac{5}{11}+4y=1$

$4y=1+\frac{5}{11}$

$4y=\frac{16}{11}$

$\frac{1}{4}(4y)=\frac{1}{4}\left(\frac{16}{11}\right)$

$y=\frac{11}{4}$

The solution set is $\left\{\left(-\frac{1}{11}, \frac{4}{11}\right)\right\}$.

15. $\begin{pmatrix} 8x-3y=13 \\ 4x+9y=3 \end{pmatrix}$

Multiply equation 1 by 3 then add to equation 2.

$$\begin{array}{r} 24x-9y=39 \\ \underline{4x+9y=3} \\ 28x=42 \end{array}$$

$x=\frac{42}{28}=\frac{3}{2}$

$8x-3y=13$

$8\left(\frac{3}{2}\right)-3y=13$

$12-3y=13$

$-3y=1$

$y=-\frac{1}{3}$

The solution set is $\left\{\left(\frac{3}{2}, -\frac{1}{3}\right)\right\}$.

17. $\begin{pmatrix} 5x+3y=-7 \\ 7x-3y=55 \end{pmatrix}$

$$\begin{array}{r} 5x+3y=-7 \\ \underline{7x-3y=55} \\ 12x=48 \end{array}$$

$x=4$

$5x+3y=-7$

$5(4)+3y=-7$

$20+3y=-7$

$3y=-27$

$y=-9$

The solution set is $\{(4,-9)\}$.

19. $\begin{pmatrix} x=5y+7 \\ 4x+9y=28 \end{pmatrix}$

$4x+9=28$

$4(5y+7)+9y=28$

$20y+28+9y=28$

$29y+28=28$

$29y=0$

$y=0$

$x=5y+7$

$x=5(0)+7$

$x=7$

The solution set is $\{(7, 0)\}$.

21. $\begin{pmatrix} x=-6y+79 \\ x=4y-41 \end{pmatrix}$

$x=-6y+79$

$4y-41=-6y+79$

$10y=120$

$y=12$

$x = -6y + 79$

$x = -6(12) + 79$

$x = -72 + 79$

$x = 7$

The solution set is $\{(7, 12)\}$.

23. $\begin{pmatrix} 4x - 3y = 2 \\ 5x - y = 3 \end{pmatrix}$

Multiply equation 2 by -3, then add to equation 1.

$$\begin{array}{r} 4x - 3y = 2 \\ -15x + 3y = -9 \\ \hline -11x = -7 \end{array}$$

$x = \frac{7}{11}$

$4x - 3y = 2$

$4\left(\frac{7}{11}\right) - 3y = 2$

$\frac{28}{11} - 3y = 2$

$-3y = 2 - \frac{28}{11}$

$-\frac{1}{3}(-3y) = -\frac{1}{3}\left(\frac{6}{11}\right)$

$y = \frac{2}{11}$

The solution set is $\left\{\left(\frac{7}{11}, \frac{2}{11}\right)\right\}$.

25. $\begin{pmatrix} 5x - 2y = 1 \\ 10x - 4y = 7 \end{pmatrix}$

Multiply equation 1 by -2, then add to equation 2.

$$\begin{array}{r} -10x + 4y = -2 \\ 10x - 4y = 7 \\ \hline 0 = 5 \end{array}$$

Since $0 \neq 5$, the system is inconsistent.

The solution set is $\emptyset$.

27. $\begin{pmatrix} 3x - 2y = 7 \\ 5x + 7y = 1 \end{pmatrix}$

Multiply equation 1 by 7 and multiply equation 2 by 2. Then add the equations.

$$\begin{array}{r} 21x - 14y = 49 \\ 10x + 14y = 2 \\ \hline 31x = 51 \end{array}$$

$x = \frac{51}{31}$

$3x - 2y = 7$

$3\left(\frac{51}{31}\right) - 2y = 7$

$\frac{153}{31} - 2y = 7$

$-2y = 7 - \frac{153}{31}$

$-2y = \frac{64}{31}$

$-\frac{1}{2}(-2y) = -\frac{1}{2}\left(\frac{64}{31}\right)$

$y = -\frac{32}{31}$

The solution set is $\left\{\left(\frac{51}{31}, -\frac{32}{31}\right)\right\}$.

29. $\begin{pmatrix} -2x + 5y = -16 \\ x = \frac{3}{4}y + 1 \end{pmatrix}$

$-2x + 5y = -16$

$-2\left(\frac{3}{4}y + 1\right) + 5y = -16$

$-\frac{3}{2}y - 2 + 5y = -16$

$\frac{7}{2}y - 2 = -16$

$\frac{7}{2}y = -14$

$\frac{2}{7}\left(\frac{7}{2}y\right) = \frac{2}{7}(-14)$

$y = -4$

$x = \frac{3}{4}y + 1$

$x = \frac{3}{4}(-4) + 1$

$x = -3 + 1$

$x = -2$

The solution set is $\{(-2, -4)\}$.

31. $\begin{pmatrix} y = \frac{2}{3}x - 4 \\ 5x - 3y = 9 \end{pmatrix}$

$5x - 3y = 9$

$5x - 3\left(\frac{2}{3}x - 4\right) = 9$

$5x - 2x + 12 = 9$

$3x = -3$

$x = -1$

$y = \frac{2}{3}x - 4$

$y = \frac{2}{3}(-1) - 4$

$y = -\frac{2}{3} - 4$

$y = -\frac{14}{3}$

The solution set is $\left\{\left(-1, -\frac{14}{3}\right)\right\}$.

33. $\begin{pmatrix} \frac{x}{6} + \frac{y}{3} = 3 \\ \frac{5x}{2} - \frac{y}{6} = -17 \end{pmatrix}$

Multiply both equations by 6.

$\begin{pmatrix} x + 2y = 18 \\ 15x - y = -102 \end{pmatrix}$

Multiply equation 2 by 2, then add to equation 1.

$$\begin{array}{r} x + 2y = 18 \\ \underline{30x - 2y = -204} \\ 31x = -186 \end{array}$$

$x = -6$

$x + 2y = 18$

$-6 + 2y = 18$

$2y = 24$

$y = 12$

The solution set is $\{(-6, 12)\}$.

35. $\begin{pmatrix} -(x-6) + 6(y+1) = 58 \\ 3(x+1) - 4(y-2) = -15 \end{pmatrix}$

$\begin{pmatrix} -x + 6 + 6y + 6 = 58 \\ 3x + 3 - 4y + 8 = -15 \end{pmatrix}$

$\begin{pmatrix} -x + 6y = 46 \\ 3x - 4y = -26 \end{pmatrix}$

Multiply equation 1 by 3, then add to equation 2.

$$\begin{array}{r} -3x + 18y = 138 \\ \underline{3x - 4y = -26} \\ 14y = 112 \end{array}$$

$y = 8$

$-x + 6y = 46$

$-x + 6(8) = 46$

$-x + 48 = 46$

$-x = -2$

$x = 2$

The solution set is $\{(2, 8)\}$.

37. $\begin{pmatrix} 5(x+1) - (y+3) = -6 \\ 2(x-2) + 3(y-1) = 0 \end{pmatrix}$

$\begin{pmatrix} 5x + 5 - y - 3 = -6 \\ 2x - 4 + 3y - 3 = 0 \end{pmatrix}$

$\begin{pmatrix} 5x - y = -8 \\ 2x + 3y = 7 \end{pmatrix}$

Multiply equation 1 by 3, then add to equation 2.

$$\begin{array}{r} 15x - 3y = -24 \\ \underline{2x + 3y = 7} \\ 17x = -17 \end{array}$$

$x = -1$

$5x - y = -8$

$5(-1) - y = -8$

$-5 - y = -8$

$-y = -3$

$y = 3$

The solution set is $\{(-1, 3)\}$.

39. $\begin{pmatrix} \frac{1}{2}x - \frac{1}{3}y = 12 \\ \frac{3}{4}x + \frac{2}{3}y = 4 \end{pmatrix}$

Multiply equation 1 by 6 and multiply equation 2 by 12.

$\begin{pmatrix} 3x - 2y = 72 \\ 9x + 8y = 48 \end{pmatrix}$

Multiply equation 1 by 4, then add to equation 2.

$$\begin{array}{r} 12x-8y=288 \\ \underline{9x+8y=\ 48} \\ 21x \qquad =336 \end{array}$$

$x=16$

$3x-2y=72$

$3(16)-2y=72$

$48-2y=72$

$-2y=24$

$y=-12$

The solution set is $\{(16,-12)\}$.

41. $\begin{pmatrix} \frac{2x}{3}-\frac{y}{2}=-\frac{5}{4} \\ \frac{x}{4}+\frac{5y}{6}=\frac{17}{16} \end{pmatrix}$

Multiply equation 1 by 12 and multiply equation 2 by 48.

$\begin{pmatrix} 8x-6y=-15 \\ 12x+40y=51 \end{pmatrix}$

Multiply equation 1 by -3 and multiply equation 2 by 2. Then add the equations.

$$\begin{array}{r} -24x+18y=\ 45 \\ \underline{24x+80y=102} \\ 98y \qquad =147 \end{array}$$

$y=\frac{147}{98}=\frac{3}{2}$

$8x-6y=-15$

$8x-6\left(\frac{3}{2}\right)=-15$

$8x-9=-15$

$8x=-6$

$x=-\frac{6}{8}=-\frac{3}{4}$

The solution set is $\left\{\left(-\frac{3}{4},\ \frac{3}{2}\right)\right\}$.

43. $\begin{pmatrix} \frac{3x+y}{2}+\frac{x-2y}{5}=8 \\ \frac{x-y}{3}-\frac{x+y}{6}=\frac{10}{3} \end{pmatrix}$

Multiply equation 1 by 10 and multiply equation 2 by 6.

$\begin{pmatrix} 5(3x+y)+2(x-2y)=80 \\ 2(x-y)-(x+y)=20 \end{pmatrix}$

$\begin{pmatrix} 15x+5y+2x-4y=80 \\ 2x-2y-x-y=20 \end{pmatrix}$

$\begin{pmatrix} 17x+y=80 \\ x-3y=20 \end{pmatrix}$

Solve equation 2 for x.

$x-3y=20$

$x=3y+20$

Substitute into equation 1.

$17x+y=80$

$17(3y+20)+y=80$

$51y+340+y=80$

$52y+340=80$

$52y=-260$

$y=-5$

$x=3y+20$

$x=3(-5)+20$

$x=-15+20$

$x=5$

The solution set is $\{(5,-5)\}$.

45. $x=$ gallons of 10% solution
$y=$ gallons of 20% solution

$\begin{pmatrix} x+y=20 \\ .10x+.20y=.175(20) \end{pmatrix}$

Solve equation 1 for x.

$x+y=20$

$x=20-y$

Substitute

$.10x+.20y=3.5$

$.10(20-y)+.20y=3.5$

$10[.10(20-y)+.20y]=10(3.5)$

$1(20-y)+2y=35$

$20-y+2y=35$

$y = 15$

$x = 20 - y$

$x = 20 - 15$

$x = 5$

There should be 5 gallons of 10% solution and 15 gallons of 20% solution.

47. Let $x =$ cost of tennis ball
$y =$ cost of golf ball

$$\begin{pmatrix} 3x + 2y = 7 \\ 6x + 3y = 12 \end{pmatrix}$$

Multiply equation 1 by -2, then add to equation 2.

$$\begin{array}{r} -6x - 4y = -14 \\ 6x + 3y = 12 \\ \hline -y = -2 \end{array}$$

$y = 2$

$3x + 2y = 7$

$3x + 2(2) = 7$

$3x + 4 = 7$

$3x = 3$

$x = 1$

Tennis balls cost \$1 and golf balls cost \$2.

49. Let $x =$ number of single rooms
$y =$ number of double rooms

$$\begin{pmatrix} x + y = 55 \\ 22x + 42y = 2010 \end{pmatrix}$$

Solve equation 1 for x.

$x + y = 55$

$x = -y + 55$

Substitute into equation 2.

$22x + 42y = 2010$

$22(-y + 55) + 42y = 2010$

$-22y + 1210 + 42y = 2010$

$20y = 800$

$y = 40$

$x = -y + 55$

$x = -40 + 55$

$x = 15$

There were 15 single rooms and 40 double rooms.

51. Let $x =$ distance the 40 pounds is from the fulcrum.
$y =$ distance the 80 pounds is from the fulcrum.

$$\begin{pmatrix} 40x = 80y \\ 60x = 80\left(y + \frac{3}{2}\right) \end{pmatrix}$$

Solve equation 1 for x.

$40x = 80y$

$x = 2y$

Substitute into equation 2.

$60x = 80\left(y + \frac{3}{2}\right)$

$60(2y) = 80y + 120$

$120y = 80y + 120$

$40y = 120$

$y = 3$

$x = 2(3)$

$x = 6$

The distance between the weights is 9 feet.

53. Let $x =$ numerator
$y =$ denominator

$$\begin{pmatrix} \frac{x+5}{y-1} = \frac{8}{3} \\ \frac{2x}{y+7} = \frac{6}{11} \end{pmatrix}$$

$$\begin{pmatrix} 3(x+5) = 8(y-1) \\ 11(2x) = 6(y+7) \end{pmatrix}$$

$$\begin{pmatrix} 3x + 15 = 8y - 8 \\ 22x = 6y + 42 \end{pmatrix}$$

$$\begin{pmatrix} 3x - 8y = -23 \\ 22x - 6y = 42 \end{pmatrix}$$

Multiply equation 1 by 3 and equation 2 by -4. Then add the equations.

$$\begin{array}{rcl} 9x-24y &=& -69 \\ -88x+24y &=& -168 \\ \hline -79x &=& -237 \end{array}$$

$x=3$

$3x-8y=-23$

$3(3)-8y=-23$

$9-8y=-23$

$-8y=-32$

$y=4$

The fraction is $\frac{3}{4}$.

55. Let $x=$ width
$y=$ length

$$\begin{pmatrix} (x+2)(y-1)=xy+28 \\ (x-1)(y+2)=xy+10 \end{pmatrix}$$

$$\begin{pmatrix} xy-x+2y-2=xy+28 \\ xy+2x-y-2=xy+10 \end{pmatrix}$$

$$\begin{pmatrix} -x+2y=30 \\ 2x-y=12 \end{pmatrix}$$

Multiply equation 1 by 2, then add to equation 2.

$$\begin{array}{rcl} -2x+4y &=& 60 \\ 2x-y &=& 12 \\ \hline 3y &=& 72 \end{array}$$

$y=24$

$-x+2y=30$

$-x+2(24)=30$

$-x+48=30$

$-x=-18$

$x=18$

The cover is 18 centimeters by 24 centimeters.

57. Let $x=$ distance from 60 pound weight to the fulcrum.

$y=$ distance from 100 pound weight to the fulcrum.

$$\begin{pmatrix} 60x=100y \\ 60(x-1)=80y \end{pmatrix}$$

Solve equation 1 for x.

$60x=100y$

$x=\frac{5}{3}y$

Substitute into equation 2.

$60(x-1)=80y$

$60\left(\frac{5}{3}y-1\right)=80y$

$100y-60=80y$

$-60=-20y$

$3=y$

$x=\frac{5}{3}y$

$x=\frac{5}{3}(3)$

$x=5$

The distance between the weights is 8 feet.

Problem Set 11.3 Systems of Three Linear Equations in Three Variables

1. $$\begin{pmatrix} x+2y-3z=2 \\ 3y-z=13 \\ 3y+5z=25 \end{pmatrix}$$

Multiply equation 2 by -1, then add to equation 3.

$$\begin{array}{rcl} -3y+z &=& -13 \\ 3y+5z &=& 25 \\ \hline 6z &=& 12 \end{array}$$

$z=2$

$3y-z=13$

$3y-(2)=13$

$3y=15$

$y=5$

$x+2y-3z=2$

$x+2(5)-3(2)=2$

$x+10-6=2$

$x+4=2$

$x=-2$

The solution set is $\{(-2, 5, 2)\}$.

3. $\begin{pmatrix} 3x+2y-2z=14 \\ x-6z=16 \\ 2x+5z=-2 \end{pmatrix}$

Multiply equation 2 by -2, then add to equation 3.

$$\begin{array}{r} -2x+12z=-32 \\ 2x+\ 5z=-\ 2 \\ \hline 17z=-34 \end{array}$$

$z=-2$

$x-6z=16$

$x-6(-2)=16$

$x=4$

$3x+2y-2z=14$

$3(4)+2y-2(-2)=14$

$12+2y+4=14$

$2y+16=14$

$2y=-2$

$y=-1$

The solution set is $\{(4,-1,-2)\}$.

5. $\begin{pmatrix} 2x-y+z=0 \\ 3x-2y+4z=11 \\ 5x+y-6z=-32 \end{pmatrix}$

Add equation 1 and equation 3.

$$\begin{array}{r} 2x-y+\ z=\ \ \ 0 \\ 5x+y-6z=-32 \\ \hline 7x\ \ \ \ -5z=-32 \end{array}$$

Multiply equation 3 by 2, then add to equation 2.

$$\begin{array}{r} 3x-2y+\ 4z=\ \ \ 11 \\ 10x+2y-12z=-64 \\ \hline 13x\ \ \ \ -\ 8z=-53 \end{array}$$

$\begin{pmatrix} 7x-5z=-32 \\ 13x-8z=-53 \end{pmatrix}$

Multiply equation 1 by 8 and multiply equation 2 by -5.

Then add the equations.

$$\begin{array}{r} 56x-40z=-256 \\ -65x+40z=\ \ \ 265 \\ \hline -9x\ \ \ \ \ \ =\ \ \ \ 9 \end{array}$$

$x=-1$

$7x-5z=-32$

$7(-1)-5z=-32$

$-7-5z=-32$

$-5z=-25$

$z=5$

$2x-y+z=0$

$2(-1)-y+5=0$

$-2-y+5=0$

$-y+3=0$

$3=y$

The solution set is $\{(-1, 3, 5)\}$.

7. $\begin{pmatrix} 4x-y+z=5 \\ 3x+y+2z=4 \\ x-2y-z=1 \end{pmatrix}$

Add equations 1 and 2.

$$\begin{array}{r} 4x-y+\ z=5 \\ 3x+y+2z=4 \\ \hline 7x\ \ \ \ +3z=9 \end{array}$$

Multiply equation 2 by 2, then add to equation 1.

$$\begin{array}{r} 6x+2y+4z=8 \\ x-2y-\ z=1 \\ \hline 7x\ \ \ \ +3z=9 \end{array}$$

$\begin{pmatrix} 7x+3z=9 \\ 7x+3z=9 \end{pmatrix}$

Multiply equation 1 by -1, then add to equation 2.

$$\begin{array}{r} -7x-3z=-9 \\ 7x+3z=\ \ \ 9 \\ \hline 0=0 \end{array}$$

Since $0=0$ is a true statement, there are infinitely many solutions.

9. $\begin{pmatrix} x-y+2z=4 \\ 2x-2y+4z=7 \\ 3x-3y+6z=1 \end{pmatrix}$

Multiply equation 1 by -2, then add to equation 2.

$$\begin{array}{r} -2x+2y-4z=-8 \\ 2x-2y+4z=\ \ 7 \\ \hline 0=-1 \end{array}$$

Since $0 \neq -1$, there is no solution.

The solution set is $\emptyset$.

11. $\begin{pmatrix} x-2y+z=-4 \\ 2x+4y-3z=-1 \\ -3x-6y+7z=4 \end{pmatrix}$

Multiply equation 1 by -2, then add to equation 2.

$$\begin{array}{r} -2x+4y-2z=\ \ 8 \\ 2x+4y-3z=-1 \\ \hline 8y-5z=\ \ 7 \end{array}$$

Multiply equation 1 by 3, then add to equation 3.

$$\begin{array}{r} 3x-6y+\ 3z=-12 \\ -3x-6y+\ 7z=\ \ 4 \\ \hline -12y+10z=\ -8 \end{array}$$

$\begin{pmatrix} 8y-5z=7 \\ -12y+10z=-8 \end{pmatrix}$

Multiply equation 1 by 2, then add to equation 2.

$$\begin{array}{r} 16y-10z=\ 14 \\ -12y+10z=-8 \\ \hline 4y\qquad=\ \ 6 \end{array}$$

$y=\frac{6}{4}=\frac{3}{2}$

$8y-5z=7$

$8\left(\frac{3}{2}\right)-5z=7$

$12-5z=7$

$-5z=-5$

$z=1$

$x-2y+z=-4$

$x-2\left(\frac{3}{2}\right)+1=-4$

$x-3+1=-4$

$x-2=-4$

$x=-2$

The solution set is $\left\{\left(-2,\ \frac{3}{2},\ 1\right)\right\}$.

13. $\begin{pmatrix} 3x-2y+4z=6 \\ 9x+4y-z=0 \\ 6x-8y-3z=3 \end{pmatrix}$

Multiply equation 1 by -3, then add to equation 2.

$$\begin{array}{r} -9x+6y-12z=-18 \\ 9x+4y-\ \ z=\ \ \ 0 \\ \hline 10y-13z=-18 \end{array}$$

Multiply equation 1 by -2, then add to equation 3.

$$\begin{array}{r} -6x+4y-8z=-12 \\ 6x-8y-3z=\ \ \ 3 \\ \hline -4y-11z=-9 \end{array}$$

$\begin{pmatrix} 10y-13z=-18 \\ -4y-11z=-9 \end{pmatrix}$

Multiply equation 1 by 2 and multiply equation 2 by 5.

$$\begin{array}{r} 20y-26z=-36 \\ -20y-55z=-45 \\ \hline -81z=-81 \end{array}$$

$z=1$

$10y-13z=-18$

$10y-13(1)=-18$

$10y-13=-18$

$10y=-5$

$y=-\frac{5}{10}=-\frac{1}{2}$

$3x-2y+4z=6$

$3x-2\left(-\frac{1}{2}\right)+4(1)=6$

$3x+1+4=6$

$3x+5=6$

$3x=1$

$x=\frac{1}{3}$

The solution set is $\left\{\left(\frac{1}{3},\ -\frac{1}{2},\ 1\right)\right\}$.

15. $\begin{pmatrix} 3x - y + 4z = 9 \\ 3x + 2y - 8z = -12 \\ 9x + 5y - 12z = -23 \end{pmatrix}$

Multiply equation 1 by -1, then add to equation 2.

$$\begin{array}{r} -3x + y - 4z = -9 \\ 3x + 2y - 8z = -12 \\ \hline 3y - 12z = -21 \end{array}$$

Multiply equation 1 by -3, then add to equation 3.

$$\begin{array}{r} -9x + 3y - 12z = -27 \\ 9x + 5y - 12z = -23 \\ \hline 8y - 24z = -50 \end{array}$$

$\begin{pmatrix} 3y - 12z = -21 \\ 8y - 24z = -50 \end{pmatrix}$

Multiply equation 1 by -2, then add to equation 2.

$$\begin{array}{r} -6y + 24z = 42 \\ 8y - 24z = -50 \\ \hline 2y \quad = -8 \end{array}$$

$y = -4$

$3y - 12z = -21$

$3(-4) - 12z = -21$

$-12 - 12z = -21$

$-12z = -9$

$z = \frac{-9}{-12} = \frac{3}{4}$

$3x - y + 4z = 9$

$3x - (-4) + 4\left(\frac{3}{4}\right) = 9$

$3x + 4 + 3 = 9$

$3x + 7 = 9$

$3x = 2$

$x = \frac{2}{3}$

The solution set is $\left\{\left(\frac{2}{3}, -4, \frac{3}{4}\right)\right\}$.

17. $\begin{pmatrix} 4x - y + 3z = -12 \\ 2x + 3y - z = 8 \\ 6x + y + 2z = -8 \end{pmatrix}$

Add equation 1 and equation 3.

$$\begin{array}{r} 4x - y + 3z = -12 \\ 6x + y + 2z = -8 \\ \hline 10x \quad + 5z = -20 \end{array}$$

Multiply equation 1 by 3, then add to equation 2.

$$\begin{array}{r} 12x - 3y + 9z = -36 \\ 2x + 3y - z = 8 \\ \hline 14x \quad + 8z = -28 \end{array}$$

$\begin{pmatrix} 10x + 5z = -20 \\ 14x + 8z = -28 \end{pmatrix}$

Multiply equation 1 by -8 and multiply equation 2 by 5. Then add the equations.

$$\begin{array}{r} -80x - 40x = 160 \\ 70x + 40x = -140 \\ \hline -10x \quad = 20 \end{array}$$

$x = -2$

$10x + 5z = -20$

$10(-2) + 5z = -20$

$-20 + 5z = -20$

$5z = 0$

$z = 0$

$4x - y + 3z = -12$

$4(-2) - y + 3(0) = -12$

$-8 - y + 0 = -12$

$-y = -4$

$y = 4$

The solution set is $\{(-2, 4, 0)\}$.

19. $\begin{pmatrix} x + y + z = 1 \\ 2x - 3y + 6z = 1 \\ -x + y + z = 0 \end{pmatrix}$

Multiply equation 1 by -1, then add to equation 3.

$$\begin{array}{r} -x - y - z = -1 \\ -x + y + z = 0 \\ \hline -2x \quad = -1 \end{array}$$

$x = \frac{-1}{-2} = \frac{1}{2}$

Substitute $x=\frac{1}{2}$ into equation 1 and equation 2.

$\frac{1}{2}+y+z=1$

$y+z=\frac{1}{2}$

$2\left(\frac{1}{2}\right)-3y+6z=1$

$1-3y+6z=1$

$-3y+6z=0$

$$\begin{pmatrix} y+z=\frac{1}{2} \\ -3y+6z=0 \end{pmatrix}$$

Multiply equation 1 by 3, then add to equation 2.

$$\begin{array}{r} 3y+3z=\frac{3}{2} \\ -3y+6z=0 \\ \hline 9z=\frac{3}{2} \end{array}$$

$\frac{1}{9}(9z)=\frac{1}{9}\left(\frac{3}{2}\right)$

$z=\frac{1}{6}$

$y+z=\frac{1}{2}$

$y+\frac{1}{6}=\frac{1}{2}$

$y=\frac{1}{2}-\frac{1}{6}$

$y=\frac{1}{3}$

The solution set is $\left\{\left(\frac{1}{2}, \frac{1}{3}, \frac{1}{6}\right)\right\}$.

21. Let $x=$ digit in hundreds place
$y=$ digit in tens place
$z=$ digit in units place

$$\begin{pmatrix} x+y+z=14 \\ 100x+10y+z=20y+14 \\ y+z=x+12 \end{pmatrix}$$

$$\begin{pmatrix} x+y+z=14 \\ 100x-10y+z=14 \\ -x+y+z=12 \end{pmatrix}$$

Add equation 1 and 3.

$$\begin{array}{r} x+y+z=14 \\ -x+y+z=12 \\ \hline 2y+2z=26 \end{array}$$

$\frac{1}{2}(2y+2z)=\frac{1}{2}(26)$

$y+z=13$

Multiply equation 3 by 100, then add to equation 2.

$$\begin{array}{r} 100x-10y+z=14 \\ -100x+100y+100z=1200 \\ \hline 90y+101z=1214 \end{array}$$

$$\begin{pmatrix} y+z=13 \\ 90y+101z=1214 \end{pmatrix}$$

Multiply equation 1 by -90, then add to equation 2.

$$\begin{array}{r} -90y-90z=1170 \\ 90y+101z=1214 \\ \hline 11z=44 \end{array}$$

$z=4$

$y+z=13$

$y+4=13$

$y=9$

$x+y+z=14$

$x+9+4=14$

$x+13=14$

$x=1$

The number is 194.

23. Let $x=$ cost of catsup
$y=$ cost of peanut butter
$z=$ cost of pickles

$$\begin{pmatrix} 2x+2y+1z=4.20 \\ 3x+4y+2z=7.70 \\ 4x+3y+5z=9.80 \end{pmatrix}$$

Multiply equation 1 by -2, then add to equation 2.

$$\begin{array}{r} -4x-4y-2z=-8.40 \\ 3x+4y+2z=7.70 \\ \hline -x=-.70 \end{array}$$

$x=.70$

Substitute $x = .70$ in equation 1 and equation 3.

$2(.70) + 2y + z = 4.20$

$1.40 + 2y + z = 4.20$

$2y + z = 2.80$

$4(.70) + 3y + 5z = 9.80$

$2.80 + 3y + 5z = 9.80$

$3y + 5z = 7.00$

$$\begin{pmatrix} 2y + z = 2.80 \\ 3y + 5z = 7.00 \end{pmatrix}$$

Multiply equation 1 by -5, then add to equation 2.

$$\begin{array}{r} -10y - 5z = -14.00 \\ \underline{3y + 5z = \quad 7.00} \\ -7y \quad = \ -7.00 \end{array}$$

$y = 1.00$

$2y + z = 2.80$

$2(1.00) + z = 2.80$

$2.00 + z = 2.80$

$z = .80$

The catsup cost \$.70, the peanut butter cost \$1.00, and the pickles cost \$.80.

25. Let $x = 1^{st}$ number
$y = 2^{nd}$ number
$z = 3^{rd}$ number

$$\begin{pmatrix} x + y + z = 20 \\ x + z = 2y + 2 \\ z - x = 3y \end{pmatrix}$$

$$\begin{pmatrix} x + y + z = 20 \\ x - 2y + z = 2 \\ -x - 3y + z = 0 \end{pmatrix}$$

Add equation 1 and equation 3.

$$\begin{array}{r} x + y + z = 20 \\ \underline{-x - 3y + z = 0} \\ -2y + 2z = 20 \end{array}$$

$\frac{1}{2}(-2y + 2z) = \frac{1}{2}(20)$

$-y + z = 10$

Add equation 2 and equation 3.

$$\begin{array}{r} x - 2y + z = 2 \\ \underline{-x - 3y + z = 0} \\ -5y + 2z = 2 \end{array}$$

$$\begin{pmatrix} -y + z = 10 \\ -5y + 2z = 2 \end{pmatrix}$$

Multiply equation 1 by -2, then add to equation 2.

$$\begin{array}{r} 2y - 2z = -20 \\ \underline{-5y + 2z = \quad 2} \\ -3y \quad = -18 \end{array}$$

$y = 6$

$-y + z = 10$

$-6 + z = 10$

$z = 16$

$x + y + z = 20$

$x + 6 + 16 = 20$

$x + 22 = 20$

$x = -2$

The numbers are -2, 6, and 16.

27. Let $x =$ largest angle
$y =$ middle angle
$z =$ smallest angle

$$\begin{pmatrix} x + y + z = 180 \\ x = 2z \\ z + x = 2y \end{pmatrix}$$

Substitute $x = 2z$ into equation 3.

$z + x = 2y$

$z + 2z = 2y$

$3z = 2y$

$\frac{3}{2}z = y$

Substitute $x = 2z$ and $y = \frac{3}{2}z$ into equation 1.

$x + y + z = 180$

$2z + \frac{3}{2}z + z = 180$

$\frac{9}{2}z = 180$

$\frac{2}{9}\left(\frac{9}{2}z\right) = \frac{2}{9}(180)$

$z = 40$

$y = \frac{3}{2}(40)$

$y = 60$

$x = 2z$

$x = 2(40)$

$x = 80$

The angles are 40°, 60°, and 80°.

29. Let $x =$ money invested at 12%
$y =$ money invested at 13%
$z =$ money invested at 14%

$$\begin{pmatrix} x+y+z = 3000 \\ .12x + .13y + .14z = 400 \\ x+y = z \end{pmatrix}$$

Substitute $x+y=z$ into equation 1.

$x+y+z = 3000$

$(x+y)+z = 3000$

$z+z = 3000$

$2z = 3000$

$z = 1500$

Substitute $z = 1500$ into equation 2.

$.12x + .13y + .14(1500) = 400$

$.12x + .13y + 210 = 400$

$.12x + .13y = 190$

$$\begin{pmatrix} x+y = 1500 \\ .12x + .13y = 190 \end{pmatrix}$$

Multiply equation 2 by 100.

$$\begin{pmatrix} x+y = 1500 \\ 12x + 13y = 19000 \end{pmatrix}$$

Multiply equation by -12, then add to equation 2.

$$\begin{array}{r} -12x - 12y = -18000 \\ 12x + 13y = 19000 \\ \hline y = 1000 \end{array}$$

$x+y = z$

$x + 1000 = 1500$

$x = 500$

There is \$500 invested at 12%, \$1000 invested at 13%, and \$1500 invested at 14%.

Problem Set 11.4 Systems Involving Nonlinear Equations and Systems of Inequalities

1. $\begin{pmatrix} y = (x+2)^2 \\ y = -2x-4 \end{pmatrix}$

$y = (x+2)^2$

$-2x-4 = (x+2)^2$

$-2x-4 = x^2+4x+4$

$0 = x^2+6x+8$

$0 = (x+4)(x+2)$

$x+4 = 0$ or $x+2 = 0$

$x = -4$ or $x = -2$

If $x = -4$

$y = -2x-4$

$y = -2(-4)-4$

$y = 8-4 = 4$

$(-4, 4)$

If $x = -2$

$y = -2x-4$

$y = -2(-2)-4$

$y = 4-4 = 0$

$(-2, 0)$

The solution set is $\{(-4, 4), (-2, 0)\}$.

3. $\begin{pmatrix} x^2+y^2 = 13 \\ 3x+2y = 0 \end{pmatrix}$

$3x+2y = 0$

$3x = -2y$

$x = -\frac{2}{3}y$

$x^2+y^2 = 13$

$\left(-\frac{2}{3}y\right)^2+y^2=13$

$\frac{4}{9}y^2+\frac{9}{9}y^2=13$

$\frac{13}{9}y^2=13$

$\frac{9}{13}\left(\frac{13}{9}y^2\right)=\frac{9}{13}(13)$

$y^2=9$

$y=\pm\sqrt{9}$

$y=\pm 3$

If $y=3$

$x=-\frac{2}{3}y$

$x=-\frac{2}{3}(3)=-2$

$(-2, 3)$

If $y=-3$

$x=-\frac{2}{3}y$

$x=-\frac{2}{3}(-3)=2$

$(2,-3)$

The solution set is $\{(-2, 3), (2,-3)\}$.

5. $\begin{pmatrix} x+y=-8 \\ x^2-y^2=16 \end{pmatrix}$

Solve equation 1 for y.

$x+y=-8$

$y=-x-8$

Substitute into equation 2.

$x^2-y^2=16$

$x^2-(-x-8)^2=16$

$x^2-(x^2+16x+64)=16$

$x^2-x^2-16x-64=16$

$-16x-64=16$

$-16x=80$

$x=-5$

Find y, when $x=-5$.

$y=-x-8$

$y=-(-5)-8$

$y=5-8$

$y=-3$

$(-5,-3)$

The solution set is $\{(-5,-3)\}$.

7. $\begin{pmatrix} y=x^2+6x+7 \\ 2x+y=-5 \end{pmatrix}$

Solve equation 2 for y.

$2x+y=-5$

$y=-2x-5$

Substitute into equation 1.

$y=x^2+6x+7$

$-2x-5=x^2+6x+7$

$0=x^2+8x+12$

$0=(x+6)(x+2)$

$x+6=0$ or $x+2=0$

$x=-6$ or $x=-2$

Find y, when $x=-6$.

$y=-2x-5$

$y=-2(-6)-5$

$y=12-5$

$y=7$

$(-6, 7)$

Find y, when $x=-2$.

$y=-2x-5$

$y=-2(-2)-5$

$y=4-5$

$y=-1$

$(-2,-1)$

The solution set is $\{(-6, 7), (-2,-1)\}$.

9. $\begin{pmatrix} xy=4 \\ y=x \end{pmatrix}$

$xy=4$

$x(x)=4$

$x^2=4$

$x = \pm\sqrt{4}$

$x = 2$ or $x = -2$

Find y, when $x = 2$.

$y = x$

$y = 2$

$(2, 2)$

Find y, when $x = -2$

$y = x$

$y = -2$

$(-2, -2)$

The solution set is $\{(2, 2), (-2, -2)\}$.

11. $\begin{pmatrix} x^2 + 2y^2 = 8 \\ x^2 - y^2 = 1 \end{pmatrix}$

Solve equation 2 for x^2.

$x^2 - y^2 = 1$

$x^2 = y^2 + 1$

Substitute into equation 1.

$x^2 + 2y^2 = 8$

$y^2 + 1 + 2y^2 = 8$

$3y^2 = 7$

$y^2 = \frac{7}{3}$

$y = \pm\sqrt{\frac{7}{3}}$

$y = \pm\frac{\sqrt{7}}{\sqrt{3}} \bullet \frac{\sqrt{3}}{\sqrt{3}} = \pm\frac{\sqrt{21}}{3}$

Find x, when $y = \frac{\sqrt{21}}{3}$

$x^2 = y^2 + 1$

$x^2 = \left(\frac{\sqrt{21}}{3}\right)^2 + 1$

$x^2 = \frac{21}{9} + 1$

$x^2 = \frac{30}{9}$

$x = \pm\sqrt{\frac{30}{9}} = \pm\frac{\sqrt{30}}{3}$

$\left(\frac{\sqrt{30}}{3}, \frac{\sqrt{21}}{3}\right), \left(-\frac{\sqrt{30}}{3}, \frac{\sqrt{21}}{3}\right)$

Find x, when $y = -\frac{\sqrt{21}}{3}$

$x^2 = y^2 + 1$

$x^2 = \left(-\frac{\sqrt{21}}{3}\right)^2 + 1$

$x^2 = \frac{21}{9} + 1$

$x^2 = \frac{30}{9}$

$x = \pm\sqrt{\frac{30}{9}} = \pm\frac{\sqrt{30}}{3}$

$\left(\frac{\sqrt{30}}{3}, -\frac{\sqrt{21}}{3}\right), \left(-\frac{\sqrt{30}}{3}, -\frac{\sqrt{21}}{3}\right)$

The solution set is

$\left\{\left(\frac{\sqrt{30}}{3}, \frac{\sqrt{21}}{3}\right), \left(-\frac{\sqrt{30}}{3}, \frac{\sqrt{21}}{3}\right), \left(\frac{\sqrt{30}}{3}, -\frac{\sqrt{21}}{3}\right), \left(-\frac{\sqrt{30}}{3}, -\frac{\sqrt{21}}{3}\right)\right\}$

13. $\begin{pmatrix} y = x^2 \\ y = x^2 - 4x + 4 \end{pmatrix}$

$y = x^2$

$x^2 - 4x + 4 = x^2$

$-4x + 4 = 0$

$-4x = -4$

$x = 1$

$y = x^2$

$y = 1^2$

$y = 1$

The solution set is $\{(1, 1)\}$.

15. $\begin{pmatrix} y = x^2 + 2x - 1 \\ y = x^2 + 4x + 5 \end{pmatrix}$

$y = x^2 + 2x - 1$

$x^2 + 4x + 5 = x^2 + 2x - 1$

$4x + 5 = 2x - 1$

$2x = -6$

$x = -3$

$y = x^2 + 2x - 1$

$y = (-3)^2 + 2(-3) - 1$

$y = 9 - 6 - 1$

$y = 2$

$(-3, 2)$

The solution set is $\{(-3, 2)\}$.

17 - 35. See back of textbook for graphs.

37. Let $x =$ one number
$y =$ other number

$$\begin{pmatrix} x^2 + y^2 = 34 \\ x^2 - y^2 = 16 \end{pmatrix}$$

$$\begin{array}{l} x^2 + y^2 = 34 \\ \underline{x^2 - y^2 = 16} \\ 2x^2 \quad\quad = 50 \end{array}$$

$x^2 = 25$

$x = \pm\sqrt{25} = \pm 5$

If $x = 5$

$x^2 + y^2 = 34$

$5^2 + y^2 = 34$

$25 = y^2 = 34$

$y^2 = 9$

$y = \pm\sqrt{9} = \pm 3$

$(5, 3), (5, -3)$

If $x = -5$

$x^2 + y^2 = 34$

$(-5)^2 + y^2 = 34$

$25 + y^2 = 34$

$y^2 = 9$

$y = \pm\sqrt{9} = \pm 3$

$(-5, 3), (-5, 3)$

The numbers are 5 and 3 or 5 and -3 or -5 and 3 or -5 and -3.

39. Let $x =$ width
$y =$ length

$$\begin{pmatrix} xy = 54 \\ 2x + 2y = 30 \end{pmatrix}$$

$2x + 2y = 30$

$2x = 30 - 2y$

$x = \dfrac{30 - 2y}{2}$

$x = 15 - y$

$xy = 54$

$(15 - y)(y) = 54$

$15y - y^2 = 54$

$0 = y^2 - 15y + 54$

$0 = (y - 9)(y - 6)$

$y - 9 = 0$ or $y - 6 = 0$

$y = 9$ or $y = 6$

If $y = 9$

$xy = 54$

$x(9) = 54$

$x = 6$

If $y = 6$

$xy = 54$

$x(6) = 54$

$x = 9$

The rectangle is 6 meters by 9 meters.

Chapter 11 Review

1a. Substitution method

$$\begin{pmatrix} 3x - 2y = -6 \\ 2x + 5y = 34 \end{pmatrix}$$

Solve equation 2 for x.

$2x + 5y = 34$

$2x = -5y + 34$

$x = -\frac{5}{2}y + 17$

Substitute into equation 1.

$3x - 2y = -6$

$3\left(-\frac{5}{2}y + 17\right) - 2y = -6$

$-\frac{15}{2}y + 51 - 2y = -6$

$-\frac{19}{2}y + 51 = -6$

$-\frac{19}{2}y = -57$

$-\frac{2}{19}\left(-\frac{19}{2}y\right) = -\frac{2}{19}(-57)$

$y = 6$

$x = -\frac{5}{2}y + 17$

$x = -\frac{5}{2}(6) + 17$

$x = -15 + 17$

$x = 2$

The solution set is $\{(2, 6)\}$.

b. Elimination method

$\begin{pmatrix} 3x - 2y = -6 \\ 2x + 5y = 34 \end{pmatrix}$

Multiply equation 1 by -2 and multiply equation 2 by 3. Then add the equations.

$$\begin{array}{r} -6x + 4y = 12 \\ 6x + 15y = 102 \\ \hline 19y = 114 \end{array}$$

$y = 6$

$3x - 2y = -6$

$3x - 2(6) = -6$

$3x - 12 = -6$

$3x = 6$

$x = 2$

The solution set is $\{(2, 6)\}$.

3. $\begin{pmatrix} x = \frac{y+9}{2} \\ 2x + 3y = 5 \end{pmatrix}$

$2x + 3y = 5$

$2\left(\frac{y+9}{2}\right) + 3y = 5$

$y + 9 + 3y = 5$

$4y + 9 = 5$

$4y = -4$

$y = -1$

$x = \frac{y+9}{2}$

$x = \frac{-1+9}{2}$

$x = \frac{8}{2} = 4$

The solution set is $\{(4, -1)\}$.

5. $\begin{pmatrix} \frac{1}{2}x + \frac{1}{3}y = -4 \\ \frac{3}{4}x - \frac{2}{3}y = 22 \end{pmatrix}$

$6\left(\frac{1}{2}x + \frac{1}{3}y\right) = 6(-4)$

$3x + 2y = -24$

$12\left(\frac{3}{4}x - \frac{2}{3}y\right) = 12(22)$

$9x - 8y = 264$

$\begin{pmatrix} 3x + 2y = -24 \\ 9x - 8y = 264 \end{pmatrix}$

Multiply equation 1 by -3 and then add to equation 2.

$$\begin{array}{r} -9x - 6y = 72 \\ 9x - 8y = 264 \\ \hline -14y = 336 \end{array}$$

$y = -24$

$3x + 2y = -24$

$3x + 2(-24) = -24$

$3x - 48 = -24$

$3x = 24$

$x = 8$

The solution set is $\{(8, -24)\}$.

7. $\begin{pmatrix} x + y - z = -2 \\ 2x - 3y + 4z = 17 \\ -3x + 2y + 5z = -7 \end{pmatrix}$

Multiply equation 1 by -2 and then add to equation 2.

$$\begin{array}{r} -2x-2y+2z = 4 \\ 2x-3y+4z = 17 \\ \hline -5y+6z = 21 \end{array}$$

Multiply equation 1 by 3 and then add to equation 3.

$$\begin{array}{r} 3x+3y-3z = -6 \\ -3x+2y+5z = -7 \\ \hline 5y+2z = -13 \end{array}$$

$$\begin{pmatrix} -5y+6z = 21 \\ 5y+2z = -13 \end{pmatrix}$$

Add the equations.

$$\begin{array}{r} -5y+6z = 21 \\ 5y+2z = -13 \\ \hline 8z = 8 \end{array}$$

$z = 1$

$5y+2z = -13$

$5y+2(1) = -13$

$5y+2 = -13$

$5y = -15$

$y = -3$

$x+y-z = -2$

$x+(-3)-1 = -2$

$x-4 = -2$

$x = 2$

The solution set is $\{(2, -3, 1)\}$.

9. $\begin{pmatrix} x+3y-2z = -7 \\ 4x+13y-7z = -21 \\ 5x+16y-8z = -23 \end{pmatrix}$

Multiply equation 1 by -4 and then add to equation 2.

$$\begin{array}{r} -4x-12y+8z = 28 \\ 4x+13y-7z = -21 \\ \hline y+z = 7 \end{array}$$

Multiply equation 1 by -5 and then add to equation 3.

$$\begin{array}{r} -5x-15y+10z = 35 \\ 5x+16y-8z = -23 \\ \hline y+2z = 12 \end{array}$$

$$\begin{pmatrix} y+z = 7 \\ y+2z = 12 \end{pmatrix}$$

Multiply equation 1 by -1 and then add to equation 2.

$$\begin{array}{r} -y-z = -7 \\ y+2z = 12 \\ \hline z = 5 \end{array}$$

$y+z = 7$

$y+5 = 7$

$y = 2$

$x+3y-2z = -7$

$x+3(2)-2(5) = -7$

$x+6-10 = -7$

$x-4 = -7$

$x = -3$

The solution set is $\{(-3, 2, 5)\}$.

11. $\begin{pmatrix} x-3y-4z = -1 \\ 2x+y-2z = 3 \\ 5x-8y-14z = 0 \end{pmatrix}$

Multiply equation 1 by -2 and then add to equation 2.

$$\begin{array}{r} -2x+6y+8z = 2 \\ 2x+y-2z = 3 \\ \hline 7y+6z = 5 \end{array}$$

Multiply equation 1 by -5 and then add to equation 2.

$$\begin{array}{r} -5x+15y+20z = 5 \\ 5x-8y-14z = 0 \\ \hline 7y+6z = 5 \end{array}$$

$$\begin{pmatrix} 7y+6z = 5 \\ 7y+6z = 5 \end{pmatrix}$$

Multiply equation 1 by -1 and then add to equation 2.

$$\begin{array}{r} -7y-6z = -5 \\ 7y+6z = 5 \\ \hline 0 = 0 \end{array}$$

Since $0 = 0$ is a true statement, the system is dependent. There are infinitely many solutions.

13. $\begin{pmatrix} x^2+y^2=7 \\ x^2-y^2=1 \end{pmatrix}$

Add the equations.

$$\begin{array}{l} x^2+y^2=7 \\ \underline{x^2-y^2=1} \\ 2x^2 \quad = 8 \end{array}$$

$x^2=4$

$x=\pm\sqrt{4}=\pm 2$

If $x=2$

$x^2+y^2=7$

$2^2+y^2=7$

$4+y^2=7$

$y^2=3$

$y=\pm\sqrt{3}$

$(2, \sqrt{3}),\ (2, -\sqrt{3})$

If $x=-2$

$x^2+y^2=7$

$(-2)^2+y^2=7$

$4+y^2=7$

$y^2=3$

$y=\pm\sqrt{3}$

The solution set is $\{(2,\sqrt{3}),\ (2,-\sqrt{3}),\ (-2,\ \sqrt{3}),\ (-2,-\sqrt{3})\}$.

15. $\begin{pmatrix} y=x^2+4x+7 \\ y=-x^2-4x+1 \end{pmatrix}$

$y=x^2+4x+7$

$-x^2-4x+1=x^2+4x+7$

$0=2x^2+8x+6$

$0=2(x^2+4x+3)$

$0=2(x+1)(x+3)$

$x+1=0 \quad \text{or} \quad x+3=0$

$x=-1 \quad \text{or} \quad x=-3$

If $x=-1$

$y=x^2+4x+7$

$y=(-1)^2+4(-1)+7$

$y=1-4+7$

$y=4$

$(-1,\ 4)$

If $x=-3$

$y=x^2+4x+7$

$y=(-3)^2+4(-3)+7$

$y=9-12+7$

$y=4$

$(-3,\ 4)$

The solution set is $\{(-1,\ 4),\ (-3,\ 4)\}$.

17 - 19. See back of textbook for graphs.

21. Let $x=$ one number
$y=$ other number

$\begin{pmatrix} x^2+y^2=13 \\ x=y+1 \end{pmatrix}$

$x^2+y^2=13$

$(y+1)^2+y^2=13$

$y^2+2y+1+y^2=13$

$2y^2+2y-12=0$

$2(y^2+y-6)=0$

$2(y+3)(y-2)=0$

$y+3=0 \quad \text{or} \quad y-2=0$

$y=-3 \quad \text{or} \quad y=2$

If $y=-3$

$x=y+1$

$x=-3+1=-2$

If $y=2$

$x=y+1$

$y=2+1=3$

The numbers are -2 and -3 or 2 and 3.

23. Let $x =$ one number
 $y =$ other number

$\begin{pmatrix} x = y^2 + 1 \\ x + y = 7 \end{pmatrix}$

$x + y = 7$

$y^2 + 1 + y = 7$

$y^2 + y - 6 = 0$

$(y + 3)(y - 2) = 0$

$y + 3 = 0$ or $y - 2 = 0$

$y = -3$ or $y = 2$

If $y = -3$

$x = y^2 + 1$

$x = (-3)^2 + 1$

$x = 9 + 1 = 10$

If $y = 2$

$x = y^2 + 1$

$x = (2)^2 + 1$

$x = 4 + 1 = 5$

The numbers are -3 and 10 or 2 and 5.

25. Let $x =$ cost of pound of cashews
 $y =$ cost of pound of peanuts

$\begin{pmatrix} 7x + 5y = 88 \\ 3x + 2y = 37 \end{pmatrix}$

Multiply equation 1 by 2.
Multiply equation 2 by -5.

$$\begin{array}{r} 14x + 10y = 176 \\ -15x - 10y = -185 \\ \hline -x \quad = -9 \end{array}$$

$x = 9$

$7x + 5y = 88$

$7(9) + 5y = 88$

$63 + 5y = 88$

$5y = 25$

$y = 5$

The cashews cost \$9 a pound and the peanuts cost \$5 a pound.

27. Let $x =$ fixed fee
 $y =$ additional fee

$\begin{pmatrix} x + 4y = 2.40 \\ x + 11y = 3.10 \end{pmatrix}$

Multiply equation 1 by -1, then add to equation 2.

$$\begin{array}{r} -x - 4y = -2.40 \\ x + 11y = 3.10 \\ \hline 7y = .70 \end{array}$$

$y = .10$

$x + 4y = 2.40$

$x + 4(.10) = 2.40$

$x = 2.00$

The fixed fee is \$2.00, the additional fee is \$.10 per pound over the initial one pound.

Chapter 11 Test

1. The graphs are parallel lines for systems III because the slopes are the same and the constants are different.

3. The solution set is $\emptyset$ for system III because the lines are parallel lines and do not intersect.

5. $\begin{pmatrix} 3x - 2y = -14 \\ 7x + 2y = -6 \end{pmatrix}$

Add the equations.

$$\begin{array}{r} 3x - 2y = -14 \\ 7x + 2y = -6 \\ \hline 10x \quad = -20 \end{array}$$

$x = -2$

$3x - 2y = -14$

$3(-2) - 2(y) = -14$

$-6 - 2y = -14$

$-2y = -8$

$y = 4$

The solution set is $\{(-2, 4)\}$.

7. $\begin{pmatrix} \frac{3}{4}x - \frac{1}{2}y = -21 \\ \frac{2}{3}x + \frac{1}{6}y = -4 \end{pmatrix}$

Multiply equation 1 by 8 and multiply equation 2 by 6.

$\begin{pmatrix} 6x - 4y = -168 \\ 4x + y = -24 \end{pmatrix}$

Multiply equation 2 by 4 then add to equation 1.

$$\begin{array}{l} 16x + 4y = -96 \\ \underline{6x - 4y = -168} \\ 22x \quad = -264 \end{array}$$

$x = -12$

9. $\begin{pmatrix} 5x + 2y = -1 \\ 3x - 4y = 3 \end{pmatrix}$

Multiply equation 1 by 2 and then add to equation 2.

$$\begin{array}{l} 10x + 4y = -2 \\ \underline{3x - 4y = 3} \\ 13x \quad = 1 \end{array}$$

$x = \frac{1}{13}$

11. $\begin{pmatrix} 2x - y + 3z = -8 \\ 3x + y - 2z = -3 \\ 5x + y + 7z = -16 \end{pmatrix}$

Add equation 1 to equation 2.

$$\begin{array}{l} 2x - y + 3z = -8 \\ \underline{3x + y - 2z = -3} \\ 5x \quad + z = -11 \end{array}$$

Add equation 1 to equation 3.

$$\begin{array}{l} 2x - y + 3z = -8 \\ \underline{5x + y + 7z = -16} \\ 7x \quad + 10z = -24 \end{array}$$

$\begin{pmatrix} 5x + z = -11 \\ 7x + 10z = -24 \end{pmatrix}$

Multiply equation 1 by -10 and then add to equation 2.

$$\begin{array}{l} -50x - 10z = 110 \\ \underline{7x + 10z = -24} \\ -43x \quad = 86 \end{array}$$

$x = -2$

13. $\begin{pmatrix} x + 2y - 3z = -7 \\ -x + y + 4z = 6 \\ x - 3y + 8z = 20 \end{pmatrix}$

Add equation 1 and 2.

$$\begin{array}{l} x + 2y - 3z = -7 \\ \underline{-x + y + 4z = 6} \\ 3y + z = -1 \end{array}$$

Add equation 2 and equation 3.

$$\begin{array}{l} -x + y + 4z = 6 \\ \underline{x - 3y + 8z = 20} \\ -2y + 12z = 26 \end{array}$$

$\begin{pmatrix} 3y + z = -1 \\ -2y + 12z = 26 \end{pmatrix}$

Multiply equation 1 by 2 and multiply equation 2 by 3. Then add the equations.

$$\begin{array}{l} 6y + 2z = -2 \\ \underline{-6y + 36z = 78} \\ 38z = 76 \end{array}$$

$z = 2$

15. $\begin{pmatrix} y = 3x - 4 \\ 9x - 3y = 12 \end{pmatrix}$

$9x - 3y = 12$

$9x - 3(3x - 4) = 12$

$9x - 9x + 12 = 12$

$12 = 12$

Since $12 = 12$ is a true statement, the system is dependent. There are infinitely many solutions.

17. $\begin{pmatrix} y = x^2 - 4x \\ y = -2x^2 + 8x - 12 \end{pmatrix}$

$y = x^2 - 4x$

$-2x^2 + 8x - 12 = x^2 - 4x$

$0 = 3x^2 - 12x - 12$

$0 = 3(x^2 - 4x - 4)$

$0 = 3(x - 2)^2$

$x - 2 = 0$

$x = 2$

There will be one solution.

19. $\begin{pmatrix} y = x^2 \\ y = -x^2 + 8 \end{pmatrix}$

$y = x^2$

$-x^2 + 8 = x^2$

$8 = 2x^2$

$4 = x^2$

$\pm\sqrt{4} = x$

$x = \pm 2$

21. Let $x =$ digit in the tens place
$y =$ digit in the units place

$\begin{pmatrix} x = 2y + 1 \\ 10y + x = 10x + y - 27 \end{pmatrix}$

$\begin{pmatrix} x = 2y + 1 \\ -9x + 9y = -27 \end{pmatrix}$

Divide equation 3 by -9.

$\begin{pmatrix} x = 2y + 1 \\ x - y = 3 \end{pmatrix}$

$x - y = 3$

$2y + 1 - y = 3$

$y + 1 = 3$

$y = 2$

$x = 2y + 1$

$x = 2(2) + 1$

$x = 5$

The original number is 52.

23. Let $x =$ number of nickels
$y =$ number of dimes
$z =$ number of quarters

$\begin{pmatrix} x + y + z = 43 \\ z = 3x + 1 \\ 5x + 10y + 25z = 725 \end{pmatrix}$

Substitute for z in both equations 1 and 3.

$x + y + 3x + 1 = 43$

$4x + y = 42$

$5x + 10y + 25(3x + 1) = 725$

$5x + 10y + 75x + 25 = 725$

$80x + 10y = 700$

$8x + y = 70$

$\begin{pmatrix} 4x + y = 42 \\ 8x + y = 70 \end{pmatrix}$

Multiply equation 1 by -1 and then add to equation 2.

$$\begin{array}{rcr} -4x - y & = & -42 \\ 8x + y & = & 70 \\ \hline 4x & = & 28 \end{array}$$

$x = 7$

$z = 3x + 1$

$z = 3(7) + 1$

$z = 22$

There are 22 quarters.

25. See back of textbook for graph.

Chapter 12 Using Matrices and Determinants to Solve Linear Systems

Problem Set 12.1 Matrix Approach to Solving Systems

1. $\begin{pmatrix} x+3y=8 \\ 7x-2y=-13 \end{pmatrix}$

$$\left[\begin{array}{cc|c} 1 & 3 & 8 \\ 7 & -2 & -13 \end{array}\right]$$

Multiply row 1 by -7 and add to row 2.

$$\left[\begin{array}{cc|c} 1 & 3 & 8 \\ 0 & -23 & -69 \end{array}\right]$$

$-23y=-69$

$y=3$

$x+3y=8$

$x+3(3)=8$

$x+9=8$

$x=-1$

The solution set is $\{(-1, 3)\}$.

3. $\begin{pmatrix} 5x+2y=-20 \\ x-4y=18 \end{pmatrix}$

$$\left[\begin{array}{cc|c} 5 & 2 & -20 \\ 1 & -4 & 18 \end{array}\right]$$

Interchange row 1 and row 2.

$$\left[\begin{array}{cc|c} 1 & -4 & 18 \\ 5 & 2 & -20 \end{array}\right]$$

Multiply row 1 by -5 and add to row 2.

$$\left[\begin{array}{cc|c} 1 & -4 & 18 \\ 0 & 22 & -110 \end{array}\right]$$

$22y=-110$

$y=-5$

$x-4y=18$

$x-4(-5)=18$

$x+20=18$

$x=-2$

The solution set is $\{(-2,-5)\}$.

5. $\begin{pmatrix} x-y=6 \\ 3x+2y=7 \end{pmatrix}$

$$\left[\begin{array}{cc|c} 1 & -1 & 6 \\ 3 & 2 & 7 \end{array}\right]$$

Multiply row 1 by -3 and add to row 2.

$$\left[\begin{array}{cc|c} 1 & -1 & 6 \\ 0 & 5 & -11 \end{array}\right]$$

$5y=-11$

$y=-\frac{11}{5}$

$x-y=6$

$x-\left(-\frac{11}{5}\right)=6$

$x+\frac{11}{5}=\frac{30}{5}$

$x=\frac{19}{5}$

The solution set is $\left\{\left(\frac{19}{5}, -\frac{11}{5}\right)\right\}$.

7. $\begin{pmatrix} 3x+7y=-45 \\ 5x-y=1 \end{pmatrix}$

$$\left[\begin{array}{cc|c} 3 & 7 & -45 \\ 5 & -1 & 1 \end{array}\right]$$

Multiply row 1 by $-\frac{5}{3}$ and add to row 2.

$$\left[\begin{array}{cc|c} 3 & 7 & -45 \\ 0 & -\frac{38}{3} & 76 \end{array}\right]$$

$-\frac{38}{3}y = 76$

$y = -\frac{3}{38}(76)$

$y = -6$

$3x + 7y = -45$

$3x + 7(-6) = -45$

$3x = -3$

$x = -1$

The solution set is $\{(-1, -6)\}$.

9. $\begin{pmatrix} 3x + 2y = 17 \\ 11x - 4y = 85 \end{pmatrix}$

$$\left[\begin{array}{cc|c} 3 & 2 & 17 \\ 11 & -4 & 85 \end{array}\right]$$

Multiply row 1 by $-\frac{11}{3}$ and add to row 2.

$$\left[\begin{array}{cc|c} 3 & 2 & 17 \\ 0 & -\frac{34}{3} & \frac{68}{3} \end{array}\right]$$

$-\frac{34}{3}y = \frac{68}{3}$

$y = -\frac{3}{34}\left(\frac{68}{3}\right)$

$y = -2$

$3x + 2y = 17$

$3x + 2(-2) = 17$

$3x - 4 = 17$

$3x = 21$

$x = 7$

The solution set is $\{(7, -2)\}$.

11. $\begin{pmatrix} 2x + 9y = 98 \\ 6x - 5y = -26 \end{pmatrix}$

$$\left[\begin{array}{cc|c} 2 & 9 & 98 \\ 6 & -5 & -26 \end{array}\right]$$

Multiply row 1 by -3 and add to row 2.

$$\left[\begin{array}{cc|c} 2 & 9 & 98 \\ 0 & -32 & -320 \end{array}\right]$$

$-32y = -320$

$y = 10$

$2x + 9y = 98$

$2x + 9(10) = 98$

$2x + 90 = 98$

$2x = 8$

$x = 4$

The solution set is $\{(4, 10)\}$.

13. $\begin{pmatrix} 9x - 2y = -6 \\ 4x + 5y = 15 \end{pmatrix}$

$$\left[\begin{array}{cc|c} 9 & -2 & -6 \\ 4 & 5 & 15 \end{array}\right]$$

Multiply row 1 by $-\frac{4}{9}$ and add to row 2.

$$\left[\begin{array}{cc|c} 9 & -2 & -6 \\ 0 & \frac{53}{9} & \frac{159}{9} \end{array}\right]$$

$\frac{53}{9}y = \frac{159}{9}$

$y = \frac{9}{53}\left(\frac{159}{9}\right)$

$y = 3$

$9x - 2y = -6$

$9x - 2(3) = -6$

$9x - 6 = -6$

$9x = 0$

$x = 0$

The solution set is $\{(0, 3)\}$.

15. $\begin{pmatrix} 2x - 9y + 4z = -27 \\ x - 5y - 2z = -11 \\ 5x + 2y - 3z = 19 \end{pmatrix}$

$$\left[\begin{array}{ccc|c} 2 & -9 & 4 & -27 \\ 1 & -5 & -2 & -11 \\ 5 & 2 & -3 & 19 \end{array}\right]$$

Interchange row 1 and row 2.

$$\left[\begin{array}{ccc|c} 1 & -5 & -2 & -11 \\ 2 & -9 & 4 & -27 \\ 5 & 2 & -3 & 19 \end{array}\right]$$

$-2(\text{row } 1) + \text{row } 2$

$-5(\text{row } 1) + \text{row } 3$

$$\left[\begin{array}{ccc|c} 1 & -5 & -2 & -11 \\ 0 & 1 & 8 & -5 \\ 0 & 27 & 7 & 74 \end{array}\right]$$

$-27(\text{row } 2) + \text{row } 3$

$$\left[\begin{array}{ccc|c} 1 & -5 & -2 & -11 \\ 0 & 1 & 8 & -5 \\ 0 & 0 & -209 & 209 \end{array}\right]$$

$-209z = 209$

$z = -1$

$y + 8z = -5$

$y + 8(-1) = -5$

$y - 8 = -5$

$y = 3$

$x - 5y - 2z = -11$

$x - 5(3) - 2(-1) = -11$

$x - 15 + 2 = -11$

$x = 2$

The solution set is $\{(2, 3, -1)\}$.

17. $\begin{pmatrix} 2x + 3y - z = -3 \\ 5x - 2y - 4z = -1 \\ 3x - 8y + 3z = 17 \end{pmatrix}$

$$\left[\begin{array}{ccc|c} 2 & 3 & -1 & -3 \\ 5 & -2 & -4 & -1 \\ 3 & -8 & 3 & 17 \end{array}\right]$$

$-\frac{5}{2}(\text{row } 1) + \text{row } 2$

$-\frac{3}{2}(\text{row } 1) + \text{row } 3$

$$\left[\begin{array}{ccc|c} 2 & 3 & -1 & -3 \\ 0 & -\frac{19}{2} & -\frac{3}{2} & \frac{13}{2} \\ 0 & -\frac{25}{2} & \frac{9}{2} & \frac{43}{2} \end{array}\right]$$

$-\frac{25}{19}(\text{row } 2) + \text{row } 3$

$$\left[\begin{array}{ccc|c} 2 & 3 & -1 & -3 \\ 0 & -\frac{19}{2} & -\frac{3}{2} & \frac{13}{2} \\ 0 & 0 & \frac{123}{19} & \frac{246}{19} \end{array}\right]$$

$\frac{123}{19}z = \frac{246}{19}$

$z = \frac{19}{123}\left(\frac{246}{19}\right)$

$z = 2$

$-\frac{19}{2}y - \frac{3}{2}z = \frac{13}{2}$

$-\frac{19}{2}y - \frac{3}{2}(2) = \frac{13}{2}$

$-\frac{19}{2}y - 3 = \frac{13}{2}$

$-\frac{19}{2}y = \frac{19}{2}$

$y = -1$

$2x + 3y - z = -3$

$2x + 3(-1) - (2) = -3$

$2x - 3 - 2 = -3$

$2x = 2$

$x = 1$

The solution set is $\{(1, -1, 2)\}$.

19. $\begin{pmatrix} 2x + 5y - 2z = -8 \\ 3x - 11y + 4z = -5 \\ 4x + 3y - z = -13 \end{pmatrix}$

$$\left[\begin{array}{ccc|c} 2 & 5 & -2 & -8 \\ 3 & -11 & 4 & -5 \\ 4 & 3 & -1 & -13 \end{array}\right]$$

$-\frac{3}{2}$(row 1) + row 2

-2(row 1) + row 3

$$\left[\begin{array}{ccc|c} 2 & 5 & -2 & -8 \\ 0 & -\frac{37}{2} & 7 & 7 \\ 0 & -7 & 3 & 3 \end{array}\right]$$

$-\frac{2}{37}$(row 2)

$$\left[\begin{array}{ccc|c} 2 & 5 & -2 & -8 \\ 0 & 1 & -\frac{14}{37} & -\frac{14}{37} \\ 0 & -7 & 3 & 3 \end{array}\right]$$

-7(row 2) + row 3

$$\left[\begin{array}{ccc|c} 2 & 5 & -2 & -8 \\ 0 & 1 & -\frac{14}{37} & -\frac{14}{37} \\ 0 & 0 & \frac{13}{37} & \frac{13}{37} \end{array}\right]$$

$\frac{13}{37}z = \frac{13}{37}$

$z = 1$

$y - \frac{14}{37}z = -\frac{14}{37}$

$y - \frac{14}{37}(1) = -\frac{14}{37}$

$y = 0$

$2x + 5y - 2z = -8$

$2x + 5(0) - 2(1) = -8$

$2x + 0 - 2 = -8$

$2x = -6$

$x = -3$

The solution set is $\{(-3, 0, 1)\}$.

21. $\begin{pmatrix} x + 5y - 2z = -5 \\ 3x - 2y - 3z = 24 \\ -2x - 9y + 5z = 3 \end{pmatrix}$

$$\left[\begin{array}{ccc|c} 1 & 5 & -2 & -5 \\ 3 & -2 & -3 & 24 \\ -2 & -9 & 5 & 3 \end{array}\right]$$

-3(row 1) + row 2

2(row 1) + row3

$$\left[\begin{array}{ccc|c} 1 & 5 & -2 & -5 \\ 0 & -17 & 3 & 39 \\ 0 & 1 & 1 & -7 \end{array}\right]$$

Interchange row 2 and row 3.

$$\left[\begin{array}{ccc|c} 1 & 5 & -2 & -5 \\ 0 & 1 & 1 & -7 \\ 0 & -17 & 3 & 39 \end{array}\right]$$

17(row 2) + row 3

$$\left[\begin{array}{ccc|c} 1 & 5 & -2 & -5 \\ 0 & 1 & 1 & -7 \\ 0 & 0 & 20 & -80 \end{array}\right]$$

$20z = -80$

$z = -4$

$y + z = -7$

$y - 4 = -7$

$y = -3$

$x + 5y - 2z = -5$

$x + 5(-3) - 2(-4) = -5$

$x - 15 + 8 = -5$

$x - 7 = -5$

$x = 2$

The solution set is $\{(2, -3, -4)\}$.

23. $\begin{pmatrix} 2x - 3y - 4z = 1 \\ x - 2y - 5z = -1 \\ -3x + 5y + 2z = 2 \end{pmatrix}$

$$\left[\begin{array}{ccc|c} 2 & -3 & -4 & 1 \\ 1 & -2 & -5 & -1 \\ -3 & 5 & 2 & 2 \end{array}\right]$$

Interchange row 1 and row 2.

$$\left[\begin{array}{ccc|c} 1 & -2 & -5 & -1 \\ 2 & -3 & -4 & 1 \\ -3 & 5 & 2 & 2 \end{array}\right]$$

-2(row 1) + row 2
3(row 1) + row 3

$$\left[\begin{array}{ccc|c} 1 & -2 & -5 & -1 \\ 0 & 1 & 6 & 3 \\ 0 & -1 & -13 & -1 \end{array}\right]$$

row 2 + row 3

$$\left[\begin{array}{ccc|c} 1 & -2 & -5 & -1 \\ 0 & 1 & 6 & 3 \\ 0 & 0 & -7 & 2 \end{array}\right]$$

$-7z = 2$

$z = -\frac{2}{7}$

$y + 6z = 3$

$y + 6\left(-\frac{2}{7}\right) = 3$

$y - \frac{12}{7} = \frac{21}{7}$

$y = \frac{33}{7}$

$x - 2y - 5z = -1$

$x - 2\left(\frac{33}{7}\right) - 5\left(-\frac{2}{7}\right) = -1$

$x - \frac{66}{7} + \frac{10}{7} = -1$

$x - 8 = -1$

$x = 7$

The solution set is $\left\{\left(7, \frac{33}{7}, -\frac{2}{7}\right)\right\}$.

25. $\begin{pmatrix} -x + y - 4z = -15 \\ -2x + 3y + 5z = 25 \\ 4x - 6y - z = -14 \end{pmatrix}$

$$\left[\begin{array}{ccc|c} -1 & 1 & -4 & -15 \\ -2 & 3 & 5 & 25 \\ 4 & -6 & -1 & -14 \end{array}\right]$$

-2(row 1) + row 2
4(row 1) + row 3

$$\left[\begin{array}{ccc|c} -1 & 1 & -4 & -15 \\ 0 & 1 & 13 & 55 \\ 0 & -2 & -17 & -74 \end{array}\right]$$

2(row 2) + row 3

$$\left[\begin{array}{ccc|c} -1 & 1 & -4 & -15 \\ 0 & 1 & 13 & 55 \\ 0 & 0 & 9 & 36 \end{array}\right]$$

$9z = 36$

$z = 4$

$y + 13z = 55$

$y + 13(4) = 55$

$y + 52 = 55$

$y = 3$

$-x + y - 4z = -15$

$-x + 3 - 4(4) = -15$

$-x + 3 - 16 = -15$

$-x - 13 = -15$

$-x = -2$

$x = 2$

The solution set is {(2, 3, 4)}.

27. $\begin{pmatrix} 4x - 3y + 2z = 17 \\ 2y + 5z = 24 \\ 3x - z = -9 \end{pmatrix}$

$$\left[\begin{array}{ccc|c} 4 & -3 & 2 & 17 \\ 0 & 2 & 5 & 24 \\ 3 & 0 & -1 & -9 \end{array}\right]$$

$-\frac{3}{4}$(row 1) + row 3

$$\left[\begin{array}{ccc|c} 4 & -3 & 2 & 17 \\ 0 & 2 & 5 & 24 \\ 0 & \frac{9}{4} & -\frac{5}{2} & -\frac{87}{4} \end{array}\right]$$

$\frac{1}{2}$(row 2)

$$\left[\begin{array}{ccc|c} 4 & -3 & 2 & 17 \\ 0 & 1 & \frac{5}{2} & 12 \\ 0 & \frac{9}{4} & -\frac{5}{2} & -\frac{87}{4} \end{array}\right]$$

$-\frac{9}{4}$(row 2) + row 3

$$\left[\begin{array}{ccc|c} 4 & -3 & 2 & 17 \\ 0 & 1 & \frac{5}{2} & 12 \\ 0 & 0 & -\frac{65}{8} & -\frac{195}{4} \end{array}\right]$$

$-\frac{65}{8}z = -\frac{195}{4}$

$z = -\frac{8}{65}\left(-\frac{195}{4}\right)$

$z = 6$

$y + \frac{5}{2}z = 12$

$y + \frac{5}{2}(6) = 12$

$y + 15 = 12$

$y = -3$

$4x - 3y + 2z = 17$

$4x - 3(-3) + 2(6) = 17$

$4x + 9 + 12 = 17$

$4x + 21 = 17$

$4x = -4$

$x = -1$

The solution set is $\{(-1, -3, 6)\}$.

29. $\begin{pmatrix} 4x - y - 2z = -7 \\ 5x + y + 3z = -20 \\ 2x - 3y - z = 9 \end{pmatrix}$

$$\left[\begin{array}{ccc|c} 4 & -1 & -2 & -7 \\ 5 & 1 & 3 & -20 \\ 2 & -3 & -1 & 9 \end{array}\right]$$

$-\frac{5}{4}$(row 1) + row 2

$-\frac{1}{2}$(row 1) + row 3

$$\left[\begin{array}{ccc|c} 4 & -1 & -2 & -7 \\ 0 & \frac{9}{4} & \frac{11}{2} & -\frac{45}{4} \\ 0 & -\frac{5}{2} & 0 & \frac{25}{2} \end{array}\right]$$

$-\frac{5}{2}y = \frac{25}{2}$

$y = -\frac{2}{5}\left(\frac{25}{2}\right)$

$y = -5$

$\frac{9}{4}y + \frac{11}{2}z = -\frac{45}{4}$

$\frac{9}{4}(-5) + \frac{11}{2}z = -\frac{45}{4}$

$-\frac{45}{4} + \frac{11}{2}z = -\frac{45}{4}$

$\frac{11}{2}z = 0$

$z = \frac{2}{11}(0)$

$z = 0$

$4x - y - 2z = -7$

$4x - (-5) - 2(0) = -7$

$4x + 5 + 0 = -7$

$4x = -12$

$x = -3$

The solution set is $\{(-3, -5, 0)\}$.

Problem Set 12.2 Reduced Echelon Form

1. $\left[\begin{array}{cc|c} 1 & 2 & 8 \\ 0 & 0 & 0 \end{array}\right]$

Yes

3. $\left[\begin{array}{ccc|c} 1 & 0 & 3 & 8 \\ 0 & 1 & 2 & -6 \\ 0 & 0 & 0 & 0 \end{array}\right]$

Yes

5. $$\left[\begin{array}{ccc|c} 1 & 0 & 0 & -7 \\ 0 & 1 & 0 & 0 \\ 0 & 0 & 1 & 9 \end{array}\right]$$

Yes

7. $$\left[\begin{array}{ccc|c} 1 & 0 & 0 & 5 \\ 0 & 3 & 0 & 8 \\ 0 & 0 & 1 & -11 \end{array}\right]$$

No

9. $$\left[\begin{array}{cccc|c} 1 & 0 & 0 & 0 & 2 \\ 0 & 0 & 1 & 0 & 4 \\ 0 & 1 & 0 & 0 & -3 \\ 0 & 0 & 0 & 1 & 9 \end{array}\right]$$

No

11. $$\begin{pmatrix} 2x+7y=-55 \\ x-4y=25 \end{pmatrix}$$

$$\left[\begin{array}{cc|c} 2 & 7 & -55 \\ 1 & -4 & 25 \end{array}\right]$$

Interchange rows.

$$\left[\begin{array}{cc|c} 1 & -4 & 25 \\ 2 & 7 & -55 \end{array}\right]$$

-2(row 1) + row 2

$$\left[\begin{array}{cc|c} 1 & -4 & 25 \\ 0 & 15 & -105 \end{array}\right]$$

$\frac{1}{15}$(row 2)

$$\left[\begin{array}{cc|c} 1 & -4 & 25 \\ 0 & 1 & -7 \end{array}\right]$$

4(row 2) + row 1

$$\left[\begin{array}{cc|c} 1 & 0 & -3 \\ 0 & 1 & -7 \end{array}\right]$$

The solution set is $\{(-3, -7)\}$.

13. $$\begin{pmatrix} x+5y=-18 \\ -2x+3y=-16 \end{pmatrix}$$

$$\left[\begin{array}{cc|c} 1 & 5 & -18 \\ -2 & 3 & -16 \end{array}\right]$$

2(row 1) + row 2

$$\left[\begin{array}{cc|c} 1 & 5 & -18 \\ 0 & 13 & -52 \end{array}\right]$$

$\frac{1}{13}$(row 2)

$$\left[\begin{array}{cc|c} 1 & 5 & -18 \\ 0 & 1 & -4 \end{array}\right]$$

-5(row 2) + row 1

$$\left[\begin{array}{cc|c} 1 & 0 & 2 \\ 0 & 1 & -4 \end{array}\right]$$

The solution set is $\{(2, -4)\}$.

15. $$\begin{pmatrix} 3x+9y=-1 \\ x+3y=10 \end{pmatrix}$$

$$\left[\begin{array}{cc|c} 3 & 9 & -1 \\ 1 & 3 & 10 \end{array}\right]$$

Interchange rows.

$$\left[\begin{array}{cc|c} 1 & 3 & 10 \\ 3 & 9 & -1 \end{array}\right]$$

-3(row 1) + row 2

$$\left[\begin{array}{cc|c} 1 & 3 & 10 \\ 0 & 0 & -31 \end{array}\right]$$

The system is inconsistent.

The solution set is $\emptyset$.

17. $\begin{pmatrix} 2x-3y=-12 \\ 3x+2y=8 \end{pmatrix}$

$$\left[\begin{array}{cc|c} 2 & -3 & -12 \\ 3 & 2 & 8 \end{array}\right]$$

$-\frac{3}{2}$(row 1) + row 2

$$\left[\begin{array}{cc|c} 2 & -3 & -12 \\ 0 & \frac{13}{2} & 26 \end{array}\right]$$

$\frac{2}{13}$(row 2)

$$\left[\begin{array}{cc|c} 2 & -3 & -12 \\ 0 & 1 & 4 \end{array}\right]$$

3(row 2) + row 1

$$\left[\begin{array}{cc|c} 2 & 0 & 0 \\ 0 & 1 & 4 \end{array}\right]$$

The solution set is {(0, 4)}.

19. $\begin{pmatrix} x+3y-4z=13 \\ 2x+7y-3z=11 \\ -2x-y+2z=-8 \end{pmatrix}$

$$\left[\begin{array}{ccc|c} 1 & 3 & -4 & 13 \\ 2 & 7 & -3 & 11 \\ -2 & -1 & 2 & -8 \end{array}\right]$$

-2(row 1) + row 2

2(row 1) + row 3

$$\left[\begin{array}{ccc|c} 1 & 3 & -4 & 13 \\ 0 & 1 & 5 & -15 \\ 0 & 5 & -6 & 18 \end{array}\right]$$

-3(row 2) + row 1

-5(row 2) + row 3

$$\left[\begin{array}{ccc|c} 1 & 0 & -19 & 58 \\ 0 & 1 & 5 & -15 \\ 0 & 0 & -31 & 93 \end{array}\right]$$

$-\frac{1}{31}$(row 3)

$$\left[\begin{array}{ccc|c} 1 & 0 & -19 & 58 \\ 0 & 1 & 5 & -15 \\ 0 & 0 & 1 & -3 \end{array}\right]$$

-5(row 3) + row 2

19(row 3) + row 1

$$\left[\begin{array}{ccc|c} 1 & 0 & 0 & 1 \\ 0 & 1 & 0 & 0 \\ 0 & 0 & 1 & -3 \end{array}\right]$$

The solution set is {(1, 0, −3)}.

21. $\begin{pmatrix} x-4y+3z=16 \\ 2x+3y-4z=-22 \\ -3x+11y-z=-36 \end{pmatrix}$

$$\left[\begin{array}{ccc|c} 1 & -4 & 3 & 16 \\ 2 & 3 & -4 & -22 \\ -3 & 11 & -1 & -36 \end{array}\right]$$

-2(row 1) + row 2

3(row 1) + row 3

$$\left[\begin{array}{ccc|c} 1 & -4 & 3 & 16 \\ 0 & 11 & -10 & -54 \\ 0 & -1 & 8 & 12 \end{array}\right]$$

Interchange row 2 and row 3.

$$\left[\begin{array}{ccc|c} 1 & -4 & 3 & 16 \\ 0 & -1 & 8 & 12 \\ 0 & 11 & -10 & -54 \end{array}\right]$$

-1(row 2)

$$\left[\begin{array}{ccc|c} 1 & -4 & 3 & 16 \\ 0 & 1 & -8 & -12 \\ 0 & 11 & -10 & -54 \end{array}\right]$$

4(row 2) + row 1
-11(row 2) + row 3

$$\left[\begin{array}{ccc|c} 1 & 0 & -29 & -32 \\ 0 & 1 & -8 & -12 \\ 0 & 0 & 78 & 78 \end{array}\right]$$

$\frac{1}{78}$(row 3)

$$\left[\begin{array}{ccc|c} 1 & 0 & -29 & -32 \\ 0 & 1 & -8 & -12 \\ 0 & 0 & 1 & 1 \end{array}\right]$$

8(row 3) + row 2
29(row 3) + row 1

$$\left[\begin{array}{ccc|c} 1 & 0 & 0 & -3 \\ 0 & 1 & 0 & -4 \\ 0 & 0 & 1 & 1 \end{array}\right]$$

The solution set is $\{(-3, -4, 1)\}$.

23. $\begin{pmatrix} -3x + 2y + z = 17 \\ x - y + 5z = -2 \\ 4x - 5y - 3z = -36 \end{pmatrix}$

$$\left[\begin{array}{ccc|c} -3 & 2 & 1 & 17 \\ 1 & -1 & 5 & -2 \\ 4 & -5 & -3 & -36 \end{array}\right]$$

Interchange row 1 and row 2.

$$\left[\begin{array}{ccc|c} 1 & -1 & 5 & -2 \\ -3 & 2 & 1 & 17 \\ 4 & -5 & -3 & -36 \end{array}\right]$$

3(row 1) + row 2
-4(row 1) + row 3

$$\left[\begin{array}{ccc|c} 1 & -1 & 5 & -2 \\ 0 & -1 & 16 & 11 \\ 0 & -1 & -23 & -28 \end{array}\right]$$

-1(row 2)

$$\left[\begin{array}{ccc|c} 1 & -1 & 5 & -2 \\ 0 & 1 & -16 & -11 \\ 0 & -1 & -23 & -28 \end{array}\right]$$

row 2 + row 1
row 2 + row 3

$$\left[\begin{array}{ccc|c} 1 & 0 & -11 & -13 \\ 0 & 1 & -16 & -11 \\ 0 & 0 & -39 & -39 \end{array}\right]$$

$-\frac{1}{39}$(row 3)

$$\left[\begin{array}{ccc|c} 1 & 0 & -11 & -13 \\ 0 & 1 & -16 & -11 \\ 0 & 0 & 1 & 1 \end{array}\right]$$

16(row 3) + row 2
11(row 3) + row 1

$$\left[\begin{array}{ccc|c} 1 & 0 & 0 & -2 \\ 0 & 1 & 0 & 5 \\ 0 & 0 & 1 & 1 \end{array}\right]$$

The solution set is $\{(-2, 5, 1)\}$.

25. $\begin{pmatrix} x + 2y - 5z = -1 \\ 2x + 3y - 2z = 2 \\ 3x + 5y - 7z = 4 \end{pmatrix}$

$$\left[\begin{array}{ccc|c} 1 & 2 & -5 & -1 \\ 2 & 3 & -2 & 2 \\ 3 & 5 & -7 & 4 \end{array}\right]$$

$-2(\text{row } 1) + \text{row } 2$

$-3(\text{row } 1) + \text{row } 3$

$$\left[\begin{array}{ccc|c} 1 & 2 & -5 & -1 \\ 0 & -1 & 8 & 4 \\ 0 & -1 & 8 & 7 \end{array}\right]$$

$-1(\text{row } 2) + \text{row } 3$

$$\left[\begin{array}{ccc|c} 1 & 2 & -5 & -1 \\ 0 & -1 & 8 & 4 \\ 0 & 0 & 0 & 3 \end{array}\right]$$

The system is inconsistent.

The solution set is $\emptyset$.

27. $\begin{pmatrix} 4x - 10y + 3z = -19 \\ 2x + 5y - z = -7 \\ x - 3y - 2z = -2 \end{pmatrix}$

$$\left[\begin{array}{ccc|c} 4 & -10 & 3 & -19 \\ 2 & 5 & -1 & -7 \\ 1 & -3 & -2 & -2 \end{array}\right]$$

Interchange row 1 and row 3.

$$\left[\begin{array}{ccc|c} 1 & -3 & -2 & -2 \\ 2 & 5 & -1 & -7 \\ 4 & -10 & 3 & -19 \end{array}\right]$$

$-2(\text{row } 1) + \text{row } 2$

$-4(\text{row } 1) + \text{row } 3$

$$\left[\begin{array}{ccc|c} 1 & -3 & -2 & -2 \\ 0 & 11 & 3 & -3 \\ 0 & 2 & 11 & -11 \end{array}\right]$$

$\frac{1}{11}(\text{row } 2)$

$$\left[\begin{array}{ccc|c} 1 & -3 & -2 & -2 \\ 0 & 1 & \frac{3}{11} & -\frac{3}{11} \\ 0 & 2 & 11 & -11 \end{array}\right]$$

$3(\text{row } 2) + \text{row } 1$

$-2(\text{row } 2) + \text{row } 3$

$$\left[\begin{array}{ccc|c} 1 & 0 & -\frac{13}{11} & -\frac{31}{11} \\ 0 & 1 & \frac{3}{11} & -\frac{3}{11} \\ 0 & 0 & \frac{115}{11} & -\frac{115}{11} \end{array}\right]$$

$\frac{11}{115}(\text{row } 3)$

$$\left[\begin{array}{ccc|c} 1 & 0 & -\frac{13}{11} & -\frac{31}{11} \\ 0 & 1 & \frac{3}{11} & -\frac{3}{11} \\ 0 & 0 & 1 & -1 \end{array}\right]$$

$-\frac{3}{11}(\text{row } 3) + \text{row } 2$

$\frac{13}{11}(\text{row } 3) + \text{row } 1$

$$\left[\begin{array}{ccc|c} 1 & 0 & 0 & -4 \\ 0 & 1 & 0 & 0 \\ 0 & 0 & 1 & -1 \end{array}\right]$$

The solution set is $\{(-4, 0, -1)\}$.

29. $\begin{pmatrix} 4x + 3y - z = 0 \\ 3x + 2y + 5z = 6 \\ 5x - y - 3z = 3 \end{pmatrix}$

$$\left[\begin{array}{ccc|c} 4 & 3 & -1 & 0 \\ 3 & 2 & 5 & 6 \\ 5 & -1 & -3 & 3 \end{array}\right]$$

$\frac{1}{4}(\text{row } 1)$

$$\left[\begin{array}{ccc|c} 1 & \frac{3}{4} & -\frac{1}{4} & 0 \\ 3 & 2 & 5 & 6 \\ 5 & -1 & -3 & 3 \end{array}\right]$$

$-3(\text{row } 1) + \text{row } 2$

$-5(\text{row } 1) + \text{row } 3$

$$\left[\begin{array}{ccc|c} 1 & \frac{3}{4} & -\frac{1}{4} & 0 \\ 0 & -\frac{1}{4} & \frac{23}{4} & 6 \\ 0 & -\frac{19}{4} & -\frac{7}{4} & 3 \end{array}\right]$$

$-4(\text{row } 2)$

$$\left[\begin{array}{ccc|c} 1 & \frac{3}{4} & -\frac{1}{4} & 0 \\ 0 & 1 & -23 & -24 \\ 0 & -\frac{19}{4} & -\frac{7}{4} & 3 \end{array}\right]$$

$-\frac{3}{4}(\text{row } 2) + \text{row } 1$

$\frac{19}{4}(\text{row } 2) + \text{row } 3$

$$\left[\begin{array}{ccc|c} 1 & 0 & 17 & 18 \\ 0 & 1 & -23 & -24 \\ 0 & 0 & -111 & -111 \end{array}\right]$$

$-\frac{1}{111}(\text{row } 3)$

$$\left[\begin{array}{ccc|c} 1 & 0 & 17 & 18 \\ 0 & 1 & -23 & -24 \\ 0 & 0 & 1 & 1 \end{array}\right]$$

$23(\text{row } 3) + \text{row } 2$

$-17(\text{row } 3) + \text{row } 1$

$$\left[\begin{array}{ccc|c} 1 & 0 & 0 & 1 \\ 0 & 1 & 0 & -1 \\ 0 & 0 & 1 & 1 \end{array}\right]$$

The solution set is $\{(1, -1, 1)\}$.

31.

$$\left[\begin{array}{cccc|c} 1 & 0 & 0 & 0 & 0 \\ 0 & 1 & 0 & 0 & -5 \\ 0 & 0 & 1 & 0 & 0 \\ 0 & 0 & 0 & 1 & 4 \end{array}\right]$$

The solution set is $\{(0, -5, 0, 4)\}$.

33.

$$\left[\begin{array}{cccc|c} 1 & 0 & 0 & 0 & 2 \\ 0 & 1 & 0 & 2 & -3 \\ 0 & 0 & 1 & 3 & 4 \\ 0 & 0 & 0 & 0 & 0 \end{array}\right]$$

The solution set is $\{(2, -2k-3, -3k+4, k)\}$.

35.

$$\left[\begin{array}{cccc|c} 1 & 3 & 0 & 0 & 0 \\ 0 & 0 & 1 & 0 & 0 \\ 0 & 0 & 0 & 0 & 1 \\ 0 & 0 & 0 & 0 & 0 \end{array}\right]$$

Row 3 would represent $0 = 1$. Since $0 \neq 1$, the system is inconsistent. The solution set is $\emptyset$.

37.

$$\left[\begin{array}{cccc|c} 1 & 0 & 0 & 0 & 7 \\ 0 & 1 & 0 & 0 & -3 \\ 0 & 0 & 1 & -2 & 5 \\ 0 & 0 & 0 & 0 & 0 \end{array}\right]$$

The solution set is $\{(7, -3, 2k+5, k)\}$.

Problem Set 12.3 Determinants and Cramer's Rule

1. $\begin{vmatrix} 6 & 2 \\ 4 & 3 \end{vmatrix} = 6(3) - 2(4)$

$= 18 - 8 = 10$

3. $\begin{vmatrix} 4 & 7 \\ 8 & 2 \end{vmatrix} = 4(2) - 7(8)$

$= 8 - 56 = -48$

5. $\begin{vmatrix} -3 & 2 \\ 7 & 5 \end{vmatrix} = -3(5)-2(7)$

$= -15-14 = -29$

7. $\begin{vmatrix} 8 & -3 \\ 6 & 4 \end{vmatrix} = 8(4)-(-3)(6)$

$= 32+18 = 50$

9. $\begin{vmatrix} -3 & 2 \\ 5 & -6 \end{vmatrix} = -3(-6)-2(5)$

$= 18-10 = 8$

11. $\begin{vmatrix} 3 & -3 \\ -6 & 8 \end{vmatrix} = 3(8)-(-3)(-6)$

$= 24-18 = 6$

13. $\begin{vmatrix} -7 & -2 \\ -2 & 4 \end{vmatrix} = -7(4)-(-2)(-2)$

$= -28-4 = -32$

15. $\begin{vmatrix} -2 & -3 \\ -4 & -5 \end{vmatrix} = -2(-5)-(-3)(-4)$

$= 10-12 = -2$

17. $\begin{vmatrix} \frac{1}{4} & -2 \\ \frac{3}{2} & 8 \end{vmatrix} = \frac{1}{4}(8)-(-2)\left(\frac{3}{2}\right)$

$= 2+3 = 5$

19. $\begin{vmatrix} \frac{3}{2} & -\frac{1}{2} \\ \frac{1}{2} & -\frac{2}{5} \end{vmatrix} = \frac{3}{2}\left(-\frac{2}{5}\right)-\left(-\frac{1}{2}\right)\left(\frac{1}{2}\right)$

$= -\frac{3}{5}+\frac{1}{4} = -\frac{7}{20}$

21. $\begin{pmatrix} 2x+y=14 \\ 3x-y=1 \end{pmatrix}$

$D = \begin{vmatrix} 2 & 1 \\ 3 & -1 \end{vmatrix} = 2(-1)-1(3)$

$= -2-3 = -5$

$D_x = \begin{vmatrix} 14 & 1 \\ 1 & -1 \end{vmatrix} = 14(-1)-1(1)$

$= -14-1 = -15$

$D_y = \begin{vmatrix} 2 & 14 \\ 3 & 1 \end{vmatrix} = 2(1)-14(3)$

$= 2-42 = -40$

$x = \frac{D_x}{D} = \frac{-15}{-5} = 3$

$y = \frac{D_y}{D} = \frac{-40}{-5} = 8$

The solution set is $\{(3, 8)\}$.

23. $\begin{pmatrix} -1x+3y=17 \\ 4x-5y=-33 \end{pmatrix}$

$D = \begin{vmatrix} -1 & 3 \\ 4 & -5 \end{vmatrix} = -1(-5)-3(4)$

$= 5-12 = -7$

$D_x = \begin{vmatrix} 17 & 3 \\ -33 & -5 \end{vmatrix} = 17(-5)-3(-33)$

$= -85+99 = 14$

$D_y = \begin{vmatrix} -1 & 17 \\ 4 & -33 \end{vmatrix} = -1(-33)-17(4)$

$= 33-68 = -35$

$x = \frac{D_x}{D} = \frac{14}{-7} = -2$

$y = \frac{D_y}{D} = \frac{-35}{-7} = 5$

The solution set is $\{(-2, 5)\}$.

25. $\begin{pmatrix} 9x + 5y = -8 \\ 7x - 4y = -22 \end{pmatrix}$

$D = \begin{vmatrix} 9 & 5 \\ 7 & -4 \end{vmatrix} = 9(-4) - 5(7)$

$= -36 - 35 = -71$

$D_x = \begin{vmatrix} -8 & 5 \\ -22 & -4 \end{vmatrix} = -8(-4) - 5(-22)$

$= 32 + 110 = 142$

$D_y = \begin{vmatrix} 9 & -8 \\ 7 & -22 \end{vmatrix} = 9(-22) - (-8)(7)$

$= -198 + 56 = -142$

$x = \frac{D_x}{D} = \frac{142}{-71} = -2$

$y = \frac{D_y}{D} = \frac{-142}{-71} = 2$

The solution set is $\{(-2, 2)\}$.

27. $\begin{pmatrix} x + 5y = 4 \\ 3x + 15y = -1 \end{pmatrix}$

$D = \begin{vmatrix} 1 & 5 \\ 3 & 15 \end{vmatrix} = 1(15) - 5(3)$

$= 15 - 15 = 0$

Since $D = 0$, Cramer's Rule can not be applied. Multiply equation 1 by -3, then add the equations.

$$\begin{array}{r} -3x - 15y = -12 \\ 3x + 15y = -\ \ 1 \\ \hline 0 = -13 \end{array}$$

Since $0 \neq -13$, the system is inconsistent. The solution set is $\emptyset$.

29. $\begin{pmatrix} 6x - y = 0 \\ 5x + 4y = 29 \end{pmatrix}$

$D = \begin{vmatrix} 6 & -1 \\ 5 & 4 \end{vmatrix} = 6(4) - (-1)(5)$

$= 24 + 5 = 29$

$D_x = \begin{vmatrix} 0 & -1 \\ 29 & 4 \end{vmatrix} = 0(4) - (-1)(29)$

$= 0 + 29 = 29$

$D_y = \begin{vmatrix} 6 & 0 \\ 5 & 29 \end{vmatrix} = 6(29) - 0(5)$

$= 174 - 0 = 174$

$x = \frac{D_x}{D} = \frac{29}{29} = 1$

$y = \frac{D_y}{D} = \frac{174}{29} = 6$

The solution set is $\{(1, 6)\}$.

31. $\begin{pmatrix} -4x + 3y = 3 \\ 4x - 6y = -5 \end{pmatrix}$

$D = \begin{vmatrix} -4 & 3 \\ 4 & -6 \end{vmatrix} = -4(-6) - 3(4)$

$= 24 - 12 = 12$

$D_x = \begin{vmatrix} 3 & 3 \\ -5 & -6 \end{vmatrix} = 3(-6) - 3(-5)$

$= -18 + 15 = -3$

$D_y = \begin{vmatrix} -4 & 3 \\ 4 & -5 \end{vmatrix} = -4(-5) - 3(4)$

$= 20 - 12 = 8$

$x = \frac{D_x}{D} = \frac{-3}{12} = -\frac{1}{4}$

$y = \frac{D_y}{D} = \frac{8}{12} = \frac{2}{3}$

The solution set is $\left\{\left(-\frac{1}{4}, \frac{2}{3}\right)\right\}$.

33. $\begin{pmatrix} 6x-5y=1 \\ 4x+7y=2 \end{pmatrix}$

$D=\begin{vmatrix} 6 & -5 \\ 4 & 7 \end{vmatrix}=6(7)-(-5)(4)$

$=42+20=62$

$D_x=\begin{vmatrix} 1 & -5 \\ 2 & 7 \end{vmatrix}=1(7)-(-5)(2)$

$=7+10=17$

$D_y=\begin{vmatrix} 6 & 1 \\ 4 & 2 \end{vmatrix}=6(2)-1(4)$

$=12-4=8$

$x=\frac{D_x}{D}=\frac{17}{62}$

$y=\frac{D_y}{D}=\frac{8}{62}=\frac{4}{31}$

The solution set is $\left\{\left(\frac{17}{62}, \frac{4}{31}\right)\right\}$.

35. $\begin{pmatrix} 7x+2y=-1 \\ y=-x+2 \end{pmatrix}$

$\begin{pmatrix} 7x+2y=-1 \\ x+y=2 \end{pmatrix}$

$D=\begin{vmatrix} 7 & 2 \\ 1 & 1 \end{vmatrix}=7(1)-2(1)$

$=7-2=5$

$D_x=\begin{vmatrix} -1 & 2 \\ 2 & 1 \end{vmatrix}=-1(1)-2(2)$

$=-1-4=-5$

$D_y=\begin{vmatrix} 7 & -1 \\ 1 & 2 \end{vmatrix}=7(2)-(-1)(1)$

$=14+1=15$

$x=\frac{D_x}{D}=\frac{-5}{5}=-1$

$y=\frac{D_y}{D}=\frac{15}{5}=3$

The solution set is $\{(-1, 3)\}$.

37. $\begin{pmatrix} -\frac{2}{3}x+\frac{1}{2}y=-7 \\ \frac{1}{3}x-\frac{3}{2}y=6 \end{pmatrix}$

Multiply equation 1 by 6 and multiply equation 2 by 6.

$\begin{pmatrix} -4x+3y=-42 \\ 2x-9y=36 \end{pmatrix}$

$D=\begin{vmatrix} -4 & 3 \\ 2 & -9 \end{vmatrix}=-4(-9)-3(2)$

$=36-6=30$

$D_x=\begin{vmatrix} -42 & 3 \\ 36 & -9 \end{vmatrix}=-42(-9)-3(36)$

$=378-108=270$

$D_y=\begin{vmatrix} -4 & -42 \\ 2 & 36 \end{vmatrix}=-4(36)-2(-42)$

$=-144+84=-60$

$x=\frac{D_x}{D}=\frac{270}{30}=9$

$y=\frac{D_y}{D}=\frac{-60}{30}=-2$

The solution set is $\{(9, -2)\}$.

39. $\begin{pmatrix} x+\frac{2}{3}y=-6 \\ -\frac{1}{4}x+3y=-8 \end{pmatrix}$

Multiply equation 1 by 3 and multiply equation 2 by 4.

$\begin{pmatrix} 3x+2y=-18 \\ -x+12y=-32 \end{pmatrix}$

$D = \begin{vmatrix} 3 & 2 \\ -1 & 12 \end{vmatrix} = 3(12) - 2(-1)$

$= 36 + 2 = 38$

$D_x = \begin{vmatrix} -18 & 2 \\ -32 & 12 \end{vmatrix} = -18(12) - 2(-32)$

$= -216 + 64 = -152$

$D_y = \begin{vmatrix} 3 & -18 \\ -1 & -32 \end{vmatrix} = 3(-32) - (-18)(-1)$

$= -96 - 18 = -114$

$x = \frac{D_x}{D} = \frac{-152}{38} = -4$

$y = \frac{D_y}{D} = \frac{-114}{38} = -3$

The solution set is $\{(-4, -3)\}$.

41. Let x = cost of cashews
y = cost of peanuts

$\begin{pmatrix} 7x + 5y = 88 \\ 3x + 2y = 37 \end{pmatrix}$

Multiply equation 1 by -2 and multiply equation 2 by 5. Then add the equations.

$$\begin{array}{r} -14x - 10y = -176 \\ 15x + 10y = 185 \\ \hline x = 9 \end{array}$$

$7x + 5y = 88$

$7(9) + 5y = 88$

$63 + 5y = 88$

$5y = 25$

$y = 5$

The cashew cost \$9 per pound and the Spanish peanuts cost \$5 per pound.

43. Let x = fixed fee
y = additional fee

$\begin{pmatrix} x + 4y = 2.40 \\ x + 11y = 3.10 \end{pmatrix}$

Multiply equation 1 by -1 and then add the equations.

$$\begin{array}{r} -x - 4y = -2.40 \\ x + 11y = 3.10 \\ \hline 7y = .70 \end{array}$$

$y = .10$

$x + 4y = 2.40$

$x + 4(.10) = 2.40$

$x + .40 = 2.40$

$x = 2.00$

The fixed fee is \$2.00 and the additional fee is \$.10 per pound.

45. Let x = larger number
y = smaller number

$\begin{pmatrix} x + y = 19 \\ x = 2y + 1 \end{pmatrix}$

$x + y = 19$

$2y + 1 + y = 19$

$3y + 1 = 19$

$3y = 18$

$y = 6$

$x = 2y + 1$

$x = 2(6) + 1$

$x = 13$

The numbers are 13 and 6.

47. Let x = amount invested at 8%
y = amount invested at 9%

$\begin{pmatrix} .08x + .09y = 101 \\ .09x + .08y = 103 \end{pmatrix}$

Multiply equation 1 by 900 and multiply equation 2 by -800. Then add the equations.

$$\begin{array}{r} 72x + 81y = 90900 \\ -72x - 64y = -82400 \\ \hline 17y = 8500 \\ y = 500 \end{array}$$

$72x+81y=90900$

$72x+81(500)=90900$

$72x+40500=90900$

$72x=50400$

$x=700$

There is \$700 invested at 8% and \$500 invested at 9%.

Problem Set 12.4 3 × 3 Determinants and Cramer's Rule

1. Expand about row 1.

$$\begin{vmatrix} 2 & 7 & 5 \\ 1 & -1 & 1 \\ -4 & 3 & 2 \end{vmatrix} = 2\begin{vmatrix} -1 & 1 \\ 3 & 2 \end{vmatrix} - 1(7)\begin{vmatrix} 1 & 1 \\ -4 & 2 \end{vmatrix} + 5\begin{vmatrix} 1 & -1 \\ -4 & 3 \end{vmatrix}$$

$=2(-2-3)-7(2+4)+5(3-4)$

$=2(-5)-7(6)+5(-1)$

$=-10-42-5=-57$

3. Expand about row 1.

$$\begin{vmatrix} 3 & -2 & 1 \\ 2 & 1 & 4 \\ -1 & 3 & 5 \end{vmatrix} = 3\begin{vmatrix} 1 & 4 \\ 3 & 5 \end{vmatrix} - 1(-2)\begin{vmatrix} 2 & 4 \\ -1 & 5 \end{vmatrix} + 1\begin{vmatrix} 2 & 1 \\ -1 & 3 \end{vmatrix}$$

$=3(5-12)+2(10+4)+1(6+1)$

$=3(-7)+2(14)+1(7)$

$=-21+28+7=14$

5. Expand about row 2.

$$\begin{vmatrix} -3 & -2 & 1 \\ 5 & 0 & 6 \\ 2 & 1 & -4 \end{vmatrix} = -1(5)\begin{vmatrix} -2 & 1 \\ 1 & -4 \end{vmatrix} + 0\begin{vmatrix} -3 & 1 \\ 2 & -4 \end{vmatrix} - 1(6)\begin{vmatrix} -3 & -2 \\ 2 & 1 \end{vmatrix}$$

$=-5(8-1)+0-6(-3+4)$

$=-5(7)-6(1)$

$=-35-6=-41$

7. Expand about row 3.

$$\begin{vmatrix} 3 & -4 & -2 \\ 5 & -2 & 1 \\ 1 & 0 & 0 \end{vmatrix} = 1\begin{vmatrix} -4 & -2 \\ -2 & 1 \end{vmatrix} - 0\begin{vmatrix} 3 & -2 \\ 5 & 1 \end{vmatrix} + 0\begin{vmatrix} 3 & -4 \\ 5 & -2 \end{vmatrix}$$

$= 1(-4-4) + 0 + 0$

$= 1(-8) = -8$

9. Expand about row 1.

$$\begin{vmatrix} 4 & -2 & 7 \\ 1 & -1 & 6 \\ 3 & 5 & -2 \end{vmatrix} = 4\begin{vmatrix} -1 & 6 \\ 5 & -2 \end{vmatrix} - 1(-2)\begin{vmatrix} 1 & 6 \\ 3 & -2 \end{vmatrix} + 7\begin{vmatrix} 1 & -1 \\ 3 & 5 \end{vmatrix}$$

$= 4(2-30) + 2(-2-18) + 7(5+3)$

$= 4(-28) + 2(-20) + 7(8)$

$= -112 - 40 + 56 = -96$

11. $\begin{pmatrix} 2x - y + 3z = -10 \\ x + 2y - 3z = 2 \\ 3x - 2y + 5z = -16 \end{pmatrix}$

For D, expand about row 1.

$$D = \begin{vmatrix} 2 & -1 & 3 \\ 1 & 2 & -3 \\ 3 & -2 & 5 \end{vmatrix} = 2\begin{vmatrix} 2 & -3 \\ -2 & 5 \end{vmatrix} - 1(-1)\begin{vmatrix} 1 & -3 \\ 3 & 5 \end{vmatrix} + 3\begin{vmatrix} 1 & 2 \\ 3 & -2 \end{vmatrix}$$

$= 2(10-6) + 1(5+9) + 3(-2-6)$

$= 2(4) + 1(14) + 3(-8)$

$= 8 + 14 - 24 = -2$

For D_x, expand about row 1.

$$D_x = \begin{vmatrix} -10 & -1 & 3 \\ 2 & 2 & -3 \\ -16 & -2 & 5 \end{vmatrix} = -10\begin{vmatrix} 2 & -3 \\ -2 & 5 \end{vmatrix} - 1(-1)\begin{vmatrix} 2 & -3 \\ -16 & 5 \end{vmatrix} + 3\begin{vmatrix} 2 & 2 \\ -16 & -2 \end{vmatrix}$$

$= -10(10-6) + 1(10-48) + 3(-4+32)$

$= -10(4) + 1(-38) + 3(28)$

$= -40 - 38 + 84 = 6$

For D_y, expand about row 1.

$$D_y = \begin{vmatrix} 2 & -10 & 3 \\ 1 & 2 & -3 \\ 3 & -16 & 5 \end{vmatrix} = 2\begin{vmatrix} 2 & -3 \\ -16 & 5 \end{vmatrix} - 1(-10)\begin{vmatrix} 1 & -3 \\ 3 & 5 \end{vmatrix} + 3\begin{vmatrix} 1 & 2 \\ 3 & -16 \end{vmatrix}$$

$= 2(10-48) + 10(5+9) + 3(-16-6)$

$= 2(-38) + 10(14) + 3(-22)$

$= -76 + 140 - 66 = -2$

For D_z, expand about row 1.

$$D_z = \begin{vmatrix} 2 & -1 & -10 \\ 1 & 2 & 2 \\ 3 & -2 & -16 \end{vmatrix} = 2\begin{vmatrix} 2 & 2 \\ -2 & -16 \end{vmatrix} - 1(-1)\begin{vmatrix} 1 & 2 \\ 3 & -16 \end{vmatrix} - 10\begin{vmatrix} 1 & 2 \\ 3 & -2 \end{vmatrix}$$

$= 2(-32+4) + 1(-16-6) - 10(-2-6)$

$= 2(-28) + 1(-22) - 10(-8)$

$= -56 - 22 + 80 = 2$

$x = \frac{D_x}{D} = \frac{6}{-2} = -3$

$y = \frac{D_y}{D} = \frac{-2}{-2} = 1$

$z = \frac{D_z}{D} = \frac{2}{-2} = -1$

The solution set is $\{(-3, 1, -1)\}$.

13. $\begin{pmatrix} x - y + 2z = -8 \\ 2x + 3y - 4z = 18 \\ -x + 2y - z = 7 \end{pmatrix}$

For D, expand about row 1.

$$D = \begin{vmatrix} 1 & -1 & 2 \\ 2 & 3 & -4 \\ -1 & 2 & -1 \end{vmatrix} = 1\begin{vmatrix} 3 & -4 \\ 2 & -1 \end{vmatrix} - 1(-1)\begin{vmatrix} 2 & -4 \\ -1 & -1 \end{vmatrix} + 2\begin{vmatrix} 2 & 3 \\ -1 & 2 \end{vmatrix}$$

$= 1(-3+8) + 1(-2-4) + 2(4+3)$

$= 1(5) + 1(-6) + 2(7)$

$= 5 - 6 + 14 = 13$

For D_x, expand about row 1.

$$D_x = \begin{vmatrix} -8 & -1 & 2 \\ 18 & 3 & -4 \\ 7 & 2 & -1 \end{vmatrix} = -8\begin{vmatrix} 3 & -4 \\ 2 & -1 \end{vmatrix} - 1(-1)\begin{vmatrix} 18 & -4 \\ 7 & -1 \end{vmatrix} + 2\begin{vmatrix} 18 & 3 \\ 7 & 2 \end{vmatrix}$$

$= -8(-3+8)+1(-18+28)+2(36-21)$
$= -8(5)+1(10)+2(15)$
$= -40+10+30=0$

For D_y, expand about row 1.

$$D_y = \begin{vmatrix} 1 & -8 & 2 \\ 2 & 18 & -4 \\ -1 & 7 & -1 \end{vmatrix} = 1\begin{vmatrix} 18 & -4 \\ 7 & -1 \end{vmatrix} - 1(-8)\begin{vmatrix} 2 & -4 \\ -1 & -1 \end{vmatrix} + 2\begin{vmatrix} 2 & 18 \\ -1 & 7 \end{vmatrix}$$

$= 1(-18+28)+8(-2-4)+2(14+18)$
$= 1(10)+8(-6)+2(32)$
$= 10-48+64=26$

For D_z, expand about row 1.

$$D_z = \begin{vmatrix} 1 & -1 & -8 \\ 2 & 3 & 18 \\ -1 & 2 & 7 \end{vmatrix} = 1\begin{vmatrix} 3 & 18 \\ 2 & 7 \end{vmatrix} - 1(-1)\begin{vmatrix} 2 & 18 \\ -1 & 7 \end{vmatrix} - 8\begin{vmatrix} 2 & 3 \\ -1 & 2 \end{vmatrix}$$

$= 1(21-36)+1(14+18)-8(4+3)$
$= 1(-15)+1(32)-8(7)$
$= -15+32-56 = -39$

$x = \frac{D_x}{D} = \frac{0}{13} = 0$

$y = \frac{D_y}{D} = \frac{26}{13} = 2$

$z = \frac{D_z}{D} = \frac{-39}{13} = -3$

The solution set is $\{(0, 2, -3)\}$.

15. $\begin{pmatrix} 3x-2y-3z=-5 \\ x+2y+3z=-3 \\ -x+4y-6z=8 \end{pmatrix}$

For D, expand about row 1.

$$D = \begin{vmatrix} 3 & -2 & -3 \\ 1 & 2 & 3 \\ -1 & 4 & -6 \end{vmatrix} = 3\begin{vmatrix} 2 & 3 \\ 4 & -6 \end{vmatrix} - 1(-2)\begin{vmatrix} 1 & 3 \\ -1 & -6 \end{vmatrix} - 3\begin{vmatrix} 1 & 2 \\ -1 & 4 \end{vmatrix}$$

$= 3(-12-12)+2(-6+3)-3(4+2)$
$= 3(-24)+2(-3)-3(6)$
$= -72-6-18 = -96$

For D_x, expand about row 1.

$$D_x = \begin{vmatrix} -5 & -2 & -3 \\ -3 & 2 & 3 \\ 8 & 4 & -6 \end{vmatrix} = -5\begin{vmatrix} 2 & 3 \\ 4 & -6 \end{vmatrix} - (-2)\begin{vmatrix} -3 & 3 \\ 8 & -6 \end{vmatrix} - 3\begin{vmatrix} -3 & 2 \\ 8 & 4 \end{vmatrix}$$

$= -5(-12-12) + 2(18-24) - 3(-12-16)$

$= -5(-24) + 2(-6) - 3(-28)$

$= 120 - 12 + 84 = 192$

For D_y, expand about row 1.

$$D_y = \begin{vmatrix} 3 & -5 & -3 \\ 1 & -3 & 3 \\ -1 & 8 & -6 \end{vmatrix} = 3\begin{vmatrix} -3 & 3 \\ 8 & -6 \end{vmatrix} - 1(-5)\begin{vmatrix} 1 & 3 \\ -1 & -6 \end{vmatrix} - 3\begin{vmatrix} 1 & -3 \\ -1 & 8 \end{vmatrix}$$

$= 3(18-24) + 5(-6+3) - 3(8-3)$

$= 3(-6) + 5(-3) - 3(5)$

$= -18 - 15 - 15 = -48$

For D_z, expand about row 1.

$$D_z = \begin{vmatrix} 3 & -2 & -5 \\ 1 & 2 & -3 \\ -1 & 4 & 8 \end{vmatrix} = 3\begin{vmatrix} 2 & -3 \\ 4 & 8 \end{vmatrix} - 1(-2)\begin{vmatrix} 1 & -3 \\ -1 & 8 \end{vmatrix} - 5\begin{vmatrix} 1 & 2 \\ -1 & 4 \end{vmatrix}$$

$= 3(16+12) + 2(8-3) - 5(4+2)$

$= 3(28) + 2(5) - 5(6)$

$= 84 + 10 - 30 = 64$

$x = \frac{D_x}{D} = \frac{192}{-96} = -2$

$y = \frac{D_y}{D} = \frac{-48}{-96} = \frac{1}{2}$

$z = \frac{D_z}{D} = \frac{64}{-96} = -\frac{2}{3}$

The solution set is $\left\{\left(-2, \frac{1}{2}, -\frac{2}{3}\right)\right\}$.

17. $\begin{pmatrix} -x+y+z=-1 \\ x-2y+5z=-4 \\ 3x+4y-6z=-1 \end{pmatrix}$

For D, expand about row 1.

$$D = \begin{vmatrix} -1 & 1 & 1 \\ 1 & -2 & 5 \\ 3 & 4 & -6 \end{vmatrix} = -1\begin{vmatrix} -2 & 5 \\ 4 & -6 \end{vmatrix} - 1(1)\begin{vmatrix} 1 & 5 \\ 3 & -6 \end{vmatrix} + 1\begin{vmatrix} 1 & -2 \\ 3 & 4 \end{vmatrix}$$

$= -1(12-20) - 1(-6-15) + 1(4+6)$

$= -1(-8) - 1(-21) + 1(10)$

$= 8 + 21 + 10 = 39$

For D_x, expand about row 1.

$$D_x = \begin{vmatrix} -1 & 1 & 1 \\ -4 & -2 & 5 \\ -1 & 4 & -6 \end{vmatrix} = -1\begin{vmatrix} -2 & 5 \\ 4 & -6 \end{vmatrix} - 1(1)\begin{vmatrix} -4 & 5 \\ -1 & -6 \end{vmatrix} + 1\begin{vmatrix} -4 & -2 \\ -1 & 4 \end{vmatrix}$$

$= -1(12-20) - 1(24+5) + 1(-16-2)$

$= -1(-8) - 1(29) + 1(-18)$

$= 8 - 29 - 18 = -39$

For D_y, expand about row 1.

$$D_y = \begin{vmatrix} -1 & -1 & 1 \\ 1 & -4 & 5 \\ 3 & -1 & -6 \end{vmatrix} = -1\begin{vmatrix} -4 & 5 \\ -1 & -6 \end{vmatrix} - 1(-1)\begin{vmatrix} 1 & 5 \\ 3 & -6 \end{vmatrix} + 1\begin{vmatrix} 1 & -4 \\ 3 & -1 \end{vmatrix}$$

$= -1(24+5) + 1(-6-15) + 1(-1+12)$

$= -1(29) + 1(-21) + 1(11)$

$= -29 - 21 + 11 = -39$

For D_z, expand about row 1.

$$D_z = \begin{vmatrix} -1 & 1 & -1 \\ 1 & -2 & -4 \\ 3 & 4 & -1 \end{vmatrix} = -1\begin{vmatrix} -2 & -4 \\ 4 & -1 \end{vmatrix} - 1(1)\begin{vmatrix} 1 & -4 \\ 3 & -1 \end{vmatrix} - 1\begin{vmatrix} 1 & -2 \\ 3 & 4 \end{vmatrix}$$

$= -1(2+16) - 1(-1+12) - 1(4+6)$

$= -1(18) - 1(11) - 1(10)$

$= -18 - 11 - 10 = -39$

$x = \frac{D_x}{D} = \frac{-39}{39} = -1$

$y = \frac{D_y}{D} = \frac{-39}{39} = -1$

$z = \frac{D_z}{D} = \frac{-39}{39} = -1$

The solution set is $\{(-1, -1, -1)\}$.

19. $\begin{pmatrix} x - y + 2z = 4 \\ 3x - 2y + 4z = 6 \\ 2x - 2y + 4z = -1 \end{pmatrix}$

For D, expand about row 1.

$$D=\begin{vmatrix}1 & -1 & 2\\ 3 & -2 & 4\\ 2 & -2 & 4\end{vmatrix}=1\begin{vmatrix}-2 & 4\\ -2 & 4\end{vmatrix}-1(-1)\begin{vmatrix}3 & 4\\ 2 & 4\end{vmatrix}+2\begin{vmatrix}3 & -2\\ 2 & -2\end{vmatrix}$$

$=1(-8+8)+1(12-8)+2(-6+4)$

$=1(0)+1(4)+2(-2)$

$0+4-4=0$

Since $D=0$, Cramer's Rule can not be used.
Multiply equation 1 by -2 and add to equation 3.

$$\begin{array}{r} -2x+2y-4z=-8\\ 2x-2y+4z=-1\\ \hline 0=-9\end{array}$$

Since $0\neq -9$, the system is inconsistent.
The solution set is $\emptyset$.

21. $\begin{pmatrix} 2x-y+3z=-5\\ 3x+4y-2z=-25\\ -x+z=6\end{pmatrix}$

For D, expand about row 3.

$$D=\begin{vmatrix}2 & -1 & 3\\ 3 & 4 & -2\\ -1 & 0 & 1\end{vmatrix}=-1\begin{vmatrix}-1 & 3\\ 4 & -2\end{vmatrix}-0\begin{vmatrix}2 & 3\\ 3 & -2\end{vmatrix}+1\begin{vmatrix}2 & -1\\ 3 & 4\end{vmatrix}$$

$=-1(2-12)+0+1(8+3)$

$=-1(-10)+0+1(11)$

$=10+0+11=21$

For D_x, expand about row 3.

$$D_x=\begin{vmatrix}-5 & -1 & 3\\ -25 & 4 & -2\\ 6 & 0 & 1\end{vmatrix}=6\begin{vmatrix}-1 & 3\\ 4 & -2\end{vmatrix}+0\begin{vmatrix}-5 & 3\\ -25 & -2\end{vmatrix}+1\begin{vmatrix}-5 & -1\\ -25 & 4\end{vmatrix}$$

$=6(2-12)+0+1(-20-25)$

$=6(-10)+0+1(-45)$

$=-60+0-45=-105$

For D_y, expand about row 1.

$$D_y=\begin{vmatrix}2 & -5 & 3\\ 3 & -25 & -2\\ -1 & 6 & 1\end{vmatrix}=2\begin{vmatrix}-25 & -2\\ 6 & 1\end{vmatrix}-1(-5)\begin{vmatrix}3 & -2\\ -1 & 1\end{vmatrix}+3\begin{vmatrix}3 & -25\\ -1 & 6\end{vmatrix}$$

$= 2(-25+12)+5(3-2)+3(18-25)$

$= 2(-13)+5(1)+3(-7)$

$= -26+5-21 = -42$

For D_z, expand about row 3.

$$D_z = \begin{vmatrix} 2 & -1 & -5 \\ 3 & 4 & -25 \\ -1 & 0 & 6 \end{vmatrix} = -1\begin{vmatrix} -1 & -5 \\ 4 & -25 \end{vmatrix} - 0\begin{vmatrix} 2 & -5 \\ 3 & -25 \end{vmatrix} + 6\begin{vmatrix} 2 & -1 \\ 3 & 4 \end{vmatrix}$$

$= -1(25+20)+0+6(8+3)$

$= -1(45)+6(11)$

$= -45+66 = 21$

$x = \frac{D_x}{D} = \frac{-105}{21} = -5$

$y = \frac{D_y}{D} = \frac{-42}{21} = -2$

$z = \frac{D_z}{D} = \frac{21}{21} = 1$

The solution set is $\{(-5, -2, 1)\}$.

23. $\begin{pmatrix} 2y - z = 10 \\ 3x + 4y = 6 \\ x - y + z = -9 \end{pmatrix}$

For D, expand about row 1.

$$D = \begin{vmatrix} 0 & 2 & -1 \\ 3 & 4 & 0 \\ 1 & -1 & 1 \end{vmatrix} = 0\begin{vmatrix} 4 & 0 \\ -1 & 1 \end{vmatrix} - 1(2)\begin{vmatrix} 3 & 0 \\ 1 & 1 \end{vmatrix} - 1\begin{vmatrix} 3 & 4 \\ 1 & -1 \end{vmatrix}$$

$= 0-2(3-0)-1(-3-4)$

$= 0-2(3)-1(-7)$

$= 0-6+7 = 1$

For D_x, expand about column 3.

$$D_x = \begin{vmatrix} 10 & 2 & -1 \\ 6 & 4 & 0 \\ -9 & -1 & 1 \end{vmatrix} = -1\begin{vmatrix} 6 & 4 \\ -9 & -1 \end{vmatrix} - 0\begin{vmatrix} 10 & 2 \\ -9 & -1 \end{vmatrix} + 1\begin{vmatrix} 10 & 2 \\ 6 & 4 \end{vmatrix}$$

$= -1(-6+36)+0+1(40-12)$

$= -1(30)+0+1(28)$

$= -30+28 = -2$

For D_y, expand about column 3.

$$D_y = \begin{vmatrix} 0 & 10 & -1 \\ 3 & 6 & 0 \\ 1 & -9 & 1 \end{vmatrix} = -1\begin{vmatrix} 3 & 6 \\ 1 & -9 \end{vmatrix} - 0\begin{vmatrix} 0 & 10 \\ 1 & -9 \end{vmatrix} + 1\begin{vmatrix} 0 & 10 \\ 3 & 6 \end{vmatrix}$$

$= -1(-27-6)+0+1(0-30)$

$= -1(-33)+0+1(-30)$

$= 33-30 = 3$

For D_z, expand about column 1.

$$D_z = \begin{vmatrix} 0 & 2 & 10 \\ 3 & 4 & 6 \\ 1 & -1 & -9 \end{vmatrix} = 0\begin{vmatrix} 4 & 6 \\ -1 & -9 \end{vmatrix} - 1(3)\begin{vmatrix} 2 & 10 \\ -1 & -9 \end{vmatrix} + 1\begin{vmatrix} 2 & 10 \\ 4 & 6 \end{vmatrix}$$

$= 0-3(-18+10)+1(12-40)$

$= 0-3(-8)+1(-28)$

$= 0+24-28 = -4$

$x = \frac{D_x}{D} = \frac{-2}{1} = -2$

$y = \frac{D_y}{D} = \frac{3}{1} = 3$

$z = \frac{D_z}{D} = \frac{-4}{1} = -4$

The solution set is $\{(-2, 3, -4)\}$.

25. $\begin{pmatrix} -2x+5y-3z = -1 \\ 2x-7y+3z = 1 \\ 4x-y-6z = -6 \end{pmatrix}$

For D, expand about row 1.

$$D = \begin{vmatrix} -2 & 5 & -3 \\ 2 & -7 & 3 \\ 4 & -1 & -6 \end{vmatrix} = -2\begin{vmatrix} -7 & 3 \\ -1 & -6 \end{vmatrix} - 1(5)\begin{vmatrix} 2 & 3 \\ 4 & -6 \end{vmatrix} - 3\begin{vmatrix} 2 & -7 \\ 4 & -1 \end{vmatrix}$$

$= -2(42+3)-5(-12-12)-3(-2+28)$

$= -2(45)-5(-24)-3(26)$

$= -90+120-78 = -48$

For D_x, expand about row 1.

$$D_x = \begin{vmatrix} -1 & 5 & -3 \\ 1 & -7 & 3 \\ -6 & -1 & -6 \end{vmatrix} = -1\begin{vmatrix} -7 & 3 \\ -1 & -6 \end{vmatrix} - 1(5)\begin{vmatrix} 1 & 3 \\ -6 & -6 \end{vmatrix} - 3\begin{vmatrix} 1 & -7 \\ -6 & -1 \end{vmatrix}$$

$= -1(42+3)-5(-6+18)-3(-1-42)$

$= -1(45)-5(12)-3(-43)$

$= -45-60+129 = 24$

For D_y, expand about row 1.

$$D_y = \begin{vmatrix} -2 & -1 & -3 \\ 2 & 1 & 3 \\ 4 & -6 & -6 \end{vmatrix} = -2\begin{vmatrix} 1 & 3 \\ -6 & -6 \end{vmatrix} - 1(-1)\begin{vmatrix} 2 & 3 \\ 4 & -6 \end{vmatrix} - 3\begin{vmatrix} 2 & 1 \\ 4 & -6 \end{vmatrix}$$

$= -2(-6+18)+1(-12-12)-3(-12-4)$

$= -2(12)+1(-24)-3(-16)$

$= -24-24+48 = 0$

For D_z, expand about row 1.

$$D_z = \begin{vmatrix} -2 & 5 & -1 \\ 2 & -7 & 1 \\ 4 & -1 & -6 \end{vmatrix} = -2\begin{vmatrix} -7 & 1 \\ -1 & -6 \end{vmatrix} - 1(5)\begin{vmatrix} 2 & 1 \\ 4 & -6 \end{vmatrix} - 1\begin{vmatrix} 2 & -7 \\ 4 & -1 \end{vmatrix}$$

$= -2(42+1)-5(-12-4)-1(-2+28)$

$= -2(43)-5(-16)-1(26)$

$= -86+80-26 = -32$

$x = \frac{D_x}{D} = \frac{24}{-48} = -\frac{1}{2}$

$y = \frac{D_y}{D} = \frac{0}{-48} = 0$

$z = \frac{D_z}{D} = \frac{-32}{-48} = \frac{2}{3}$

The solution set is $\left\{\left(-\frac{1}{2},\ 0,\ \frac{2}{3}\right)\right\}$.

27. $\begin{pmatrix} -x-y+5z=4 \\ x+y-7z=-6 \\ 2x+3y+4z=13 \end{pmatrix}$

For D, expand about row 1.

$$D = \begin{vmatrix} -1 & -1 & 5 \\ 1 & 1 & -7 \\ 2 & 3 & 4 \end{vmatrix} = -1\begin{vmatrix} 1 & -7 \\ 3 & 4 \end{vmatrix} - 1(-1)\begin{vmatrix} 1 & -7 \\ 2 & 4 \end{vmatrix} + 5\begin{vmatrix} 1 & 1 \\ 2 & 3 \end{vmatrix}$$

$= -1(4+21)+1(4+14)+5(3-2)$

$= -1(25)+1(18)+5(1)$

$= -25+18+5 = -2$

For D_x, expand about row 1.

$$D_x = \begin{vmatrix} 4 & -1 & 5 \\ -6 & 1 & -7 \\ 13 & 3 & 4 \end{vmatrix} = 4\begin{vmatrix} 1 & -7 \\ 3 & 4 \end{vmatrix} - 1(-1)\begin{vmatrix} -6 & -7 \\ 13 & 4 \end{vmatrix} + 5\begin{vmatrix} -6 & 1 \\ 13 & 3 \end{vmatrix}$$

$= 4(4+21)+1(-24+91)+5(-18-13)$

$= 4(25)+1(67)+5(-31)$

$= 100+67-155 = 12$

For D_y, expand about row 1.

$$D_y = \begin{vmatrix} -1 & 4 & 5 \\ 1 & -6 & -7 \\ 2 & 13 & 4 \end{vmatrix} = -1\begin{vmatrix} -6 & -7 \\ 13 & 4 \end{vmatrix} - 1(4)\begin{vmatrix} 1 & -7 \\ 2 & 4 \end{vmatrix} + 5\begin{vmatrix} 1 & -6 \\ 2 & 13 \end{vmatrix}$$

$= -1(-24+91)-4(4+14)+5(13+12)$

$= -1(67)-4(18)+5(25)$

$= -67-72+125 = -14$

For D_z, expand about row 1.

$$D_z = \begin{vmatrix} -1 & -1 & 4 \\ 1 & 1 & -6 \\ 2 & 3 & 13 \end{vmatrix} = -1\begin{vmatrix} 1 & -6 \\ 3 & 13 \end{vmatrix} - (-1)\begin{vmatrix} 1 & -6 \\ 2 & 13 \end{vmatrix} + 4\begin{vmatrix} 1 & 1 \\ 2 & 3 \end{vmatrix}$$

$= -1(13+18)+1(13+12)+4(3-2)$

$= -1(31)+1(25)+4(1)$

$= -31+25+4 = -2$

$x = \frac{D_x}{D} = \frac{12}{-2} = -6$

$y = \frac{D_y}{D} = \frac{-14}{-2} = 7$

$z = \frac{D_z}{D} = \frac{-2}{-2} = 1$

The solution set is $\{(-6, 7, 1)\}$.

29. $\begin{pmatrix} 5x - y + 2z = 10 \\ 7x + 2y - 2z = -4 \\ -3x - y + 4z = 1 \end{pmatrix}$

For D, expand about row 1.

$$D = \begin{vmatrix} 5 & -1 & 2 \\ 7 & 2 & -2 \\ -3 & -1 & 4 \end{vmatrix} = 5\begin{vmatrix} 2 & -2 \\ -1 & 4 \end{vmatrix} - 1(-1)\begin{vmatrix} 7 & -2 \\ -3 & 4 \end{vmatrix} + 2\begin{vmatrix} 7 & 2 \\ -3 & -1 \end{vmatrix}$$

$= 5(8-2) + 1(28-6) + 2(-7+6)$
$= 5(6) + 1(22) + 2(-1)$
$= 30 + 22 - 2 = 50$

For D_x, expand about row 1.

$$D_x = \begin{vmatrix} 10 & -1 & 2 \\ -4 & 2 & -2 \\ 1 & -1 & 4 \end{vmatrix} = 10\begin{vmatrix} 2 & -2 \\ -1 & 4 \end{vmatrix} - 1(-1)\begin{vmatrix} -4 & -2 \\ 1 & 4 \end{vmatrix} + 2\begin{vmatrix} -4 & 2 \\ 1 & -1 \end{vmatrix}$$

$= 10(8-2) + 1(-16+2) + 2(4-2)$
$= 10(6) + 1(-14) + 2(2)$
$= 60 - 14 + 4 = 50$

For D_y, expand about row 1.

$$D_y = \begin{vmatrix} 5 & 10 & 2 \\ 7 & -4 & -2 \\ -3 & 1 & 4 \end{vmatrix} = 5\begin{vmatrix} -4 & -2 \\ 1 & 4 \end{vmatrix} - 1(10)\begin{vmatrix} 7 & -2 \\ -3 & 4 \end{vmatrix} + 2\begin{vmatrix} 7 & -4 \\ -3 & 1 \end{vmatrix}$$

$= 5(-16+2) - 10(28-6) + 2(7-12)$
$= 5(-14) - 10(22) + 2(-5)$
$= -70 - 220 - 10 = -300$

For D_z, expand about row 1.

$$D_z = \begin{vmatrix} 5 & -1 & 10 \\ 7 & 2 & -4 \\ -3 & -1 & 1 \end{vmatrix} = 5\begin{vmatrix} 2 & -4 \\ -1 & 1 \end{vmatrix} - 1(-1)\begin{vmatrix} 7 & -4 \\ -3 & 1 \end{vmatrix} + 10\begin{vmatrix} 7 & 2 \\ -3 & -1 \end{vmatrix}$$

$= 5(2-4) + 1(7-12) + 10(-7+6)$
$= 5(-2) + 1(-5) + 10(-1)$
$= -10 - 5 - 10 = -25$

$x = \frac{D_x}{D} = \frac{50}{50} = 1$

$y = \frac{D_y}{D} = \frac{-300}{50} = -6$

$z = \frac{D_z}{D} = \frac{-25}{50} = -\frac{1}{2}$

The solution set is $\left\{\left(1, -6, -\frac{1}{2}\right)\right\}$.

Chapter 12 Review

1. $\begin{pmatrix} 2x+5y=2 \\ 4x-3y=30 \end{pmatrix}$

$$\left[\begin{array}{cc|c} 2 & 5 & 2 \\ 4 & -3 & 30 \end{array}\right]$$

$-2(\text{row } 1)+\text{row } 2$

$$\left[\begin{array}{cc|c} 2 & 5 & 2 \\ 0 & -13 & 26 \end{array}\right]$$

$-13y=26$

$y=-2$

$2x+5y=2$

$2x+5(-2)=2$

$2x-10=2$

$2x=12$

$x=6$

The solution set is $\{(6,-2)\}$.

3. $\begin{pmatrix} 4x-3y=34 \\ 3x+2y=0 \end{pmatrix}$

$$\left[\begin{array}{cc|c} 4 & -3 & 34 \\ 3 & 2 & 0 \end{array}\right]$$

$-\frac{3}{4}(\text{row } 1)+\text{row } 2$

$$\left[\begin{array}{cc|c} 4 & -3 & 34 \\ 0 & \frac{17}{4} & -\frac{102}{4} \end{array}\right]$$

$\frac{17}{4}y=-\frac{102}{4}$

$y=\frac{4}{17}\left(-\frac{102}{4}\right)$

$y=-6$

$4x-3y=34$

$4x-3(-6)=34$

$4x+18=34$

$4x=16$

$x=4$

The solution set is $\{(4,-6)\}$.

5. $\begin{pmatrix} x+y-2z=-7 \\ 3x+4y+z=6 \\ 5x-y-3z=-1 \end{pmatrix}$

$$\left[\begin{array}{ccc|c} 1 & 1 & -2 & -7 \\ 3 & 4 & 1 & 6 \\ 5 & -1 & -3 & -1 \end{array}\right]$$

$-3(\text{row } 1)+\text{row } 2$

$-5(\text{row } 1)+\text{row } 3$

$$\left[\begin{array}{ccc|c} 1 & 1 & -2 & -7 \\ 0 & 1 & 7 & 27 \\ 0 & -6 & 7 & 34 \end{array}\right]$$

$-1(\text{row } 2)+\text{row } 1$

$6(\text{row } 2)+\text{row } 3$

$$\left[\begin{array}{ccc|c} 1 & 0 & -9 & -34 \\ 0 & 1 & 7 & 27 \\ 0 & 0 & 49 & 196 \end{array}\right]$$

$\frac{1}{49}(\text{row } 3)$

$$\left[\begin{array}{ccc|c} 1 & 0 & -9 & -34 \\ 0 & 1 & 7 & 27 \\ 0 & 0 & 1 & 4 \end{array}\right]$$

$-7(\text{row } 3)+\text{row } 2$

$9(\text{row } 3)+\text{row } 1$

$$\left[\begin{array}{ccc|c} 1 & 0 & 0 & 2 \\ 0 & 1 & 0 & -1 \\ 0 & 0 & 1 & 4 \end{array}\right]$$

The solution set is $\{(2,-1,4)\}$.

7. $\begin{pmatrix} 3x-4y-3z=10 \\ -2x+9y-2z=14 \\ x-5y+z=-8 \end{pmatrix}$

$$\left[\begin{array}{ccc|c} 3 & -4 & -3 & 10 \\ -2 & 9 & -2 & 14 \\ 1 & -5 & 1 & -8 \end{array}\right]$$

Interchange row 1 and 3.

$$\left[\begin{array}{ccc|c} 1 & -5 & 1 & -8 \\ -2 & 9 & -2 & 14 \\ 3 & -4 & -3 & 10 \end{array}\right]$$

2(row 1) + row 2

−3(row 1) + row 3

$$\left[\begin{array}{ccc|c} 1 & -5 & 1 & -8 \\ 0 & -1 & 0 & -2 \\ 0 & 11 & -6 & 34 \end{array}\right]$$

−1(row 2)

$$\left[\begin{array}{ccc|c} 1 & -5 & 1 & -8 \\ 0 & 1 & 0 & 2 \\ 0 & 11 & -6 & 34 \end{array}\right]$$

5(row 2) + row 1

−11(row 2) + row 3

$$\left[\begin{array}{ccc|c} 1 & 0 & 1 & 2 \\ 0 & 1 & 0 & 2 \\ 0 & 0 & -6 & 12 \end{array}\right]$$

$-\frac{1}{6}$(row 3)

$$\left[\begin{array}{ccc|c} 1 & 0 & 1 & 2 \\ 0 & 1 & 0 & 2 \\ 0 & 0 & 1 & -2 \end{array}\right]$$

−1(row 3) + row 1

$$\left[\begin{array}{ccc|c} 1 & 0 & 0 & 4 \\ 0 & 1 & 0 & 2 \\ 0 & 0 & 1 & -2 \end{array}\right]$$

The solution set is $\{(4, 2, -2)\}$.

9. $\begin{pmatrix} 3x-2y-5z=2 \\ -4x+3y+11z=3 \\ 2x-y+z=-1 \end{pmatrix}$

$$\left[\begin{array}{ccc|c} 3 & -2 & -5 & 2 \\ -4 & 3 & 11 & 3 \\ 2 & -1 & 1 & -1 \end{array}\right]$$

$\frac{1}{3}$(row 1)

$$\left[\begin{array}{ccc|c} 1 & -\frac{2}{3} & -\frac{5}{3} & \frac{2}{3} \\ -4 & 3 & 11 & 3 \\ 2 & -1 & 1 & -1 \end{array}\right]$$

4(row 1) + row 2

−2(row 1) + row 3

$$\left[\begin{array}{ccc|c} 1 & -\frac{2}{3} & -\frac{5}{3} & \frac{2}{3} \\ 0 & \frac{1}{3} & \frac{13}{3} & \frac{17}{3} \\ 0 & \frac{1}{3} & \frac{13}{3} & -\frac{7}{3} \end{array}\right]$$

3(row 2)

3(row 3)

$$\left[\begin{array}{ccc|c} 1 & -\frac{2}{3} & -\frac{5}{3} & \frac{2}{3} \\ 0 & 1 & 13 & 17 \\ 0 & 1 & 13 & -7 \end{array}\right]$$

−1(row 2) + row 3

$$\left[\begin{array}{ccc|c} 1 & -\frac{2}{3} & -\frac{5}{3} & \frac{2}{3} \\ 0 & 1 & 13 & 17 \\ 0 & 0 & 0 & -24 \end{array}\right]$$

The system is inconsistent.
The solution set is $\emptyset$.

11. $\begin{vmatrix} -2 & 6 \\ 3 & 8 \end{vmatrix} = -2(8) - 6(3) = -16 - 18 = -34$

13. $\begin{vmatrix} 5 & -3 \\ -4 & -2 \end{vmatrix} = 5(-2) - (-3)(-4) = -10 - 12 = -22$

15. Expand about row 1.

$$\begin{vmatrix} 2 & 3 & -1 \\ 3 & 4 & -5 \\ 6 & 4 & 2 \end{vmatrix} = 2\begin{vmatrix} 4 & -5 \\ 4 & 2 \end{vmatrix} - 1(3)\begin{vmatrix} 3 & -5 \\ 6 & 2 \end{vmatrix} - 1\begin{vmatrix} 3 & 4 \\ 6 & 4 \end{vmatrix}$$

$= 2[8 - (-20)] - 3[6 - (-30)] - 1(12 - 24)$

$= 2(28) - 3(36) - 1(-12)$

$= 56 - 108 + 12 = -40$

17. Expand about column 3.

$$\begin{vmatrix} 5 & 4 & 3 \\ 2 & -7 & 0 \\ 3 & -2 & 0 \end{vmatrix} = 3\begin{vmatrix} 2 & -7 \\ 3 & -2 \end{vmatrix} - 0\begin{vmatrix} 5 & 4 \\ 3 & -2 \end{vmatrix} + 0\begin{vmatrix} 5 & 4 \\ 2 & -7 \end{vmatrix}$$

$= 3[-4 - (-21)] + 0 + 0$

$= 3(17) = 51$

19. $\begin{pmatrix} 2x - 3y = 12 \\ 3x + 5y = -20 \end{pmatrix}$

$D = \begin{vmatrix} 2 & -3 \\ 3 & 5 \end{vmatrix} = 10 - (-9) = 19$

$D_x = \begin{vmatrix} 12 & -3 \\ -20 & 5 \end{vmatrix} = 60 - 60 = 0$

$D_y = \begin{vmatrix} 2 & 12 \\ 3 & -20 \end{vmatrix} = -40 - 36 = -76$

$x = \frac{D_x}{D} = \frac{0}{19} = 0$

$y = \frac{D_y}{D} = \frac{-76}{19} = -4$

The solution set is $\{(0, -4)\}$.

21. $\begin{pmatrix} \frac{3}{4}x - \frac{1}{2}y = -15 \\ \frac{2}{3}x + \frac{1}{4}y = -5 \end{pmatrix}$

Multiply equation 1 by 4
and multiply equation 2 by 12.

$\begin{pmatrix} 3x - 2y = -60 \\ 8x + 3y = -60 \end{pmatrix}$

$$D = \begin{vmatrix} 3 & -2 \\ 8 & 3 \end{vmatrix} = 9 - (-16) = 25$$

$$D_x = \begin{vmatrix} -60 & -2 \\ -60 & 3 \end{vmatrix} = -180 - (120) = -300$$

$$D_y = \begin{vmatrix} 3 & -60 \\ 8 & -60 \end{vmatrix} = -180 - (-480) = 300$$

$$x = \frac{D_x}{D} = \frac{-300}{25} = -12$$

$$y = \frac{D_y}{D} = \frac{300}{25} = 12$$

The solution set is $\{(-12, 12)\}$.

25. $\begin{pmatrix} 3x + y - z = -6 \\ 3x + 2y + 3z = 9 \\ 6x - 3y + 2z = 9 \end{pmatrix}$

$$D = \begin{vmatrix} 3 & 1 & -1 \\ 3 & 2 & 3 \\ 6 & -3 & 2 \end{vmatrix}$$

Expand about row 1.

$$D = 3\begin{vmatrix} 2 & 3 \\ -3 & 2 \end{vmatrix} - 1(1)\begin{vmatrix} 3 & 3 \\ 6 & 2 \end{vmatrix} - 1\begin{vmatrix} 3 & 2 \\ 6 & -3 \end{vmatrix}$$

$D = 3[4 - (-9)] - 1(6 - 18) - 1(-9 - 12)$

$D = 3(13) - 1(-12) - 1(-21)$

$D = 39 + 12 + 21$

$D = 72$

$$D_x = \begin{vmatrix} -6 & 1 & -1 \\ 9 & 2 & 3 \\ 9 & -3 & 2 \end{vmatrix}$$

Expand about row 1.

$$D_x = -6\begin{vmatrix} 2 & 3 \\ -3 & 2 \end{vmatrix} - 1(1)\begin{vmatrix} 9 & 3 \\ 9 & 2 \end{vmatrix} - 1\begin{vmatrix} 9 & 2 \\ 9 & -3 \end{vmatrix}$$

$D_x = -6[4-(-9)] - 1(18-27) - 1(-27-18)$

$D_x = -6(13) - 1(-9) - 1(-45)$

$D_x = -78 + 9 + 45$

$D_x = -24$

$$D_y = \begin{vmatrix} 3 & -6 & -1 \\ 3 & 9 & 3 \\ 6 & 9 & 2 \end{vmatrix}$$

Expand about row 1.

$$D_y = 3\begin{vmatrix} 9 & 3 \\ 9 & 2 \end{vmatrix} - 1(-6)\begin{vmatrix} 3 & 3 \\ 6 & 2 \end{vmatrix} - 1\begin{vmatrix} 3 & 9 \\ 6 & 9 \end{vmatrix}$$

$D_y = 3(18-27) + 6(6-18) - 1(27-54)$

$D_y = 3(-9) + 6(-12) - 1(-27)$

$D_y = -27 - 72 + 27$

$D_y = -72$

$$D_z = \begin{vmatrix} 3 & 1 & -6 \\ 3 & 2 & 9 \\ 6 & -3 & 9 \end{vmatrix}$$

Expand about row 1.

$$D_z = 3\begin{vmatrix} 2 & 9 \\ -3 & 9 \end{vmatrix} - 1(1)\begin{vmatrix} 3 & 9 \\ 6 & 9 \end{vmatrix} - 6\begin{vmatrix} 3 & 2 \\ 6 & -3 \end{vmatrix}$$

$D_z = 3[18-(-27)] - 1(27-54) - 6(-9-12)$

$D_z = 3(45) - 1(-27) - 6(-21)$

$D_z = 135 + 27 + 126$

$D_z = 288$

$x = \frac{D_x}{D} = \frac{-24}{72} = -\frac{1}{3}$

$y = \frac{D_y}{D} = \frac{-72}{72} = -1$

$z = \frac{D_z}{D} = \frac{288}{72} = 4$

The solution set is $\left\{\left(-\frac{1}{3}, -1, 4\right)\right\}$.

27. $\begin{pmatrix} -2x - 7y + z = 9 \\ x + 3y - 4z = -11 \\ 4x + 5y - 3z = -11 \end{pmatrix}$

$$D = \begin{vmatrix} -2 & -7 & 1 \\ 1 & 3 & -4 \\ 4 & 5 & -3 \end{vmatrix}$$

Expand about row 1.

$$D = -2\begin{vmatrix} 3 & -4 \\ 5 & -3 \end{vmatrix} - 1(-7)\begin{vmatrix} 1 & -4 \\ 4 & -3 \end{vmatrix} + 1\begin{vmatrix} 1 & 3 \\ 4 & 5 \end{vmatrix}$$

$D = -2[-9 - (-20)] + 7[-3 - (-16)] + 1(5 - 12)$

$D = -2(11) + 7(13) + 1(-7)$

$D = -22 + 91 - 7$

$D = 62$

$$D_x = \begin{vmatrix} 9 & -7 & 1 \\ -11 & 3 & -4 \\ -11 & 5 & -3 \end{vmatrix}$$

Expand about row 1.

$$D_x = 9\begin{vmatrix} 3 & -4 \\ 5 & -3 \end{vmatrix} - 1(-7)\begin{vmatrix} -11 & -4 \\ -11 & -3 \end{vmatrix} + 1\begin{vmatrix} -11 & 3 \\ -11 & 5 \end{vmatrix}$$

$D_x = 9[-9 - (-20)] + 7(33 - 44) + 1[-55 - (-33)]$

$D_x = 9(11) + 7(-11) + 1(-22)$

$D_x = 99 - 77 - 22$

$D_x = 0$

$$D_y = \begin{vmatrix} -2 & 9 & 1 \\ 1 & -11 & -4 \\ 4 & -11 & -3 \end{vmatrix}$$

Expand about row 1.

$$D_y = -2\begin{vmatrix} -11 & -4 \\ -11 & -3 \end{vmatrix} - 1(9)\begin{vmatrix} 1 & -4 \\ 4 & -3 \end{vmatrix} + 1\begin{vmatrix} 1 & -11 \\ 4 & -11 \end{vmatrix}$$

$D_y = -2(33-44) - 9[-3-(-16)] + 1[-11-(-44)]$

$D_y = -2(-11) - 9(13) + 1(33)$

$D_y = 22 - 117 + 33$

$D_y = -62$

$$D_z = \begin{vmatrix} -2 & -7 & 9 \\ 1 & 3 & -11 \\ 4 & 5 & -11 \end{vmatrix}$$

Expand about row 1.

$$D_z = -2\begin{vmatrix} 3 & -11 \\ 5 & -11 \end{vmatrix} - 1(-7)\begin{vmatrix} 1 & -11 \\ 4 & -11 \end{vmatrix} + 9\begin{vmatrix} 1 & 3 \\ 4 & 5 \end{vmatrix}$$

$D_z = -2[-33-(-55)] + 7[-11-(-44)] + 9(5-12)$

$D_z = -2(22) + 7(33) + 9(-7)$

$D_z = -44 + 231 - 63$

$D_z = 124$

$x = \frac{D_x}{D} = \frac{0}{62} = 0$

$y = \frac{D_y}{D} = \frac{-62}{62} = -1$

$z = \frac{D_z}{D} = \frac{124}{62} = 2$

The solution set is $\{(0, -1, 2)\}$.

29. $\begin{pmatrix} 3x - y + z = -10 \\ 6x - 2y + 5z = -35 \\ 7x + 3y - 4z = 19 \end{pmatrix}$

$$D = \begin{vmatrix} 3 & -1 & 1 \\ 6 & -2 & 5 \\ 7 & 3 & -4 \end{vmatrix}$$

Expand about row 1.

$$D = 3\begin{vmatrix} -2 & 5 \\ 3 & -4 \end{vmatrix} - 1(-1)\begin{vmatrix} 6 & 5 \\ 7 & -4 \end{vmatrix} + 1\begin{vmatrix} 6 & -2 \\ 7 & 3 \end{vmatrix}$$

$D = 3(8 - 15) + 1(-24 - 35) + 1[18 - (-14)]$

$D = 3(-7) + 1(-59) + 1(32)$

$D = -21 - 59 + 32$

$D = -48$

$$D_x = \begin{vmatrix} -10 & -1 & 1 \\ -35 & -2 & 5 \\ 19 & 3 & -4 \end{vmatrix}$$

Expand about row 1.

$$D_x = -10\begin{vmatrix} -2 & 5 \\ 3 & -4 \end{vmatrix} - 1(-1)\begin{vmatrix} -35 & 5 \\ 19 & -4 \end{vmatrix} + 1\begin{vmatrix} -35 & -2 \\ 19 & 3 \end{vmatrix}$$

$D_x = -10(8 - 15) + 1(140 - 95) + 1[-105 - (-38)]$

$D_x = -10(-7) + 1(45) + 1(-67)$

$D_x = 70 + 45 - 67$

$D_x = 48$

$$D_y = \begin{vmatrix} 3 & -10 & 1 \\ 6 & -35 & 5 \\ 7 & 19 & -4 \end{vmatrix}$$

Expand about row 1.

$$D_y = 3\begin{vmatrix} -35 & 5 \\ 19 & -4 \end{vmatrix} - 1(-10)\begin{vmatrix} 6 & 5 \\ 7 & -4 \end{vmatrix} + 1\begin{vmatrix} 6 & -35 \\ 7 & 19 \end{vmatrix}$$

$D_y = 3(140 - 95) + 10(-24 - 35) + 1[114 - (-245)]$

$D_y = 3(45) + 10(-59) + 1(359)$

$D_y = 135 - 590 + 359$

$D_y = -96$

$$D_z = \begin{vmatrix} 3 & -1 & -10 \\ 6 & -2 & -35 \\ 7 & 3 & 19 \end{vmatrix}$$

Expand about row 1.

$$D_z = 3\begin{vmatrix} -2 & -35 \\ 3 & 19 \end{vmatrix} - 1(-1)\begin{vmatrix} 6 & -35 \\ 7 & 19 \end{vmatrix} - 10\begin{vmatrix} 6 & -2 \\ 7 & 3 \end{vmatrix}$$

$D_z = 3[-38-(-105)]+[114-(-245)]-10[18-(-14)]$

$D_z = 3(67)+1(359)-10(32)$

$D_z = 201+359-320$

$D_z = 240$

$x = \frac{D_x}{D} = \frac{48}{-48} = -1$

$y = \frac{D_y}{D} = \frac{-96}{-48} = 2$

$z = \frac{D_z}{D} = \frac{240}{-48} = -5$

The solution set is $\{(-1, 2, -5)\}$.

31. Let x = amount invested at 10%
y = amount invested at 12%

$$\begin{pmatrix} x+y=2500 \\ .12y=.10x+102 \end{pmatrix}$$

$$\begin{pmatrix} x+y=2500 \\ 12y=10x+10200 \end{pmatrix}$$

$x+y=2500$

$x=2500-y$

$12y=10x+10200$

$12y=10(2500-y)+10200$

$12y=25000-10y+10200$

$22y=35200$

$y=1600$

$x=2500-y$

$x=2500-1600$

$x=900$

There is \$900 invested at 10% and \$1600 invested at 12%.

33. Let x = largest angle
y = smallest angle
z = other angle

$$\begin{pmatrix} x+y+z=180 \\ x=4y+10 \\ y+x=3z \end{pmatrix}$$

$$\begin{pmatrix} x+y+z=180 \\ x-4y=10 \\ x+y-3z=0 \end{pmatrix}$$

Multiply equation 1 by 3 and then add to equation 3.

$$\begin{array}{l} 3x+3y+3z=540 \\ \underline{x+\ y-3z=\ \ 0} \\ 4x+4y \qquad =540 \end{array}$$

$$\begin{pmatrix} 4x+4y=540 \\ x-4y=10 \end{pmatrix}$$

Add the equations.

$$\begin{array}{l} 4x+4y=540 \\ \underline{x-4y=\ 10} \\ 5x \qquad =550 \end{array}$$

$x=110$

$x-4y=10$

$110-4y=10$

$-4y=-100$

$y=25$

$x+y+z=180$

$110+25+z=180$

$z=45$

The angles are 110°, 25°, and 45°.

Chapter 12 Test

1. $\begin{pmatrix} 2x-3y=-7 \\ 3x+4y=-2 \end{pmatrix}$

$$\left[\begin{array}{cc|c} 2 & -3 & -7 \\ 3 & 4 & -2 \end{array}\right]$$

$\frac{3}{4}(\text{row } 2)+\text{row } 1$

$$\left[\begin{array}{cc|c} \frac{17}{4} & 0 & -\frac{17}{2} \\ 3 & 4 & -2 \end{array}\right]$$

$\frac{17}{4}x=-\frac{17}{2}$

$\frac{4}{17}\left(\frac{17}{4}x\right)=\frac{4}{17}\left(-\frac{17}{2}\right)$

$x=-2$

3. $\begin{pmatrix} 2x+7y=-33 \\ 3x-5y=28 \end{pmatrix}$

$$\left[\begin{array}{cc|c} 2 & 7 & -33 \\ 3 & -5 & 28 \end{array}\right]$$

$\frac{7}{5}(\text{row } 2)+\text{row } 1$

$$\left[\begin{array}{cc|c} \frac{31}{5} & 0 & \frac{31}{5} \\ 3 & -5 & 28 \end{array}\right]$$

$\frac{31}{5}x=\frac{31}{5}$

$x=1$

5. $\begin{pmatrix} x+y+z=-1 \\ -2x-y+2z=3 \\ 3x+4y-z=-2 \end{pmatrix}$

$$\left[\begin{array}{ccc|c} 1 & 1 & 1 & -1 \\ -2 & -1 & 2 & 3 \\ 3 & 4 & -1 & -2 \end{array}\right]$$

$2(\text{row } 3)+\text{row } 2$

$\text{row } 3+\text{row } 1$

$$\left[\begin{array}{ccc|c} 4 & 5 & 0 & -3 \\ 4 & 7 & 0 & -1 \\ 3 & 4 & -1 & -2 \end{array}\right]$$

$-\frac{5}{7}(\text{row } 2)+\text{row } 1$

$$\left[\begin{array}{ccc|c} \frac{8}{7} & 0 & 0 & -\frac{16}{7} \\ 4 & 7 & 0 & -1 \\ 3 & 4 & -1 & -2 \end{array}\right]$$

$\frac{8}{7}x=-\frac{16}{7}$

$\frac{7}{8}\left(\frac{8}{7}x\right)=\frac{7}{8}\left(-\frac{16}{7}\right)$

$x=-2$

7. $\begin{pmatrix} x+2y+z=5 \\ 3x+5y-2z=-15 \\ -2x-y-z=-4 \end{pmatrix}$

$$\left[\begin{array}{ccc|c} 1 & 2 & 1 & 5 \\ 3 & 5 & -2 & -15 \\ -2 & -1 & -1 & -4 \end{array}\right]$$

$-3(\text{row } 1)+\text{row } 2$

$2(\text{row } 1)+\text{row } 3$

$$\left[\begin{array}{ccc|c} 1 & 2 & 1 & 5 \\ 0 & -1 & -5 & -30 \\ 0 & 3 & 1 & 6 \end{array}\right]$$

$3(\text{row } 2)+\text{row } 3$

$$\left[\begin{array}{ccc|c} 1 & 2 & 1 & 5 \\ 0 & -1 & -5 & -30 \\ 0 & 0 & -14 & -84 \end{array}\right]$$

$-14z=-84$

$z=6$

9. $$\left[\begin{array}{ccc|c} 1 & 2 & -3 & 4 \\ 0 & 1 & 2 & 5 \\ 0 & 0 & 2 & -8 \end{array}\right]$$

$-1(\text{row } 3)+\text{row } 2$

$$\left[\begin{array}{ccc|c} 1 & 2 & -3 & 4 \\ 0 & 1 & 0 & 13 \\ 0 & 0 & 2 & -8 \end{array}\right]$$

$y=13$

11. $\begin{pmatrix} 3x-y-2z=1 \\ 4x+2y+z=5 \\ 6x-2y-4z=9 \end{pmatrix}$

Multiply equation 1 by -2
and then add to equation 3.

$$\begin{array}{r} -6x+2y+4z=-2 \\ 6x-2y-4z=\ \ 9 \\ \hline 0\neq 7 \end{array}$$

This is a contradiction. The system of equations is inconsistent. There are no solutions.

13. $\begin{vmatrix} 3 & -2 \\ -5 & 4 \end{vmatrix} = 12-10=2$

15. $\begin{vmatrix} \frac{1}{2} & \frac{1}{3} \\ \frac{3}{4} & -\frac{2}{3} \end{vmatrix} = -\frac{1}{3}-\frac{1}{4}=-\frac{7}{12}$

17. Expand about row 2.

$$\begin{vmatrix} 2 & 4 & -5 \\ -4 & 3 & 0 \\ -2 & 6 & 1 \end{vmatrix} = -1(-4)\begin{vmatrix} 4 & -5 \\ 6 & 1 \end{vmatrix} + 3\begin{vmatrix} 2 & -5 \\ -2 & 1 \end{vmatrix} + 0\begin{vmatrix} 2 & 4 \\ -2 & 6 \end{vmatrix}$$

$=4[4-(-30)]+3(2-10)+0$

$=4(34)+3(-8)+0$

$=112$

19. $\begin{pmatrix} 5x-2y=-41 \\ 3x+4y=-9 \end{pmatrix}$

$D=\begin{vmatrix} 5 & -2 \\ 3 & 4 \end{vmatrix} = 20-(-6)=26$

$D_x=\begin{vmatrix} -41 & -2 \\ -9 & 4 \end{vmatrix} = -164-18=-182$

$x=\frac{D_x}{D}=\frac{-182}{26}=-7$

21. $\begin{pmatrix} \frac{3}{2}x-\frac{1}{3}y=-22 \\ \frac{2}{3}x+\frac{1}{4}y=-5 \end{pmatrix}$

Multiply equation 1 by 6 and
multiply equation 2 by 12.

$$\begin{pmatrix} 9x-2y=-132 \\ 8x+3y=-60 \end{pmatrix}$$

$$D=\begin{vmatrix} 9 & -2 \\ 8 & 3 \end{vmatrix}=27-(-16)=43$$

$$D_x=\begin{vmatrix} -132 & -2 \\ -60 & 3 \end{vmatrix}$$

$$D_x=-396-120=-516$$

$$x=\frac{D_x}{D}=\frac{-516}{43}=-12$$

23. $$\begin{pmatrix} x-4y+z=12 \\ -2x+3y-z=-11 \\ 5x-3y+2z=17 \end{pmatrix}$$

$$D=\begin{vmatrix} 1 & -4 & 1 \\ -2 & 3 & -1 \\ 5 & -3 & 2 \end{vmatrix}$$

Expand about row 1.

$$D=1\begin{vmatrix} 3 & -1 \\ -3 & 2 \end{vmatrix}-1(-4)\begin{vmatrix} -2 & -1 \\ 5 & 2 \end{vmatrix}+1\begin{vmatrix} -2 & 3 \\ 5 & -3 \end{vmatrix}$$

$$D=1(6-3)+4[-4-(-5)]+1(6-15)$$

$$D=1(3)+4(1)+1(-9)$$

$$D=3+4-9=-2$$

$$D_x=\begin{vmatrix} 12 & -4 & 1 \\ -11 & 3 & -1 \\ 17 & -3 & 2 \end{vmatrix}$$

Expand about row 1.

$$D_x=12\begin{vmatrix} 3 & -1 \\ -3 & 2 \end{vmatrix}-1(-4)\begin{vmatrix} -11 & -1 \\ 17 & 2 \end{vmatrix}+1\begin{vmatrix} -11 & 3 \\ 17 & -3 \end{vmatrix}$$

$$D_x=12(6-3)+4[-22-(-17)]+1(33-51)$$

$$D_x=12(3)+4(-5)+1(-18)$$

$D_x = 36 - 20 - 18 = -2$

$x = \frac{D_x}{D} = \frac{-2}{-2} = 1$

25. $\begin{pmatrix} 3x - 2z = -6 \\ x - 3y + 5z = -11 \\ 2y + 3z = 6 \end{pmatrix}$

$$D = \begin{vmatrix} 3 & 0 & -2 \\ 1 & -3 & 5 \\ 0 & 2 & 3 \end{vmatrix}$$

Expand about row 1.

$$D = 3\begin{vmatrix} -3 & 5 \\ 2 & 3 \end{vmatrix} - 1(0)\begin{vmatrix} 1 & 5 \\ 0 & 3 \end{vmatrix} - 2\begin{vmatrix} 1 & -3 \\ 0 & 2 \end{vmatrix}$$

$D = 3(-9 - 10) + 0 - 2(2 - 0)$

$D = 3(-19) + 0 - 2(2)$

$D = -57 + 0 - 4 = -61$

$$D_z = \begin{vmatrix} 3 & 0 & -6 \\ 1 & -3 & -11 \\ 0 & 2 & 6 \end{vmatrix}$$

Expand about row 1.

$$D_z = 3\begin{vmatrix} -3 & -11 \\ 2 & 6 \end{vmatrix} - 1(0)\begin{vmatrix} 1 & -11 \\ 0 & 6 \end{vmatrix} - 6\begin{vmatrix} 1 & -3 \\ 0 & 2 \end{vmatrix}$$

$D_z = 3[-18 - (-22)] + 0 - 6(2 - 0)$

$D_z = 3(4) + 0 - 6(2)$

$D_z = 12 + 0 - 12 = 0$

$z = \frac{D_z}{D} = \frac{0}{-61} = 0$

Chapter 12 Cumulative Review

1. $-5(x - 1) - 3(2x+4) + 3(3x - 1)$; for $x = -2$

$-5x + 5 - 6x - 12 + 9x - 3$

$-2x - 10$

$-2(-2)-10$

$4-10$

-6

3. $\frac{2}{n}-\frac{3}{2n}+\frac{5}{3n}$; for $n=4$

$\frac{2}{4}-\frac{3}{2(4)}+\frac{5}{3(4)}$

$\frac{1}{2}-\frac{3}{8}+\frac{5}{12}$; LCD $=24$

$\frac{1}{2}\bullet\frac{12}{12}-\frac{3}{8}\bullet\frac{3}{3}+\frac{5}{12}\bullet\frac{2}{2}$

$\frac{12}{24}-\frac{9}{24}+\frac{10}{24}$

$\frac{13}{24}$

5. $\frac{3}{x-2}-\frac{5}{x+3}$; for $x=3$

$\frac{3}{3-2}-\frac{5}{3+3}$

$\frac{3}{1}-\frac{5}{6}$; LCD $=6$

$\frac{18}{6}-\frac{5}{6}$

$\frac{13}{6}$

7. $(2\sqrt{x}-3)(\sqrt{x}+4)$

$2\sqrt{x^2}+8\sqrt{x}-3\sqrt{x}-12$

$2x+5\sqrt{x}-12$

9. $(2x-1)(x^2+6x-4)$

$2x^3+12x^2-8x-x^2-6x+4$

$2x^3+11x^2-14x+4$

11. $\frac{16x^2y}{24xy^3}\div\frac{9xy}{8x^2y^2}$

$\frac{16x^2y}{24xy^3}\bullet\frac{8x^2y^2}{9xy}$

$\frac{16(8)x^4y^3}{24(9)x^2y^4}$

$\frac{16x^2}{27y}$

13. $\frac{7}{12ab}-\frac{11}{15a^2}$; LCD $=60a^2b$

$\frac{7}{12ab}\bullet\frac{5a}{5a}-\frac{11}{15a^2}\bullet\frac{4b}{4b}$

$\frac{35a}{60a^2b}-\frac{44b}{60a^2b}$

$\frac{35a-44b}{60a^2b}$

15.

$$\begin{array}{r|l} & 2x^2 - x - 4 \\ \hline 4x-1 & 8x^3-6x^2-15x+4 \\ & \underline{8x^3-2x^2} \\ & \quad -4x^2-15x \\ & \quad \underline{-4x^2+x} \\ & \qquad -16x+4 \\ & \qquad \underline{-16x+4} \\ & \qquad\qquad 0 \end{array}$$

$2x^2-x-4$

17. $\frac{\frac{2}{x}-3}{\frac{3}{y}+4}$

$\frac{xy}{xy}\bullet\frac{\left(\frac{2}{x}-3\right)}{\left(\frac{3}{y}+4\right)}$

$\frac{xy\left(\frac{2}{x}\right)-3xy}{xy\left(\frac{3}{y}\right)+4xy}$

$\frac{2y-3xy}{3x+4xy}$

19. $\frac{3a}{2-\frac{1}{a}}-1$

$\frac{a}{a}\bullet\frac{3a}{\left(2-\frac{1}{a}\right)}-1$

$\frac{3a^2}{2a-1}-1$

$\frac{3a^2}{2a-1}-1\bullet\frac{(2a-1)}{2a-1}$

$\frac{3a^2-1(2a-1)}{2a-1}$

$\frac{3a^2-2a+1}{2a-1}$

21. $16x^3+54$

$2(8x^3+27)$

$2(2x+3)(4x^2-6x+9)$

23. $12x^3-52x^2-40x$

$4x(3x^2-13x-10)$

$4x(3x+2)(x-5)$

25. $10+9x-9x^2$

$(5-3x)(2+3x)$

27. $\dfrac{3}{\left(\frac{4}{3}\right)^{-1}}$

$3\left(\frac{4}{3}\right)^1=4$

29. $-\sqrt{.09}=-.3$

31. $4^0+4^{-1}+4^{-2}$

$1+\frac{1}{4}+\frac{1}{4^2}$

$1+\frac{1}{4}+\frac{1}{16}$

$\frac{16}{16}+\frac{4}{16}+\frac{1}{16}$

$\frac{21}{16}$

33. $(2^{-3}-3^{-2})^{-1}$

$\left(\frac{1}{2^3}-\frac{1}{3^2}\right)^{-1}$

$\left(\frac{1}{8}-\frac{1}{9}\right)^{-1}$

$\left(\frac{9}{72}-\frac{8}{72}\right)^{-1}$

$\left(\frac{1}{72}\right)^{-1}$

72

35. $\log_3\left(\frac{1}{9}\right)=x$

$3^x=\frac{1}{9}$

$3^x=\frac{1}{3^2}$

$3^x=3^{-2}$

$x=-2$

37. $\dfrac{48x^{-4}y^2}{6xy}$

$8x^{-4-1}y^{2-1}$

$8x^{-5}y^1$

$\dfrac{8y}{x^5}$

39. $\sqrt{80}$

$\sqrt{16}\sqrt{5}$

$4\sqrt{5}$

41. $\sqrt{\frac{75}{81}}=\frac{\sqrt{75}}{\sqrt{81}}=\frac{\sqrt{25}\sqrt{3}}{9}=\frac{5\sqrt{3}}{9}$

43. $\sqrt[3]{56}$

$\sqrt[3]{8}\sqrt[3]{7}$

$2\sqrt[3]{7}$

45. $4\sqrt{52x^3y^2}$

$4\sqrt{4x^2y^2}\sqrt{13x}$

$4(2xy)\sqrt{13x}$

$8xy\sqrt{13x}$

47. $-3\sqrt{24}+6\sqrt{54}-\sqrt{6}$

$-3\sqrt{4}\sqrt{6}+6\sqrt{9}\sqrt{6}-\sqrt{6}$

$-3(2)\sqrt{6}+6(3)\sqrt{6}-\sqrt{6}$

$-6\sqrt{6}+18\sqrt{6}-\sqrt{6}$

$11\sqrt{6}$

49. $8\sqrt[3]{3}-6\sqrt[3]{24}-4\sqrt[3]{81}$

$8\sqrt[3]{3}-6\sqrt[3]{8}\sqrt[3]{3}-4\sqrt[3]{27}\sqrt[3]{3}$

$8\sqrt[3]{3}-6(2)\sqrt[3]{3}-4(3)\sqrt[3]{3}$

$8\sqrt[3]{3}-12\sqrt[3]{3}-12\sqrt[3]{3}$

$-16\sqrt[3]{3}$

51. $\frac{3\sqrt{5}-\sqrt{3}}{2\sqrt{3}+\sqrt{7}}$

$\frac{(3\sqrt{5}-\sqrt{3})}{(2\sqrt{3}+\sqrt{7})} \bullet \frac{(2\sqrt{3}-\sqrt{7})}{(2\sqrt{3}-\sqrt{7})}$

$\frac{6\sqrt{15}-3\sqrt{35}-2\sqrt{9}+\sqrt{21}}{4\sqrt{9}-\sqrt{49}}$

$\frac{6\sqrt{15}-3\sqrt{35}-2(3)+\sqrt{21}}{4(3)-7}$

$\frac{6\sqrt{15}-3\sqrt{35}-6+\sqrt{21}}{5}$

53. $\frac{.00072}{.0000024}$

$\frac{(7.2)(10)^{-4}}{(2.4)(10)^{-6}} = 3(10)^{-4-(-6)}$

$3(10)^2 = 300$

55. $(5-2i)(4+6i)$

$20+30i-8i-12i^2$

$20+22i+12$

$32+22i$

57. $\frac{5}{4i} = \frac{5}{4i} \bullet \frac{(-i)}{(-i)} = \frac{-5i}{-4i^2} = \frac{-5i}{4}$

$0-\frac{5}{4}i$

59. $(2,-3)$ and $(-1,\ 7)$

$m = \frac{7-(-3)}{-1-2} = \frac{10}{-3} = -\frac{10}{3}$

61. $(4,\ 5)$ and $(-2,\ 1)$

$d = \sqrt{(-2-4)^2+(1-5)^2}$

$d = \sqrt{(-6)^2+(-4)^2}$

$d = \sqrt{36+16} = \sqrt{52}$

$d = \sqrt{4}\sqrt{13} = 2\sqrt{13}$

63. $3x-4y=6;\ (-3,-2)$

$-4y = -3x+6$

$y = \frac{-3}{-4}x + \frac{6}{-4}$

$y = \frac{3}{4}x - \frac{3}{2}$

Perpendicular lines have slopes that are negative reciprocals.

$m = -\frac{4}{3};\ (-3,-2)$

$y-(-2) = -\frac{4}{3}[x-(-3)]$

$y+2 = -\frac{4}{3}(x+3)$

$3(y+2) = 3\left[-\frac{4}{3}(x+3)\right]$

$3y+6 = -4(x+3)$

$3y+6 = -4x-12$

$4x+3y+6 = -12$

$4x+3y = -18$

65. $y = x^2+10x+21$

$y = x^2+10x+25-25+21$

$y = (x+5)^2-4$

vertex $(-5,-4)$

67 - 81. See back of textbook for graphs.

83. $f(x) = x-3$ and $g(x) = 2x^2-x-1$

$(g \circ f)(x) = 2(x-3)^2-(x-3)-1$

$(g \circ f)(x) = 2(x^2-6x+9)-x+3-1$

$(g \circ f)(x) = 2x^2-12x+18-x+2$

$(g \circ f)(x) = 2x^2-13x+20$

$(f \circ g)(x) = (2x^2-x-1)-3$

$(f \circ g)(x) = 2x^2-x-4$

85. $f(x) = -\frac{1}{2}x+\frac{2}{3}$

$y = -\frac{1}{2}x+\frac{2}{3}$

Interchange x and y.

$x = -\frac{1}{2}y+\frac{2}{3}$

$x-\frac{2}{3} = -\frac{1}{2}y$

$-2\left(x-\frac{2}{3}\right)=-2\left(-\frac{1}{2}y\right)$

$-2x+\frac{4}{3}=y$

$f^{-1}(x)=-2x+\frac{4}{3}$

87. $y=\frac{k}{x^2}$

$4=\frac{k}{3^2}$

$4=\frac{k}{9}$

$36=k$

$y=\frac{36}{x^2}$

$y=\frac{36}{6^2}=\frac{36}{36}=1$

89. $\begin{vmatrix} -2 & 4 \\ 7 & 6 \end{vmatrix}=-12-28=-40$

91. $3(2x-1)-2(5x+1)=4(3x+4)$

$6x-3-10x-2=12x+16$

$-4x-5=12x+16$

$-16x=21$

$x=-\frac{21}{16}$

The solution set is $\left\{-\frac{21}{16}\right\}$.

93. $.92+.9(x-.3)=2x-5.95$

$100[.92+.9(x-.3)]=100(2x-5.95)$

$92+90(x-.3)=200x-595$

$92+90x-27=200x-595$

$-110x=-660$

$x=6$

The solution set is $\{6\}$.

95. $3x^2=7x$

$3x^2-7x=0$

$x(3x-7)=0$

$x=0$ or $3x-7=0$

$x=0$ or $3x=7$

$x=0$ or $x=\frac{7}{3}$

The solution set is $\left\{0, \frac{7}{3}\right\}$.

97. $30x^2+13x-10=0$

$(6x+5)(5x-2)=0$

$6x+5=0$ or $5x-2=0$

$6x=-5$ or $5x=2$

$x=-\frac{5}{6}$ or $x=\frac{2}{5}$

The solution set is $\left\{-\frac{5}{6}, \frac{2}{5}\right\}$.

99. $x^4+8x^2-9=0$

$(x^2+9)(x^2-1)=0$

$x^2+9=0$ or $x^2-1=0$

$x^2=-9$ or $x^2=1$

$x=\pm\sqrt{-9}$ or $x=\pm\sqrt{1}$

$x=\pm 3i$ or $x=\pm 1$

The solution set is $\{\pm 3i, \pm 1\}$.

101. $2-\frac{3x}{x-4}=\frac{14}{x+7}$; $x\neq 4$, $x\neq -7$

$(x-4)(x+7)\left(2-\frac{3x}{x-4}\right)=(x-4)(x+7)\left(\frac{14}{x+7}\right)$

$2(x-4)(x+7)-3x(x+7)=14(x-4)$

$2(x^2+3x-28)-3x^2-21x=14x-56$

$2x^2+6x-56-3x^2-21x=14x-56$

$-x^2-15x-56=14x-56$

$0=x^2+29x$

$0=x(x+29)$

$x=0$ or $x+29=0$

$x=0$ or $x=-29$

The solution set is $\{-29, 0\}$.

103. $\sqrt{3y}-y=-6$

$\sqrt{3y}=y-6$

$(\sqrt{3y})^2=(y-6)^2$

$3y=y^2-12y+36$

$0 = y^2 - 15y + 36$

$0 = (y - 12)(y - 3)$

$y - 12 = 0$ or $y - 3 = 0$

$y = 12$ or $y = 3$

Checking $y = 12$

$\sqrt{3(12)} - 12 \stackrel{?}{=} -6$

$\sqrt{36} - 12 \stackrel{?}{=} -6$

$6 - 12 \stackrel{?}{=} -6$

$-6 = -6$

Checking $y = 3$

$\sqrt{3(3)} - 3 \stackrel{?}{=} -6$

$\sqrt{9} - 3 \stackrel{?}{=} -6$

$3 - 3 \stackrel{?}{=} -6$

$0 \neq -6$

The solution set is $\{12\}$.

105. $(3x - 1)^2 = 45$

$3x - 1 = \pm\sqrt{45}$

$3x - 1 = \pm 3\sqrt{5}$

$3x = 1 \pm 3\sqrt{5}$

$x = \frac{1 \pm 3\sqrt{5}}{3}$

The solution set is $\left\{\frac{1 \pm 3\sqrt{5}}{3}\right\}$.

107. $2x^2 - 3x + 4 = 0$

$x = \frac{-(-3) \pm \sqrt{(-3)^2 - 4(2)(4)}}{2(2)}$

$x = \frac{3 \pm \sqrt{9 - 32}}{4}$

$x = \frac{3 \pm \sqrt{-23}}{4}$

$x = \frac{3 \pm i\sqrt{23}}{4}$

The solution set is $\left\{\frac{3 \pm i\sqrt{23}}{4}\right\}$.

109. $\frac{5}{n-3} - \frac{3}{n+3} = 1;\ n \neq 3,\ n \neq -3$

$(n-3)(n+3)\left(\frac{5}{n-3} - \frac{3}{n+3}\right) = (n-3)(n+3)(1)$

$5(n+3) - 3(n-3) = n^2 - 9$

$5n + 15 - 3n + 9 = n^2 - 9$

$2n + 24 = n^2 - 9$

$0 = n^2 - 2n - 33$

$n = \frac{-(-2) \pm \sqrt{(-2)^2 - 4(1)(-33)}}{2(1)}$

$n = \frac{2 \pm \sqrt{4 + 132}}{2}$

$n = \frac{2 \pm \sqrt{136}}{2}$

$n = \frac{2 \pm 2\sqrt{34}}{2}$

$n = \frac{2(1 \pm \sqrt{34})}{2}$

$n = 1 \pm \sqrt{34}$

The solution set is $\{1 \pm \sqrt{34}\}$.

111. $2x^2 + 5x + 5 = 0$

$x = \frac{-5 \pm \sqrt{(5)^2 - 4(2)(5)}}{2(2)}$

$x = \frac{-5 \pm \sqrt{25 - 40}}{4}$

$x = \frac{-5 \pm \sqrt{-15}}{4}$

$x = \frac{-5 \pm i\sqrt{15}}{4}$

The solution set is $\left\{\frac{-5 \pm i\sqrt{15}}{4}\right\}$.

113. $6x^3 - 19x^2 + 9x + 10 = 0$

Factors of 6: $\pm 1, \pm 2, \pm 3, \pm 6$

Factors of 10: $\pm 1, \pm 2, \pm 5, \pm 10$

Possible rational roots:

$\pm 1, \pm 2, \pm 5, \pm 10 \pm \frac{5}{3}, \pm \frac{10}{3}, \pm \frac{5}{6},$

$\pm \frac{5}{2}, \pm \frac{1}{3}, \pm \frac{2}{3}, \pm \frac{1}{6}, \pm \frac{1}{2}$

$$\begin{array}{r|rrrr} 2 & 6 & -19 & 9 & 10 \\ & & 12 & -14 & -10 \\ \hline & 6 & -7 & -5 & 0 \end{array}$$

$(x-2)(6x^2-7x-5)=0$

$(x-2)(3x-5)(2x+1)=0$

$x-2=0$ or $3x-5=0$ or $2x+1=0$

$x=2$ or $3x=5$ or $2x=-1$

$x=2$ or $x=\frac{5}{3}$ or $x=-\frac{1}{2}$

The solution set is $\left\{-\frac{1}{2}, \frac{5}{3}, 2\right\}$.

115. $\log_3 x=4$

$3^4=x$

$81=x$

The solution set is {81}.

117. $\ln(3x-4)-\ln(x+1)=\ln 2$

$\ln\frac{3x-4}{x+1}=\ln 2$

$\frac{3x-4}{x+1}=2;\ x\neq -1$

$3x-4=2(x+1)$

$3x-4=2x+2$

$x-4=2$

$x=6$

The solution set is {6}.

119. $-5(y-1)+3>3y-4-4y$

$-5y+5+3>-y-4$

$-5y+8>-y-4$

$-4y>-12$

$-\frac{1}{4}(-4y)<-\frac{1}{4}(-12)$

$y<3$

The solution set is $\{(-\infty, 3)\}$.

121. $|5x-2|>13$

$5x-2<-13$ or $5x-2>13$

$5x<-11$ or $5x>15$

$x<-\frac{11}{5}$ or $x>3$

The solution set is

$\{(-\infty,-\frac{11}{5})\cup(3, \infty)\}$.

123. $\frac{x-2}{5}-\frac{3x-1}{4}\leq\frac{3}{10}$

$20\left(\frac{x-2}{5}-\frac{3x-1}{4}\right)\leq 20\left(\frac{3}{10}\right)$

$4(x-2)-5(3x-1)\leq 6$

$4x-8-15x+5\leq 6$

$-11x-3\leq 6$

$-11x\leq 9$

$-\frac{1}{11}(-11x)\geq -\frac{1}{11}(9)$

$x\geq -\frac{9}{11}$

The solution set is $\{[-\frac{9}{11},\infty)\}$.

125. $(3x-1)(x-4)>0$

$3x-1=0$ or $x-4=0$

$3x=1$ or $x=4$

$x=\frac{1}{3}$ or $x=4$

Critical points on the number line: $\frac{1}{3}$, 4

Test Point:	-1	1	5
$(3x-1)$:	negative	positive	positive
$(x-4)$:	negative	negative	positive
product:	positive	negative	positive

The solution set is

$\{(-\infty, \frac{1}{3})\cup(4, \infty)\}$.

127. $\frac{x-3}{x-7}\geq 0;\ x\neq 7$

$x-3=0$ or $x-7=0$

$x=3$ or $x=7$

Critical points on the number line: 3, 7

Test Point:	0	5	8
$(x-3)$:	negative	positive	positive
$(x-7)$:	negative	negative	positive
quotient:	positive	negative	positive

The solution set is

$\{(-\infty, 3]\cup(7, \infty)\}$.

129. $\begin{pmatrix}4x-3y=18\\3x-2y=15\end{pmatrix}$

$$D=\begin{vmatrix}4 & -3\\3 & -2\end{vmatrix}=-8+9=1$$

$$D_x = \begin{vmatrix} 18 & -3 \\ 15 & -2 \end{vmatrix} = -36 + 45 = 9$$

$$D_y = \begin{vmatrix} 4 & 18 \\ 3 & 15 \end{vmatrix} = 60 - 54 = 6$$

$$x = \frac{D_x}{D} = \frac{9}{1} = 9$$

$$y = \frac{D_y}{D} = \frac{6}{1} = 6$$

The solution set is {(9, 6)}.

131. $\begin{pmatrix} \frac{x}{2} - \frac{y}{3} = 1 \\ \frac{2x}{5} + \frac{y}{2} = 2 \end{pmatrix}$

Multiply equation 1 by 6 and multiply equation 2 by 10.

$$\begin{pmatrix} 3x - 2y = 6 \\ 4x + 5y = 20 \end{pmatrix}$$

$$D = \begin{vmatrix} 3 & -2 \\ 4 & 5 \end{vmatrix} = 15 + 8 = 23$$

$$D_x = \begin{vmatrix} 6 & -2 \\ 20 & 5 \end{vmatrix} = 30 + 40 = 70$$

$$D_y = \begin{vmatrix} 3 & 6 \\ 4 & 20 \end{vmatrix} = 60 - 24 = 36$$

$$x = \frac{D_x}{D} = \frac{70}{23}$$

$$y = \frac{D_y}{D} = \frac{36}{23}$$

The solution set is $\{(\frac{70}{23}, \frac{36}{23})\}$.

133. $\begin{pmatrix} x - y + 5z = -10 \\ 5x + 2y - 3z = 6 \\ -3x + 2y - z = 12 \end{pmatrix}$

$$\left[\begin{array}{ccc|c} 1 & -1 & 5 & -10 \\ 5 & 2 & -3 & 6 \\ -3 & 2 & -1 & 12 \end{array}\right]$$

$-5(\text{row } 1) + \text{row } 2$
$3(\text{row } 1) + \text{row } 3$

$$\left[\begin{array}{ccc|c} 1 & -1 & 5 & -10 \\ 0 & 7 & -28 & 56 \\ 0 & -1 & 14 & -18 \end{array}\right]$$

Interchange row 2 and row 3.

$$\left[\begin{array}{ccc|c} 1 & -1 & 5 & -10 \\ 0 & -1 & 14 & -18 \\ 0 & 7 & -28 & 56 \end{array}\right]$$

$7(\text{row } 2) + \text{row } 3$

$$\left[\begin{array}{ccc|c} 1 & -1 & 5 & -10 \\ 0 & -1 & 14 & -18 \\ 0 & 0 & 70 & -70 \end{array}\right]$$

$70z = -70$

$z = -1$

$-y + 14z = -18$

$-y + 14(-1) = -18$

$-y - 14 = -18$

$-y = -4$

$y = 4$

$x - y + 5z = -10$

$x - 4 + 5(-1) = -10$

$x - 4 - 5 = -10$

$x = -1$

The solution set is $\{(-1, 4, -1)\}$.

135. Let x = number of nickels
y = number of dimes
z = number of quarters

$$\begin{pmatrix} x + y + z = 63 \\ y = x + 6 \\ z = 2x + 1 \end{pmatrix}$$

$x + y + z = 63$

$x + x + 6 + 2x + 1 = 63$

$4x + 7 = 63$

$4x = 56$

$x = 14$

$y = x + 6$

$y = 14 + 6$

$y = 20$

$z = 2x + 1$

$z = 2(14) + 1$

$z = 28 + 1$

$z = 29$

There are 14 nickels, 20 dimes, and 29 quarters.

137. Let $x =$ selling price

$300 + (50\%)(x) = x$

$300 + .5x = x$

$10(300 + .5x) = 10x$

$3000 + 5x = 10x$

$3000 = 5x$

$600 = x$

The selling price should be $600.

139.

	rate	time	distance
east	x+10	4.5	4.5(x+10)
west	x	4.5	4.5x

$4.5(x + 10) + 4.5x = 639$

$10[4.5(x + 10) + 4.5x] = 10(639)$

$45(x + 10) + 45x = 6390$

$45x + 450 + 45x = 6390$

$90x = 5940$

$x = 66$

The westbound train is traveling at 66 mph and the eastbound train is traveling at 76 mph.

141. Let $x =$ score on the 4th day

$\frac{70 + 73 + 76 + x}{4} \leq 72$

$\frac{219 + x}{4} \leq 72$

$219 + x \leq 288$

$x \leq 69$

The score must be 69 or less.

143. Let $x =$ width of strip
width $= 8 - 2x$
length $= 14 - 2x$

$(8 - 2x)(14 - 2x) = 72$

$112 - 16x - 28x + 4x^2 = 72$

$4x^2 - 44x + 40 = 0$

$4(x^2 - 11x + 10) = 0$

$4(x - 10)(x - 1) = 0$

$x - 10 = 0$ or $x - 1 = 0$

$x = 10$ or $x = 1$

Discard the root $x = 10$.
The width of the strip is 1 inch.

145.

	Time in hours	Rate
Sue	x	$\frac{1}{x}$
Dean	2	$\frac{1}{2}$
Together	$\frac{6}{5}$	$\frac{5}{6}$

$\frac{1}{x} + \frac{1}{2} = \frac{5}{6}$

$6x\left(\frac{1}{x} + \frac{1}{2}\right) = 6x\left(\frac{5}{6}\right)$

$6 + 3x = 5x$

$6 = 2x$

$3 = x$

It would take Sue 3 hours.

147. Let $x =$ units digit
$y =$ tens digit

$\begin{pmatrix} x = 2y + 1 \\ x + y = 10 \end{pmatrix}$

$x + y = 10$

$2y + 1 + y = 10$

$3y + 1 = 10$

$3y = 9$

$y = 3$

$x = 2(3) + 1$

$x = 6 + 1$

$x = 7$

The number is 37.

Chapter 13 Sequences and Series

Problem Set 13.1 Arithmetic Sequences

1. $a_n = 3n - 4$
   $a_1 = 3(1) - 4 = -1$
   $a_2 = 3(2) - 4 = 2$
   $a_3 = 3(3) - 4 = 5$
   $a_4 = 3(4) - 4 = 8$
   $a_5 = 3(5) - 4 = 11$

3. $a_n = -2n + 5$
   $a_1 = -2(1) + 5 = 3$
   $a_2 = -2(2) + 5 = 1$
   $a_3 = -2(3) + 5 = -1$
   $a_4 = -2(4) + 5 = -3$
   $a_5 = -2(5) + 5 = -5$

5. $a_n = n^2 - 2$
   $a_1 = 1^2 - 2 = -1$
   $a_2 = 2^2 - 2 = 2$
   $a_3 = 3^2 - 2 = 7$
   $a_4 = 4^2 - 2 = 14$
   $a_5 = 5^2 - 2 = 23$

7. $a_n = -n^2 + 1$
   $a_1 = -(1)^2 + 1 = 0$
   $a_2 = -(2)^2 + 1 = -3$
   $a_3 = -(3)^2 + 1 = -8$
   $a_4 = -(4)^2 + 1 = -15$
   $a_5 = -(5)^2 + 1 = -24$

9. $a_n = 2n^2 - 3$
   $a_1 = 2(1)^2 - 3 = 2(1) - 3 = -1$
   $a_2 = 2(2)^2 - 3 = 2(4) - 3 = 5$
   $a_3 = 2(3)^2 - 3 = 2(9) - 3 = 15$
   $a_4 = 2(4)^2 - 3 = 2(16) - 3 = 29$
   $a_5 = 2(5)^2 - 3 = 2(25) - 3 = 47$

11. $a_n = 2^{n-2}$
   $a_1 = 2^{1-2} = 2^{-1} = \frac{1}{2}$
   $a_2 = 2^{2-2} = 2^0 = 1$
   $a_3 = 2^{3-2} = 2^1 = 2$
   $a_4 = 2^{4-2} = 2^2 = 4$
   $a_5 = 2^{5-2} = 2^3 = 8$

13. $a_n = -2(3)^{n-2}$
   $a_1 = -2(3)^{1-2} = -2(3)^{-1} = -2\left(\frac{1}{3}\right) = -\frac{2}{3}$
   $a_2 = -2(3)^{2-2} = -2(3)^0 = -2(1) = -2$
   $a_3 = -2(3)^{3-2} = -2(3)^1 = -2(3) = -6$
   $a_4 = -2(3)^{4-2} = -2(3)^2 = -2(9) = -18$
   $a_5 = -2(3)^{5-2} = -2(3)^3 = -2(27) = -54$

15. $a_n = n^2 - n - 2$
   $a_8 = 8^2 - 8 - 2$
   $a_8 = 64 - 8 - 2 = 54$

   $a_{12} = 12^2 - 12 - 2$
   $a_{12} = 144 - 12 - 2 = 130$

17. $a_n = (-2)^{n-2}$
   $a_7 = (-2)^{7-2} = (-2)^5 = -32$
   $a_8 = (-2)^{8-2} = (-2)^6 = 64$

19. 1, 3, 5, 7, 9, ...
   $d = 3 - 1 = 2$
   $a_n = a_1 + (n-1)d$
   $a_n = 1 + (n-1)(2)$
   $a_n = 1 + 2n - 2$
   $a_n = 2n - 1$

21. −2, 2, 6, 10, 14, ...
   $d = 2 - (-2) = 4$
   $a_n = a_1 + (n-1)d$
   $a_n = -2 + (n-1)(4)$

$a_n = -2+4n-4$

$a_n = 4n-6$

23. 5, 3, 1, −1, −3, ...

$d = 3-5 = -2$

$a_n = a_1 + (n-1)d$

$a_n = 5+(n-1)(-2)$

$a_n = 5-2n+2$

$a_n = -2n+7$

25. −7, −10, −13, −16, −19, ...

$d = -10-(-7) = -3$

$a_n = a_1 + (n-1)d$

$a_n = -7+(n-1)(-3)$

$a_n = -7-3n+3$

$a_n = -3n-4$

27. 1, $\frac{3}{2}$, 2, $\frac{5}{2}$, 3, ...

$d = \frac{3}{2}-1 = \frac{1}{2}$

$a_n = a_1 + (n-1)d$

$a_n = 1+(n-1)\left(\frac{1}{2}\right)$

$a_n = 1+\frac{1}{2}n-\frac{1}{2}$

$a_n = \frac{1}{2}n+\frac{1}{2}$

29. 7, 10, 13, 16, ...

$d = 10-7 = 3$

$a_n = a_1 + (n-1)d$

$a_n = 7+(n-1)(3)$

$a_n = 7+3n-3$

$a_n = 3n+4$

$a_{10} = 3(10)+4 = 34$

31. 2, 6, 10, 14, ...

$d = 6-2 = 4$

$a_n = a_1 + (n-1)d$

$a_{20} = 2+(n-1)d$

$a_{20} = 2+19(4)$

$a_{20} = 2+76 = 78$

33. −7, −9, −11, −13, ...

$d = -9-(-7) = -2$

$a_n = a_1 + (n-1)d$

$a_{75} = -7+(75-1)(-2)$

$a_{75} = -7+74(-2)$

$a_{75} = -7-148 = -155$

35. 1, 3, 5, 7, ..., 211

$d = 3-1 = 2$

$a_n = a_1 + (n-1)d$

$211 = 1+(n-1)(2)$

$210 = 2n-2$

$212 = 2n$

$106 = n$

37. 10, 13, 16, 19, ..., 157

$d = 13-10 = 3$

$a_n = a_1 + (n-1)d$

$157 = 10+(n-1)3$

$147 = 3n-3$

$150 = 3n$

$50 = n$

39. −7, −9, −11, −13, ..., −345

$d = -9-(-7) = -2$

$a_n = a_1 + (n-1)d$

$-345 = -7+(n-1)(-2)$

$-338 = -2n+2$

$-340 = -2n$

$170 = n$

41. $a_6 = 24$

$a_{10} = 44$

$24 = a_1 + (6-1)d$

$24 = a_1 + 5d$

$44 = a_1 + (10 - 1)d$

$44 = a_1 + 9d$

$\begin{pmatrix} a_1 + 5d = 24 \\ a_1 + 9d = 44 \end{pmatrix}$

Multiply equation 1 by -1, then add to equation 2.

$$\begin{array}{r} -a_1 - 5d = -24 \\ a_1 + 9d = \ \ 44 \\ \hline 4d = 20 \end{array}$$

$d = 5$

$a_1 + 5d = 24$

$a_1 + 5(5) = 24$

$a_1 + 25 = 24$

$a_1 = -1$

43. $a_4 = -9$

$a_9 = -29$

$-9 = a_1 + (4 - 1)d$

$-9 = a_1 + 3d$

$-29 = a_1 + (9 - 1)d$

$-29 = a_1 + 8d$

$\begin{pmatrix} a_1 + 3d = -9 \\ a_1 + 8d = -29 \end{pmatrix}$

Multiply equation 1 by -1, then add to eqaution 2.

$$\begin{array}{r} -a_1 - 3d = \ \ 9 \\ a_1 + 8d = -29 \\ \hline 5d = -20 \end{array}$$

$d = -4$

$a_1 + 3d = -9$

$a_1 + 3(-4) = -9$

$a_1 - 12 = -9$

$a_1 = 3$

$a_5 = a_1 + (5 - 1)d$

$a_5 = 3 + 4(-4)$

$a_5 = 3 - 16 = -13$

45. $.97,\ 1.00,\ 1.03,\ 1.06,\ \ldots$

$d = 1.00 - .97 = .03$

$a_n = a_1 + (n - 1)d$

$5.02 = .97 + (n - 1)(.03)$

$4.05 = .03n - .03$

$4.08 = .03n$

$136 = n$

47. $a_n = a_1 + (n - 1)d$

$a_1 = 12,500$

$d = 900$

The salary for 1992 means $n = 18$.

$a_n = 12500 + (18 - 1)(900)$

$a_n = 12500 + 17(900)$

$a_n = 12500 + 15300$

$a_n = \$27,800$

49. $a_n = a_1 + (n - 1)d$

$a_1 = 900$

$d = 30$

$n = 2(5) = 10$

$a_n = 900 + (10 - 1)30$

$a_n = 900 + 9(30)$

$a_n = 900 + 270$

$a_n = \$1170$

Problem Set 13.2 Arithmetic Series

1. $2 + 4 + 6 + 8 + \ldots$

$a_1 = 2 \qquad d = 2$

$a_{50} = 2 + (50 - 1)(2)$

$a_{50} = 2 + 49(2)$

$a_{50} = 2 + 98 = 100$

$S_{50} = \dfrac{50(2 + 100)}{2}$

$S_{50} = 25(102) = 2550$

3. $3+8+13+18+\ldots$

$a_1 = 3 \qquad d = 5$

$a_{60} = 3+(60-1)(5)$

$a_{60} = 3+59(5)$

$a_{60} = 3+295 = 298$

$S_{60} = \frac{60(3+298)}{2}$

$S_{60} = 30(301) = 9030$

5. $(-1)+(-3)+(-5)+(-7)+\ldots$

$a_1 = -1 \qquad d = -2$

$a_{65} = -1+(65-1)(-2)$

$a_{65} = -1+64(-2)$

$a_{65} = -1-128 = -129$

$S_{65} = \frac{65(-1-129)}{2}$

$S_{65} = \frac{65(-130)}{2}$

$S_{65} = -4225$

7. $\frac{1}{2}+1+\frac{3}{2}+2+\ldots$

$a_1 = \frac{1}{2} \qquad d = \frac{1}{2}$

$a_{40} = \frac{1}{2}+(40-1)\left(\frac{1}{2}\right)$

$a_{40} = \frac{1}{2}+\frac{39}{2} = 20$

$S_{40} = \frac{40\left(\frac{1}{2}+20\right)}{2}$

$S_{40} = 20(20.5) = 410$

9. $7+10+13+16+\ldots$

$a_1 = 7 \qquad d = 3$

$a_{75} = 7+(75-1)(3)$

$a_{75} = 7+74(3)$

$a_{75} = 7+222 = 229$

$S_{75} = \frac{75(7+229)}{2}$

$S_{75} = \frac{75(236)}{2}$

$S_{75} = 75(118) = 8850$

11. $4+8+12+16+\ldots+212$

$a_1 = 4 \quad a_n = 212 \quad d = 4$

$a_n = a_1+(n-1)d$

$212 = 4+(n-1)4$

$212 = 4+4n-4$

$212 = 4n$

$53 = n$

$S = \frac{53(4+212)}{2}$

$S = \frac{53(216)}{2}$

$S = 53(108)$

$S = 5724$

13. $(-4)+(-1)+2+5+\ldots+173$

$a_1 = -4 \quad a_n = 173 \quad d = 3$

$a_n = a_1+(n-1)d$

$173 = -4+(n-1)(3)$

$177 = 3n-3$

$180 = 3n$

$60 = n$

$S = \frac{60(-4+173)}{2}$

$S = 30(169) = 5070$

15. $2.5+3.0+3.5+4.0+\ldots+18.5$

$a_1 = 2.5 \quad a_n = 18.5 \quad d = .5$

$a_n = a_1+(n-1)d$

$18.5 = 2.5+(n-1)(.5)$

$16.0 = .5n-.5$

$16.5 = .5n$

$33 = n$

$S = \frac{33(2.5+18.5)}{2}$

$S = \frac{33(21)}{2}$

$S = 346.5$

17. $a_n = 3n - 1$

$a_1 = 3(1) - 1 = 2$

$a_{50} = 3(50) - 1 = 149$

$S = \frac{50(2+149)}{2}$

$S = 25(151)$

$S = 3775$

19. $a_n = 5n + 1$

$a_1 = 5(1) + 1 = 6$

$a_{125} = 5(125) + 1 = 626$

$S = \frac{125(6+626)}{2}$

$S = \frac{125(632)}{2}$

$S = 125(316) = 39{,}500$

21. $a_n = -4n - 1$

$a_1 = -4(1) - 1 = -5$

$a_{65} = -4(65) - 1 = -261$

$S = \frac{65(-5-261)}{2}$

$S = \frac{65(-266)}{2}$

$S = 65(-133)$

$S = -8645$

23. $2 + 4 + 6 + 8 + \ldots$

$a_1 = 2 \quad n = 350 \quad d = 2$

$a_n = 2 + (350 - 1)(2)$

$a_n = 2 + 349(2)$

$a_n = 2 + 698 = 700$

$S = \frac{350(2+700)}{2}$

$S = 175(702)$

$S = 122{,}850$

25. $a_1 = 15 \quad a_n = 397 \quad d = 2$

$397 = 15 + (n - 1)(2)$

$382 = 2n - 2$

$384 = 2n$

$192 = n$

$S = \frac{192(15+397)}{2}$

$S = 96(412) = 39{,}552$

27. $a_1 = 20 \quad d = 4 \quad n = 15$

$a_n = 20 + (15 - 1)(4)$

$a_n = 20 + 14(4) = 76$

$S = \frac{15(20+76)}{2}$

$S = \frac{15(96)}{2}$

$S = 15(48) = 720$

There are 76 seats in the 15th row and 720 seats in the auditorium.

29. $a_1 = 1 \quad d = 2 \quad n = 1000$

$a_n = a_1 + (n - 1)d$

$a_{1000} = 1 + (1000 - 1)(2)$

$a_{1000} = 1 + 999(2) = 1999$

$S = \frac{1000(1+1999)}{2}$

$S = 500(2000) = 1{,}000{,}000$ cents

$S = \$10{,}000$

31. $a_1 = 18500 \quad d = 1500 \quad n = 13$

$a_n = a_1 + (n - 1)d$

$a_{13} = 18500 + (13 - 1)(1500)$

$a_{13} = 18500 + 12(1500)$

$a_{13} = 36500$

$S = \frac{13(18500+36500)}{2}$

$S = \$357{,}500$

33. $a_1 = 25 \quad d = -2 \quad a_n = 1$

$a_n = a_1 + (n - 1)d$

$1 = 25 + (n - 1)(-2)$

$-24 = -2n + 2$

$-26 = -2n$

$13 = n$

$S = \frac{13(25+1)}{2}$

$S = 169$

There are 169 cans in the display.

Problem Set 13.3 Geometric Sequences and Series

1. 1, 3, 9, 27, ...

$a_1 = 1 \qquad r = \frac{3}{1} = 3$

$a_n = a_1 r^{n-1}$

$a_n = 1(3)^{n-1}$

$a_n = 3^{n-1}$

3. 2, 8, 32, 128, ...

$a_1 = 2 \qquad r = \frac{8}{2} = 4$

$a_n = 2(4)^{n-1}$

$a_n = 2(2^2)^{n-1}$

$a_n = 2^1(2^{2n-2})$

$a_n = 2^{2n-2+1}$

$a_n = 2^{2n-1}$

5. $1, \frac{1}{3}, \frac{1}{9}, \frac{1}{27}, \ldots$

$a_1 = 1 \qquad r = \frac{\frac{1}{3}}{1} = \frac{1}{3}$

$a_n = 1\left(\frac{1}{3}\right)^{n-1}$

$a_n = \left(\frac{1}{3}\right)^{n-1}$

7. .2, .04, .008, .0016, ...

$a_1 = .2 \qquad r = \frac{.04}{.2} = .2$

$a_n = .2(.2)^{n-1}$

$a_n = (.2)^n$

9. $9, 6, 4, \frac{8}{3}, \ldots$

$a_1 = 9 \qquad r = \frac{6}{9} = \frac{2}{3}$

$a_n = 9\left(\frac{2}{3}\right)^{n-1}$

$a_n = 3^2\left(\frac{2^{n-1}}{3^{n-1}}\right)$

$a_n = \frac{3^2}{3^{n-1}}(2^{n-1})$

$a_n = 3^{2-(n-1)}2^{n-1}$

$a_n = 3^{-n+3}(2^{n-1})$

11. $1, -4, 16, -64, \ldots$

$a_1 = 1 \qquad r = \frac{-4}{1} = -4$

$a_n = 1(-4)^{n-1}$

$a_n = (-4)^{n-1}$

13. $\frac{1}{9}, \frac{1}{3}, 1, 3, \ldots$

$a_1 = \frac{1}{9} \quad r = \frac{3}{1} = 3$

$a_{12} = \frac{1}{9}(3)^{12-1}$

$a_{12} = \frac{1}{9}(3)^{11} = \frac{3^{11}}{3^2} = 3^9 = 19{,}683$

15. $1, -2, 4, -8, \ldots$

$a_1 = 1 \qquad r = \frac{-2}{1} = -2$

$a_{10} = 1(-2)^{10-1}$

$a_{10} = (-2)^9 = -512$

17. $-1, -\frac{3}{2}, -\frac{9}{4}, -\frac{27}{8}, \ldots$

$a_1 = -1 \qquad r = \frac{-\frac{3}{2}}{-1} = \frac{3}{2}$

$a_9 = -1\left(\frac{3}{2}\right)^{9-1}$

$a_9 = -1\left(\frac{3}{2}\right)^8 = -\frac{6561}{256}$

19. $\frac{1}{2} + \frac{3}{2} + \frac{9}{2} + \frac{27}{2} + \ldots$

$a_1 = \frac{1}{2} \quad r = \frac{\frac{3}{2}}{\frac{1}{2}} = 3 \quad n = 10$

$S = \frac{a_1 r^n - a_1}{r - 1}$

$S = \frac{\frac{1}{2}(3)^{10} - \frac{1}{2}}{3 - 1}$

$S = \frac{\frac{1}{2}(3^{10} - 1)}{2}$

$S = \frac{\frac{1}{2}(59049 - 1)}{2}$

$S = \frac{1}{4}(59048) = 14,762$

21. $-2 + 6 + (-18) + 54 + \ldots$

$a_1 = -2 \quad r = \frac{6}{-2} = -3 \quad n = 9$

$S = \frac{-2(-3)^9 - (-2)}{-3 - 1}$

$S = \frac{-2(-19683) + 2}{-4}$

$S = -9842$

23. $1 + 3 + 9 + 27 + \ldots$

$a_1 = 1 \quad r = \frac{3}{1} = 3 \quad n = 7$

$S = \frac{1(3)^7 - 1}{3 - 1}$

$S = \frac{2187 - 1}{2} = 1093$

25. $a_n = 2^{n-1}$

$a_1 = 2^{1-1} = 2^0 = 1$

$a_2 = 2^{2-1} = 2^1 = 2$

$r = \frac{2}{1} = 2$

$n = 9$

$S = \frac{1(2)^9 - 1}{2 - 1}$

$S = \frac{512 - 1}{1} = 511$

27. $a_n = 2(3)^n$

$a_1 = 2(3)^1 = 6$

$a_2 = 2(3)^2 = 18$

$r = \frac{18}{6} = 3$

$n = 8$

$S = \frac{6(3)^8 - 6}{3 - 1}$

$S = \frac{39366 - 6}{2} = 19,680$

29. $a_n = (-2)^n$

$a_1 = (-2)^1 = -2$

$a_2 = (-2)^2 = 4$

$r = \frac{4}{-2} = -2$

$n = 12$

$S = \frac{-2(-2)^{12} - (-2)}{-2 - 1}$

$S = \frac{-2(4096) + 2}{-3} = 2730$

31. $1 + 3 + 9 + \ldots + 729$

$a_1 = 1 \qquad r = \frac{3}{1} = 3$

$a_n = a_1 r^{n-1}$

$729 = 1(3)^{n-1}$

$729 = 3^{n-1}$

$3^6 = 3^{n-1}$

$6 = n - 1$

$7 = n$

$S = \frac{1(3)^7 - 1}{3 - 1}$

$S = \frac{2187 - 1}{2} = 1093$

33. $1 + \frac{1}{2} + \frac{1}{4} + \ldots + \frac{1}{1024}$

$a_1 = 1 \qquad r = \frac{\frac{1}{2}}{1} = \frac{1}{2}$

$a_n = a_1 r^{n-1}$

$\frac{1}{1024} = 1\left(\frac{1}{2}\right)^{n-1}$

$\left(\frac{1}{2}\right)^{10} = \left(\frac{1}{2}\right)^{n-1}$

$10 = n - 1$

$11 = n$

$S = \frac{1\left(\frac{1}{2}\right)^{11} - 1}{\frac{1}{2} - 1}$

$S = \frac{\frac{1}{2048} - 1}{-\frac{1}{2}}$

$S = \frac{-\frac{2047}{2048}}{-\frac{1}{2}}$

$S = -\frac{2}{1} \bullet -\frac{2047}{2048}$

$S = \frac{2047}{1024} = 1\frac{1023}{1024}$

35. $8 + 4 + 2 + \ldots + \frac{1}{32}$

$a_1 = 8 \qquad r = \frac{4}{8} = \frac{1}{2}$

$a_n = a_1 r^{n-1}$

$\frac{1}{32} = 8\left(\frac{1}{2}\right)^{n-1}$

$\frac{1}{8}\left(\frac{1}{32}\right) = \frac{1}{8}\left[8\left(\frac{1}{2}\right)^{n-1}\right]$

$\frac{1}{256} = \left(\frac{1}{2}\right)^{n-1}$

$\left(\frac{1}{2}\right)^{8} = \left(\frac{1}{2}\right)^{n-1}$

$8 = n - 1$

$9 = n$

$S = \frac{8\left(\frac{1}{2}\right)^{9} - 8}{\frac{1}{2} - 1}$

$S = \frac{\frac{1}{64} - 8}{-\frac{1}{2}} = \frac{\frac{1}{64} - \frac{512}{64}}{-\frac{1}{2}}$

$S = \frac{-\frac{511}{64}}{-\frac{1}{2}}$

$S = -\frac{2}{1} \bullet \frac{-511}{64} = \frac{511}{32} = 15\frac{31}{32}$

37. $a_2 = \frac{1}{6} \qquad a_5 = \frac{1}{48}$

$a_2 = a_1 r^1 \qquad a_5 = a_1 r^4$

$\frac{a_5}{a_2} = \frac{a_1 r^4}{a_1 r^1} = r^3$

$\frac{\frac{1}{48}}{\frac{1}{6}} = r^3$

$\frac{6}{48} = r^3$

$\frac{1}{8} = r^3$

$\frac{1}{2} = r$

39. $a_n = (-1)^n$

$a_1 = (-1)^1 = -1$

$a_2 = (-1)^2 = 1$

$r = \frac{1}{-1} = -1$

$n = 16$

$S_{16} = \frac{-1(-1)^{16} - (-1)}{-1 - 1}$

$S_{16} = \frac{-1(1) + 1}{-2} = \frac{0}{-2} = 0$

$S_{19} = \frac{-1(-1)^{19} - (-1)}{-1 - 1}$

$S_{19} = \frac{-1(-1) + 1}{-2} = \frac{2}{-2} = -1$

41. $a_n = 16000\left(\frac{1}{2}\right)^{n-1}$

$a_8 = 16000\left(\frac{1}{2}\right)^{8-1}$

$a_8 = 16000\left(\frac{1}{2}\right)^{7}$

$a_8 = 125$

There will be
125 liters remaining.

43. $a_n = .05(2)^{n-1}$

$a_{12} = .05(2)^{12-1}$

$a_{12} = .05(2)^{11} = \$102.40$

$S_{12} = \frac{.05(2)^{12} - .05}{2-1}$

$S_{12} = \frac{204.80 - .05}{1} = \$204.75$

The savings on the 12th day will be \$102.40. The total savings is \$204.75.

45. Traveling down

$a_1 = 486 \quad r = \frac{1}{3} \quad n = 7$

$S_d = \frac{486\left(\frac{1}{3}\right)^7 - 486}{\frac{1}{3} - 1}$

$S_d = \frac{486\left[\left(\frac{1}{3}\right)^7 - 1\right]}{-\frac{2}{3}}$

$S_d = \frac{486\left(\frac{1}{2187} - \frac{2187}{2187}\right)}{-\frac{2}{3}}$

$S_d = -\frac{3}{2} \bullet (486)\left(-\frac{2186}{2187}\right)$

$S_d = (486)\left(\frac{1093}{729}\right)$

$S_d = (2)\left(\frac{1093}{3}\right) = \frac{2186}{3}$

Traveling up

$a_1 = \frac{1}{3}(486) = 162$

$r = \frac{1}{3} \qquad n = 6$

$S_u = \frac{162\left(\frac{1}{3}\right)^6 - 162}{\frac{1}{3} - 1}$

$S_u = \frac{162\left[\left(\frac{1}{3}\right)^6 - 1\right]}{-\frac{2}{3}}$

$S_u = \frac{162\left(\frac{1}{729} - \frac{729}{729}\right)}{-\frac{2}{3}}$

$S_u = -\frac{3}{2}(162)\left(-\frac{728}{729}\right)$

$S_u = -\frac{3}{2}(2)\left(-\frac{728}{9}\right)$

$S_u = \frac{728}{3}$

Total travel $= \frac{2186}{3} + \frac{728}{3} = \frac{2914}{3}$

971.3 meters

The ball travels 971.3 meters.

47. $a_1 = 9500 \quad r = .9 \quad n = 6$

$a_6 = 9500(.9)^{6-1}$

$a_6 = 9500(.59049)$

$a_6 = \$5609.66$

The car is worth \$5609.66 in 5 years.

Problem Set 13.4 Infinite Geometric Series

1. $1 + \frac{3}{4} + \frac{9}{16} + \frac{27}{64} + \ldots$

$a_1 = 1 \qquad r = \frac{3}{4}$

$S = \frac{1}{1 - \frac{3}{4}} = \frac{1}{\frac{1}{4}} = 4$

3. $\frac{1}{2} + \frac{1}{4} + \frac{1}{8} + \frac{1}{16} + \ldots$

$a_1 = \frac{1}{2} \qquad r = \frac{1}{2}$

$S = \frac{\frac{1}{2}}{1 - \frac{1}{2}} = \frac{\frac{1}{2}}{\frac{1}{2}} = 1$

5. $\frac{2}{3} + \frac{4}{9} + \frac{8}{27} + \frac{16}{81} + \ldots$

$a_1 = \frac{2}{3} \qquad r = \frac{2}{3}$

$S = \frac{\frac{2}{3}}{1 - \frac{2}{3}} = \frac{\frac{2}{3}}{\frac{1}{3}} = \frac{3}{1} \bullet \frac{2}{3} = 2$

7. $1 - \frac{1}{2} + \frac{1}{4} - \frac{1}{8} + \ldots$

$a_1 = 1 \qquad r = -\frac{1}{2}$

$S = \frac{1}{1 - \left(-\frac{1}{2}\right)} = \frac{1}{1 + \frac{1}{2}} = \frac{1}{\frac{3}{2}} = \frac{2}{3}$

9. $6+2+\frac{2}{3}+\frac{2}{9}+\ldots$

$a_1=6 \qquad r=\frac{1}{3}$

$S=\frac{6}{1-\frac{1}{3}}=\frac{6}{\frac{2}{3}}=\frac{3}{2}(6)=9$

11. $2+(-6)+18+(-54)+\ldots$

$a_1=2 \qquad r=-3$

No sum exists.

13. $1+\left(-\frac{3}{4}\right)+\frac{9}{16}+\left(-\frac{27}{64}\right)+\ldots$

$a_1=1 \qquad r=-\frac{3}{4}$

$S=\frac{1}{1-\left(-\frac{3}{4}\right)}=\frac{1}{\frac{7}{4}}=\frac{4}{7}$

15. $8-4+2-1+\ldots$

$a_1=8 \qquad r=-\frac{1}{2}$

$S=\frac{8}{1-\left(-\frac{1}{2}\right)}=\frac{8}{\frac{3}{2}}=\frac{2}{3}(8)=\frac{16}{3}$

17. $1+\frac{3}{2}+\frac{9}{4}+\frac{27}{8}+\ldots$

$a_1=1 \qquad r=\frac{3}{2}$

No sum exists.

19. $27+9+3+1+\ldots$

$a_1=27 \qquad r=\frac{1}{3}$

$S=\frac{27}{1-\frac{1}{3}}=\frac{27}{\frac{2}{3}}=\frac{3}{2}(27)=\frac{81}{2}$

21. $.\overline{4}=.4+.04+.004+\ldots$

$S=\frac{.4}{1-.1}=\frac{.4}{.9}=\frac{4}{9}$

23. $.\overline{47}=.47+.0047+.000047+\ldots$

$S=\frac{.47}{1-.01}=\frac{.47}{.99}=\frac{47}{99}$

25. $.\overline{45}=.45+.0045+.000045+\ldots$

$S=\frac{.45}{1-.01}=\frac{.45}{.99}=\frac{45}{99}=\frac{5}{11}$

27. $.\overline{427}=.427+.000427+.000000427+\ldots$

$S=\frac{.427}{1-.001}=\frac{.427}{.999}=\frac{427}{999}$

29. $.4\overline{6}=.4+.06+.006+.0006+\ldots$

Find the sum for

$.06+.006+.0006+\ldots$

$S=\frac{.06}{1-.1}=\frac{.06}{.90}=\frac{6}{90}=\frac{1}{15}$

The total sum

$S=.4+\frac{1}{15}=\frac{4}{10}+\frac{1}{15}=\frac{7}{15}$

31. $2.\overline{18}$

$[2]+[.18+.0018+.000018+\ldots]$

Find the sum for

$.18+.0018+.000018+\ldots$

$S=\frac{.18}{1-.01}=\frac{.18}{.99}=\frac{18}{99}=\frac{2}{11}$

The total sum

$S=2+\frac{2}{11}=\frac{24}{11}$

33. $.4\overline{27}$

$.4+.027+.00027+.0000027+\ldots$

Find the sum for

$.027+.00027+.0000027+\ldots$

$S=\frac{.027}{1-.01}=\frac{.027}{.990}=\frac{27}{990}=\frac{3}{110}$

The total sum

$S=.4+\frac{3}{110}=\frac{4}{10}+\frac{3}{110}=\frac{47}{110}$

Problem Set 13.5 Binomial Expansions

1. $(x+y)^8$

$x^8+8x^7y+28x^6y^2+56x^5y^3+70x^4y^4+56x^3y^5+28x^2y^6+8xy^7+y^8$

3. $(3x+y)^4$

$(3x)^4+4(3x)^3(y)+6(3x)^2(y)^2+4(3x)(y)^3+y^4$

$81x^4+108x^3y+54x^2y^2+12xy^3+y^4$

5. $(x-y)^5$

$(x)^5+5(x)^4(-y)+10(x)^3(-y)^2+10(x)^2(-y)^3+5(x)(-y)^4+(-y)^5$

$x^5-5x^4y+10x^3y^2-10x^2y^3+5xy^4-y^5$

7. $(x+y)^{10}$

$$x^{10}+10x^9y+\frac{10\bullet 9}{2!}x^8y^2+\frac{10\bullet 9\bullet 8}{3!}x^7y^3+\frac{10\bullet 9\bullet 8\bullet 7}{4!}x^6y^4+\frac{10\bullet 9\bullet 8\bullet 7\bullet 6}{5!}x^5y^5$$

$$+\frac{10\bullet 9\bullet 8\bullet 7\bullet 6\bullet 5}{6!}x^4y^6+\frac{10\bullet 9\bullet 8\bullet 7\bullet 6\bullet 5\bullet 4}{7!}x^3y^7+\frac{10\bullet 9\bullet 8\bullet 7\bullet 6\bullet 5\bullet 4\bullet 3}{8!}x^2y^8$$

$$+10xy^9+y^{10}$$

$x^{10}+10x^9y+45x^8y^2+120x^7y^3+210x^6y^4+252x^5y^5+210x^4y^6+120x^3y^7+45x^2y^8+10xy^9+y^{10}$

9. $(2x+y)^6=(2x)^6+6(2x)^5(y)+\frac{6\bullet 5}{2!}(2x)^4(y)^2+\frac{6\bullet 5\bullet 4}{3!}(2x)^3(y)^3$

$$+\frac{6\bullet 5\bullet 4\bullet 3}{4!}(2x)^2(y)^4+\frac{6\bullet 5\bullet 4\bullet 3\bullet 2}{5!}(2x)(y)^5+y^6$$

$64x^6+192x^5y+240x^4y^2+160x^3y^3+60x^2y^4+12xy^5+y^6$

11. $(x-3y)^5$

$x^5+5x^4(-3y)+\frac{5\bullet 4}{2!}x^3(-3y)^2+\frac{5\bullet 4\bullet 3}{3!}x^2(-3y)^3+5x(-3y)^4+(-3y)^5$

$x^5-15x^4y+90x^3y^2-270x^2y^3+405xy^4-243y^5$

13. $(3a-2b)^5$

$(3a)^5+5(3a)^4(-2b)+\frac{5\bullet 4}{2!}(3a)^3(-2b)^2+\frac{5\bullet 4\bullet 3}{3!}(3a)^2(-2b)^3+5(3a)(-2b)^4+(-2b)^5$

$243a^5-810a^4b+1080a^3b^2-720a^2b^3+240ab^4-32b^5$

15. $(x+y^3)^6$

$x^6+6x^5y^3+\frac{6\bullet 5}{2!}x^4(y^3)^2+\frac{6\bullet 5\bullet 4}{3!}x^3(y^3)^3+\frac{6\bullet 5\bullet 4\bullet 3}{4!}x^2(y^3)^4+6x(y^3)^5+(y^3)^6$

$x^6+6x^5y^3+15x^4y^6+20x^3y^9+15x^2y^{12}+6xy^{15}+y^{18}$

17. $(x+2)^7$

$x^7+7x^6(2)+\frac{7\bullet 6}{2!}x^5(2)^2+\frac{7\bullet 6\bullet 5}{3!}x^4(2)^3+\frac{7\bullet 6\bullet 5\bullet 4}{4!}x^3(2)^4+\frac{7\bullet 6\bullet 5\bullet 4\bullet 3}{5!}x^2(2)^5+7x(2)^6+(2)^7$

$x^7+14x^6+84x^5+280x^4+560x^3+672x^2+448x+128$

19. $(x-3)^4$

$x^4+4x^3(-3)+6x^2(-3)^2+4x(-3)^3+(-3)^4$

$x^4-12x^3+54x^2-108x+81$

21. $(x+y)^{15}$

$x^{15}+15x^{14}y+\frac{15\bullet 14}{2!}x^{13}y^2+\frac{15\bullet 14\bullet 13}{3!}x^{12}y^3$

$x^{15}+15x^{14}y+105x^{13}y^2+455x^{12}y^3$

23. $(a-2b)^{13}$

$a^{13}+13a^{12}(-2b)+\frac{13\bullet 12}{2!}a^{11}(-2b)^2+\frac{13\bullet 12\bullet 11}{3!}a^{10}(-2b)^3$

$a^{13}-26a^{12}b+312a^{11}b^2-2288a^{10}b^3$

25. $(x+y)^{11}$ 7th term

$\frac{11\bullet 10\bullet 9\bullet 8\bullet 7\bullet 6}{6!}x^5y^6$

$462x^5y^6$

27. $(x-2y)^6$ 4th term

$\frac{6\bullet 5\bullet 4}{3!}x^3(-2y)^3$

$-160x^3y^3$

29. $(2x-5y)^5$ 3rd term

$\frac{5\bullet 4}{2!}(2x)^3(-5y)^2$

$2000x^3y^2$

Chapter 13 Review

1. 3, 9, 15, 21, ...

$a_1=3,\ d=6$

$a_n=3+(n-1)(6)$

$a_n=3+6n-6$

$a_n=6n-3$

3. $10, 20, 40, 80, \ldots$

$a_1 = 10,\ r = 2$

$a_n = 10(2)^{n-1}$

$a_n = 10(2)^n(2)^{-1}$

$a_n = 10(2)^n\left(\frac{1}{2}\right)$

$a_n = 5(2)^n$

5. $-5, -3, -1, 1, \ldots$

$a_1 = -5,\ d = 2$

$a_n = -5 + (n-1)(2)$

$a_n = -5 + 2n - 2$

$a_n = 2n - 7$

7. $-1, 2, -4, 8, \ldots$

$a_1 = -1,\ r = -2$

$a_n = -1(-2)^{n-1}$

9. $\frac{2}{3}, 1, \frac{4}{3}, \frac{5}{3}, \ldots$

$a_1 = \frac{2}{3},\ d = \frac{1}{3}$

$a_n = \frac{2}{3} + (n-1)\left(\frac{1}{3}\right)$

$a_n = \frac{2}{3} + \frac{1}{3}n - \frac{1}{3}$

$a_n = \frac{1}{3}n + \frac{1}{3}$

$a_n = \frac{n+1}{3}$

11. $1, 5, 9, 13, \ldots$

$a_1 = 1,\ d = 4,\ n = 19$

$a_{19} = 1 + (19-1)(4)$

$a_{19} = 1 + (18)(4)$

$a_{19} = 1 + 72 = 73$

13. $8, 4, 2, 1, \ldots$

$a_1 = 8,\ r = \frac{1}{2},\ n = 9$

$a_9 = 8\left(\frac{1}{2}\right)^{9-1}$

$a_9 = 8\left(\frac{1}{2}\right)^8$

$a_9 = 8\left(\frac{1}{256}\right)$

$a_9 = \frac{1}{32}$

15. $7, 4, 1, -2, \ldots$

$a_1 = 7,\ d = -3,\ n = 34$

$a_{34} = 7 + (34-1)(-3)$

$a_{34} = 7 + (33)(-3)$

$a_{34} = 7 - 99 = -92$

17. $a_5 = -19,\ a_8 = -34$

$a_5 = a_1 + (5-1)d$

$-19 = a_1 + 4d$

$a_8 = a_1 + (8-1)d$

$-34 = a_1 + 7d$

$$\begin{pmatrix} a_1 + 4d = -19 \\ a_1 + 7d = -34 \end{pmatrix}$$

Multiply equation 1 by -1 and then add the equations.

$$\begin{array}{r} -a_1 - 4d = \quad 19 \\ a_1 + 7d = -34 \\ \hline 3d = -15 \end{array}$$

$d = -5$

19. $a_3 = 5,\ a_6 = 135$

$a_3 = a_1 r^{3-1}$

$5 = a_1 r^2$

$a_6 = a_1 r^{6-1}$

$135 = a_1 r^5$

$$\begin{pmatrix} 5 = a_1 r^2 \\ 135 = a_1 r^5 \end{pmatrix}$$

$\frac{5}{135} = \frac{a_1 r^2}{a_1 r^5}$

$\frac{1}{27} = r^{-3}$

$\frac{1}{27} = \frac{1}{r^3}$

$27 = r^3$

$3 = r$

$5 = a_1r^2$

$5 = a_1(3)^2$

$5 = a_1(9)$

$\frac{5}{9} = a_1$

21. $81,\ 27,\ 9,\ 3,\ \ldots$

$a_1 = 81,\ r = \frac{1}{3},\ n = 9$

$$S_9 = \frac{81\left(\frac{1}{3}\right)^9 - 81}{\frac{1}{3} - 1}$$

$$S_9 = \frac{\frac{1}{243} - 81}{-\frac{2}{3}}$$

$$S_9 = -\frac{3}{2}\left(\frac{1}{243} - 81\right)$$

$$S_9 = -\frac{1}{162} + \frac{243}{2}$$

$$S_9 = -\frac{1}{162} + \frac{19683}{162}$$

$$S_9 = \frac{19682}{162} = 121\frac{80}{162}$$

$$S_9 = 121\frac{40}{81}$$

23. $5,\ 1,\ -3,\ -7,\ \ldots$

$a_1 = 5,\ d = -4,\ n = 75$

$a_{75} = 5 + (75-1)(-4)$

$a_{75} = 5 + (74)(-4) = -291$

$$S_{75} = \frac{75(5-291)}{2}$$

$$S_{75} = \frac{75(-286)}{2} = -10,725$$

25. $a_n = 7n + 1$

$a_1 = 7(1) + 1 = 8$

$a_{95} = 7(95) + 1 = 666$

$$S_{95} = \frac{95(8+666)}{2}$$

$$S_{95} = \frac{95(674)}{2} = 32,015$$

27. $64 + 16 + 4 + \ldots + \frac{1}{64}$

$a_1 = 64,\ r = \frac{1}{4}$

$a_n = a_1r^{n-1}$

$\frac{1}{64} = 64\left(\frac{1}{4}\right)^{n-1}$

$\frac{1}{64}\left(\frac{1}{64}\right) = \left(\frac{1}{4}\right)^{n-1}$

$\frac{1}{4^3}\left(\frac{1}{4^3}\right) = \left(\frac{1}{4}\right)^{n-1}$

$\left(\frac{1}{4^6}\right) = \left(\frac{1}{4}\right)^{n-1}$

$\left(\frac{1}{4}\right)^6 = \left(\frac{1}{4}\right)^{n-1}$

$6 = n - 1$

$7 = n$

$$S_7 = \frac{64\left(\frac{1}{4}\right)^7 - 64}{\frac{1}{4} - 1}$$

$$S_7 = \frac{4^3\left(\frac{1}{4}\right)^7 - 64}{-\frac{3}{4}}$$

$$S_7 = \frac{\frac{1}{4^4} - 64}{-\frac{3}{4}}$$

$$S_7 = -\frac{4}{3}\left(\frac{1}{4^4} - 64\right)$$

$$S_7 = -\frac{1}{192} + \frac{256}{3}$$

$$S_7 = -\frac{1}{192} + \frac{16384}{192}$$

$$S_7 = \frac{16383}{192} = 85\frac{63}{192} = 85\frac{21}{64}$$

29. $a_1 = 27,\ a_n = 276,\ d = 3$

$a_n = a_1 + (n-1)d$

$276 = 27 + (n-1)(3)$

$276 = 27 + 3n - 3$

$252 = 3n$

$84 = n$

$S = \frac{84(27+276)}{2} = 12,726$

31. $.\overline{36} = .36 + .0036 + .000036 + \ldots$

$a_1 = .36,\ r = .01$

$S = \frac{.36}{1-.01} = \frac{.36}{.99} = \frac{36}{99} = \frac{4}{11}$

$.\overline{36} = \frac{4}{11}$

33. $a_1 = 3750,\ d = -250,\ n = 13$

$a_{12} = 3750 + (13-1)(-250)$

$a_{12} = 3750 + (12)(-250)$

$a_{12} = 750$

There will be \$750 in the account.

35. $a_1 = .10,\ r = 2,\ n = 15$

$S = \frac{.10(2)^{15} - .10}{2-1}$

$S = \frac{.10(32768) - .10}{1}$

$S = 3276.70$

She will save \$3,276.70.

37. $16, 48, 80, 112, \ldots$

$a_1 = 16,\ d = 32,\ n = 15$

$a_{15} = 16 + (15-1)(32)$

$a_{15} = 16 + (14)(32) = 464$

$S = \frac{15(16+464)}{2} = 3600$

The object will fall 3600 feet.

39. $(x-3y)^4$

$x^4 + 4x^3(-3y) + 6x^2(-3y)^2 + 4x(-3y)^3 + (-3y)^4$

$x^4 - 12x^3y + 54x^2y^2 - 108xy^3 + 81y^4$

Chapter 13 Test

1. $a_n = -3n - 1$

$a_{15} = -3(15) - 1$

$a_{15} = -46$

3. $-3, 1, 5, 9, \ldots$

$a_1 = -3 \quad d = 4$

$a_n = -3 + (n-1)(4)$

$a_n = -3 + 4n - 4$

$a_n = 4n - 7$

5. $6, 3, 0, -3, \ldots$

$a_1 = 6 \quad d = -3$

$a_n = 6 + (n-1)(-3)$

$a_n = 6 - 3n + 3$

$a_n = -3n + 9$

7. $1, 4, 7, 10, \ldots$

$a_1 = 1 \quad d = 3$

$a_{75} = 1 + (75-1)(3)$

$a_{75} = 1 + 74(3)$

$a_{75} = 223$

9. $a_4 = 13 \quad a_7 = 22$

$13 = a_1 + (4-1)d$

$13 = a_1 + 3d$

$22 = a_1 + (7-1)d$

$22 = a_1 + 6d$

$\begin{pmatrix} a_1 + 3d = 13 \\ a_1 + 6d = 22 \end{pmatrix}$

Multiply equation 1 by -1, then add to equation 2.

$$\begin{array}{r} -a_1 - 3d = -13 \\ a_1 + 6d = \ \ 22 \\ \hline 3d = 9 \end{array}$$

$d = 3$

$a_1 + 3d = 13$

$a_1 + 3(3) = 13$

$a_1 = 4$

$a_{15} = 4 + (15-1)(3)$

$a_{15} = 4 + 42$

$a_{15} = 46$

11. $3+6+12+24+\ldots$

$a_1=3 \qquad r=2$

$S=\frac{3(2)^8-3}{2-1}$

$S=\frac{768-3}{1}=765$

13. $3+9+27+\ldots+2187$

$a_1=3 \qquad r=3$

$2187=3(3)^{n-1}$

$3^7=3^n$

$7=n$

$S=\frac{3(3)^7-3}{3-1}$

$S=\frac{6561-3}{2}=3279$

15. $a_n=3(2)^n$

$a_1=3(2)^1=6$

$a_2=3(2)^2=12$

$r=\frac{12}{6}=2$

$S=\frac{6(2)^{10}-6}{2-1}$

$S=\frac{6144-6}{1}=6138$

17. $a_1=11 \qquad d=2$

$193=11+(n-1)(2)$

$182=2n-2$

$184=2n$

$92=n$

$S=\frac{92(11+193)}{2}$

$S=\frac{92(204)}{2}=9384$

19. $a_1=.10 \qquad r=2$

$S=\frac{.10(2)^{15}-.10}{2-1}$

$S = \frac{3276.8 - .10}{1}$

$S = 3276.70$

The total savings is \$3276.70.

21. $.\overline{37} = .37 + .0037 + .000037 + \ldots$

$a_1 = .37 \quad r = .01$

$S = \frac{.37}{1 - .01} = \frac{.37}{.99} = \frac{37}{99}$

23. $(x - 3y)^5$

$x^5 + 5x^4(-3y) + \frac{5 \bullet 4}{2!}x^3(-3y)^2 + \frac{5 \bullet 4 \bullet 3}{3!}x^2(-3y)^3 + 5x(-3y)^4 + (-3y)^5$

$x^5 - 15x^4y + 90x^3y^2 - 270x^2y^3 + 405xy^4 - 243y^5$

25. $(a + b)^{12}$ 5th term

$\frac{12 \bullet 11 \bullet 10 \bullet 9}{4!}a^8b^4$

$495a^8b^4$

Chapter 14 Counting Techniques and Probability

Problem Set 14.1 Fundamental Principle of Counting

1. $2 \times 10 = 20$

3. $4 \times 3 \times 2 \times 1 = 24$

5. $7 \times 6 \times 4 = 168$

7. $6 \times 2 \times 4 = 48$

9. $2 \times 3 \times 6 = 36$

11. $20 \times 19 \times 18 = 6840$

13. $6 \times 5 \times 4 \times 3 \times 2 \times 1 = 720$

15. $6 \times 5 \times 4 \times 3 \times 2 \times 1 = 720$

17. $\underline{3} \times \underline{3} \times \underline{2} \times \underline{1} \times \underline{2} = 36$

19. $\underline{4} \times \underline{3} \times \underline{1} \times \underline{2} \times \underline{1} = 24$

21. There are three choices of mailboxes for the first letter, three choices of mailboxes for the second letter, etc.

    $3 \times 3 \times 3 \times 3 \times 3 = 3^5 = 243$

23. Impossible

25. $6 \times 6 \times 6 = 216$

27. Count the ways of getting 5 or less.

1st die	2nd die
1	1
1	2
1	3
1	4
2	1
2	2
2	3
3	1
3	2
4	1

There are ten ways to get a sum of 5 or less. There are $6 \times 6 = 36$ possible outcomes. Therefore there are $36 - 10 = 26$ outcomes of greater than 5.

29. For 3 digit numbers $2 \times 3 \times 2 = 12$

    For 4 digit number $4 \times 3 \times 2 \times 1 = 24$

    $12 + 24 = 36$ numbers

31. $\underline{4} \times \underline{3} \times \underline{3} \times \underline{2} \times \underline{2} \times \underline{1} \times \underline{1} = 144$

33. There are two choices for each of the ten questions.

    $2 \times 2 \times 2 \times 2 \times 2 \times 2 \times 2 \times 2 \times 2 \times 2$

    $= 2^{10} = 1024$

35. If the first digit is 4 $\underline{1} \times \underline{3} \times \underline{2} \times \underline{1} \times \underline{3} = 18$

    If the first digit is 5 $\underline{1} \times \underline{3} \times \underline{2} \times \underline{1} \times \underline{2} = 12$

    $18 + 12 = 30$ numbers

37a. $26 \times 26 \times 9 \times 10 \times 10 \times 10 = 6{,}084{,}000$

b. $26 \times 25 \times 9 \times 10 \times 10 \times 10 = 5{,}850{,}000$

c. $26 \times 26 \times 9 \times 9 \times 8 \times 7 = 3{,}066{,}336$

d. $26 \times 25 \times 9 \times 9 \times 8 \times 7 = 2{,}948{,}400$

Problem Set 14.2 Permutations and Combinations

1. $P(5,\ 3) = 5 \bullet 4 \bullet 3 = 60$

3. $P(6,\ 4) = 6 \bullet 5 \bullet 4 \bullet 3 = 360$

5. $C(7,\ 2) = \frac{P(7,\ 2)}{2!} = \frac{7 \bullet 6}{2 \bullet 1} = 21$

7. $C(10, 5) = \frac{P(10, 5)}{5!} = \frac{10 \bullet 9 \bullet 8 \bullet 7 \bullet 6}{5 \bullet 4 \bullet 3 \bullet 2 \bullet 1} = 252$

9. $C(15, 2) = \frac{P(15, 2)}{2!} = \frac{15 \bullet 14}{2 \bullet 1} = 105$

11. $C(5, 5) = \frac{P(5, 5)}{5!} = \frac{5 \bullet 4 \bullet 3 \bullet 2 \bullet 1}{5 \bullet 4 \bullet 3 \bullet 2 \bullet 1} = 1$

13. $P(4, 4) = 4 \bullet 3 \bullet 2 \bullet 1 = 24$

15. $C(9, 3) = \frac{P(9, 3)}{3!} = \frac{9 \bullet 8 \bullet 7}{3 \bullet 2 \bullet 1} = 84$

17a. $P(8, 3) = 8 \bullet 7 \bullet 6 = 336$

b. $8 \bullet 8 \bullet 8 = 512$

19. $P(4, 4) \times P(5, 5)$

$4 \bullet 3 \bullet 2 \bullet 1 \times 5 \bullet 4 \bullet 3 \bullet 2 \bullet 1 = 2880$

21. $C(7, 4) \times C(8, 4)$

$\frac{P(7, 4)}{4!} \times \frac{P(8, 4)}{4!}$

$\frac{7 \bullet 6 \bullet 5 \bullet 4}{4 \bullet 3 \bullet 2 \bullet 1} \times \frac{8 \bullet 7 \bullet 6 \bullet 5}{4 \bullet 3 \bullet 2 \bullet 1} = 2450$

23. $C(5, 3) = \frac{P(5, 3)}{3!} = \frac{5 \bullet 4 \bullet 3}{3 \bullet 2 \bullet 1} = 10$

25. $C(5, 2) = \frac{P(5, 2)}{2!} = \frac{5 \bullet 4}{2 \bullet 1} = 10$

27. $\frac{7!}{4!3!} = \frac{7 \bullet 6 \bullet 5 \bullet 4 \bullet 3 \bullet 2 \bullet 1}{4 \bullet 3 \bullet 2 \bullet 1 \bullet 3 \bullet 2 \bullet 1} = 35$

29. $\frac{9!}{3!4!2!} = \frac{9 \bullet 8 \bullet 7 \bullet 6 \bullet 5 \bullet 4 \bullet 3 \bullet 2 \bullet 1}{3 \bullet 2 \bullet 1 \bullet 4 \bullet 3 \bullet 2 \bullet 1 \bullet 2 \bullet 1}$

$= 1260$

31. $\frac{7!}{2!1!1!1!1!1!} = 2520$

33. $\frac{6!}{4!2!} = \frac{6 \bullet 5 \bullet 4 \bullet 3 \bullet 2 \bullet 1}{4 \bullet 3 \bullet 2 \bullet 1 \bullet 2 \bullet 1} = 15$

35. $C(10, 5) = \frac{P(10, 5)}{5!}$

$= \frac{10 \bullet 9 \bullet 8 \bullet 7 \bullet 6 \bullet 5}{5 \bullet 4 \bullet 3 \bullet 2 \bullet 1} = 252$

The order of the two teams is not important so the number of teams is $\frac{252}{2} = 126$.

37. One defective bulb

$C(4, 1) \times C(9, 2)$

$\frac{P(4, 1)}{1!} \times \frac{P(9, 2)}{2!}$

$4 \times \frac{9 \bullet 8}{2 \bullet 1}$

$4 \times 36 = 144$

At least one defective bulb
First find the number of all possible samples and subtract the number of samples that have no defective bulbs.

$C(13, 3) = \frac{P(13, 3)}{3!}$

$= \frac{13 \bullet 12 \bullet 11}{3 \bullet 2 \bullet 1} = 286$

$C(9, 3) = \frac{P(9, 3)}{3!}$

$= \frac{9 \bullet 8 \bullet 7}{3 \bullet 2 \bullet 1} = 84$

$286 - 84 = 202$

39. $C(6, 4) \times C(2, 2)$

$\frac{P(6, 4)}{4!} \times \frac{P(2, 2)}{2!}$

$\frac{6 \bullet 5 \bullet 4 \bullet 3}{4 \bullet 3 \bullet 2 \bullet 1} \times \frac{2 \bullet 1}{2 \bullet 1} = 15$

$C(6, 3) \times C(3, 3)$

$\frac{P(6, 3)}{3!} \times \frac{P(3, 3)}{3!}$

$\frac{6 \bullet 5 \bullet 4}{3 \bullet 2 \bullet 1} \times \frac{3 \bullet 2 \bullet 1}{3 \bullet 2 \bullet 1} = 20$

41. Subsets containing A and not B

$C(5, 3) = \frac{P(5, 3)}{3!} = \frac{5 \bullet 4 \bullet 3}{3 \bullet 2 \bullet 1} = 10$

Subsets conyaining B and not A

$C(5,\ 3) = \frac{P(5,\ 3)}{3!} = \frac{5 \bullet 4 \bullet 3}{3 \bullet 2 \bullet 1} = 10$

$10 + 10 = 20$

43. 5 points

$C(5,\ 2) = \frac{P(5,\ 2)}{2!} = \frac{5 \bullet 4}{2 \bullet 1} = 10$

6 points

$C(6,\ 2) = \frac{P(6,\ 2)}{2!} = \frac{6 \bullet 5}{2 \bullet 1} = 15$

7 points

$C(7,\ 2) = \frac{P(7,\ 2)}{2!} = \frac{7 \bullet 6}{2 \bullet 1} = 21$

n points

$C(n,\ 2) = \frac{P(n,\ 2)}{2!} = \frac{n(n-1)}{2}$

Problem Set 14.3 Probability

1. $P(E) = \frac{n(E)}{n(S)} = \frac{2}{4} = \frac{1}{2}$

3. $P(E) = \frac{n(E)}{n(S)} = \frac{3}{4}$

5. $P(E) = \frac{n(E)}{n(S)} = \frac{1}{8}$

7. $P(E) = \frac{n(E)}{n(S)} = \frac{7}{8}$

9. $P(E) = \frac{n(E)}{n(S)} = \frac{1}{16}$

11. $P(E) = \frac{n(E)}{n(S)} = \frac{6}{16} = \frac{3}{8}$

13. $P(E) = \frac{n(E)}{n(S)} = \frac{2}{6} = \frac{1}{3}$

15. $P(E) = \frac{n(E)}{n(S)} = \frac{3}{6} = \frac{1}{2}$

17. $P(E) = \frac{n(E)}{n(S)} = \frac{5}{36}$

19. $P(E) = \frac{n(E)}{n(S)} = \frac{6}{36} = \frac{1}{6}$

21. $P(E) = \frac{n(E)}{n(S)} = \frac{11}{36}$

23. $P(E) = \frac{n(E)}{n(S)} = \frac{13}{52} = \frac{1}{4}$

25. $P(E) = \frac{n(E)}{n(S)} = \frac{26}{52} = \frac{1}{2}$

27. $P(E) = \frac{n(E)}{n(S)} = \frac{1}{25}$

29. $P(E) = \frac{n(E)}{n(S)} = \frac{9}{25}$

31. $n(S) = C(5,\ 2) = 10$

$n(E) = C(4,\ 1) = 4$

$P(E) = \frac{n(E)}{n(S)} = \frac{4}{10} = \frac{2}{5}$

33. $n(S) = C(5,\ 2) = 10$

Only one committee with Bill and Carl.

$n(E) = 9$

$P(E) = \frac{n(E)}{n(S)} = \frac{9}{10}$

35. $n(S) = C(8,\ 5) = 56$

$n(E) = C(6,\ 3) = 20$

$P(E) = \frac{n(E)}{n(S)} = \frac{20}{56} = \frac{5}{14}$

37. $n(S) = C(8,\ 5) = 56$

Number of committees with either Chad or Devon.

$C(7,\ 4) + C(7,\ 4) - C(6,\ 3)$

$35+35-20$

50

There are 50 committees with Chad and Devon but now we need to find the number of committees with Chad or Devon but not both Chad and Devon.

$n(E)=50-C(6,\ 3)=50-20=30$

$P(E)=\frac{n(E)}{n(S)}=\frac{30}{56}=\frac{15}{28}$

39. $n(S)=C(10,\ 3)=120$

$n(E)=C(8,\ 3)=56$

$P(E)=\frac{n(E)}{n(S)}=\frac{56}{120}=\frac{7}{15}$

41. $n(S)=C(10,\ 3)=120$

$n(E)=C(2,\ 2)\bullet C(8,\ 1)=1\bullet 8=8$

$P(E)=\frac{n(E)}{n(S)}=\frac{8}{120}=\frac{1}{15}$

43. $n(S)=P(3,\ 3)=6$

$n(E)=4$

$P(E)=\frac{n(E)}{n(S)}=\frac{4}{6}=\frac{2}{3}$

45. $n(S)=C(5,\ 4)=5$

$n(E)=C(4,\ 4)=1$

$P(E)=\frac{n(E)}{n(S)}=\frac{1}{5}$

47. $n(S)=P(9,\ 9)=362{,}880$

There are two ways, either math books first or history books first.

$n(E)=2\times P(4,\ 4)\times P(5,\ 5)$

$n(E)=2\times 24\times 120=5760$

$P(E)=\frac{n(E)}{n(S)}=\frac{5760}{362{,}880}=\frac{1}{63}$

49. $n(S)=P(4,\ 4)=24$

$n(E)=2\times P(3,\ 3)$

$n(E)=2\times 6=12$

$P(E)=\frac{n(E)}{n(S)}=\frac{12}{24}=\frac{1}{2}$

51. $n(S)=C(11,\ 4)=330$

$n(E)=C(6,\ 2)\times C(5,\ 2)$

$n(E)=15\times 10=150$

$P(E)=\frac{n(E)}{n(S)}=\frac{150}{330}=\frac{5}{11}$

53. $n(S)=C(10,\ 5)=252$

There are two teams that Al, Bob, and Carl could be on together.

$n(E)=2\times C(7,\ 2)$

$n(E)=2\times 21=42$

$P(E)=\frac{n(E)}{n(S)}=\frac{42}{252}=\frac{1}{6}$

55. $n(S)=2^9=512$

$n(E)=C(9,\ 3)=84$

$P(E)=\frac{n(E)}{n(S)}=\frac{84}{512}=\frac{21}{128}$

57. $n(S)=2^5=32$

Getting 0 heads

$n(E_0)=C(5,\ 0)=1$

Getting 1 head

$n(E_1)=C(5,\ 1)=5$

Getting 2 heads

$n(E_2)=C(5,\ 2)=10$

Getting 3 heads

$n(E_3)=C(5,\ 3)=10$

For not getting more than 3 heads

$n(E)=1+5+10+10=26$

$P(E)=\frac{n(E)}{n(S)}=\frac{26}{32}=\frac{13}{16}$

59. $n(S) = \frac{7!}{2!3!1!1!} = 420$

$n(E) = \frac{5!}{3!1!1!} = 20$

$P(E) = \frac{20}{420} = \frac{1}{21}$

Problem Set 14.4 Some Properties of Probability and Tree Diagrams

1a. $n(S) = 36$

$n(E) = 5$

$P(E) = \frac{n(E)}{n(S)} = \frac{5}{36}$

b. $n(S) = 36$

$n(E') = 1$

$n(E) = 36 - 1 = 35$

$P(E) = \frac{n(E)}{n(S)} = \frac{35}{36}$

c. $n(S) = 36$

$n(E) = 21$

$P(E) = \frac{n(E)}{n(S)} = \frac{21}{36} = \frac{7}{12}$

d. $n(S) = 36$

$n(E') = 0$

$n(E) = 36 - 0 = 36$

$P(E) = \frac{n(E)}{n(S)} = \frac{36}{36} = 1$

3. $n(S) = 36$

$n(E') = 6$

$n(E) = 36 - 6 = 30$

$P(E) = \frac{n(E)}{n(S)} = \frac{30}{36} = \frac{5}{6}$

5a. $n(S) = 2^5 = 32$

$n(E) = 1$

$P(E) = \frac{n(E)}{n(S)} = \frac{1}{32}$

b. $n(S) = 2^5 = 32$

$n(E) = 5$

$P(E) = \frac{n(E)}{n(S)} = \frac{5}{32}$

c. $n(S) = 32$

$n(E') = 1$

$n(E) = 32 - 1 = 31$

$P(E) = \frac{n(E)}{n(S)} = \frac{31}{32}$

7. $n(S) = 52$

$n(E') = 4$

$n(E) = 52 - 4 = 48$

$P(E) = \frac{n(E)}{n(S)} = \frac{48}{52} = \frac{12}{13}$

9. $n(S) = C(9, 2) = 36$

$n(E') = C(6, 2) = 15$

$n(E) = 36 - 15 = 21$

$P(E) = \frac{n(E)}{n(S)} = \frac{21}{36} = \frac{7}{12}$

11. $n(S) = C(12, 3) = 220$

$n(E') = C(7, 3) = 35$

$n(E) = 220 - 35 = 185$

$P(E) = \frac{n(E)}{n(S)} = \frac{185}{220} = \frac{37}{44}$

13a. $P(BB) = \frac{5}{11} \times \frac{4}{10} = \frac{2}{11}$

b. $P(BG) = \frac{5}{11} \times \frac{6}{10} = \frac{3}{11}$

$P(GB) = \frac{6}{11} \times \frac{5}{10} = \frac{3}{11}$

P(one blue, other gold) $= \frac{3}{11} + \frac{3}{11} = \frac{6}{11}$

c. P(no gold) = P(B, B) $= \frac{5}{11} \times \frac{4}{10} = \frac{2}{11}$

P(at least one gold) $= 1 - \frac{2}{11} = \frac{9}{11}$

d. $P(GG) = \frac{6}{11} \times \frac{5}{10} = \frac{3}{11}$

15a. $P(RR) = \frac{5}{17} \times \frac{4}{16} = \frac{5}{68}$

b. $P(WW) = \frac{12}{17} \times \frac{11}{16} = \frac{33}{68}$

c. $P(RW) = \frac{5}{17} \times \frac{12}{16} = \frac{15}{68}$

$P(WR) = \frac{12}{17} \times \frac{5}{16} = \frac{15}{68}$

P(one white, other red) $= \frac{15}{68} + \frac{15}{68} = \frac{30}{68} = \frac{15}{34}$

d. P(no red) = P(WW) $= \frac{12}{17} \times \frac{11}{16} = \frac{33}{68}$

P(at least one red) $= 1 - \frac{33}{68} = \frac{35}{68}$

17a. $P(WW) = \frac{4}{7} \times \frac{1}{3} = \frac{4}{21}$

b. $P(RR) = \frac{3}{7} \times \frac{2}{3} = \frac{2}{7}$

c. $P(RW) = \frac{3}{7} \times \frac{1}{3} = \frac{1}{7}$

$P(WR) = \frac{4}{7} \times \frac{2}{3} = \frac{8}{21}$

P(one white, other red) $= \frac{1}{7} + \frac{8}{21} = \frac{11}{21}$

Chapter 14 Review

1. $P(6, 6) = 6! = 720$

3. $6 \times 5 \times 4 = 120$

5. $C(6, 3) = \frac{P(6, 3)}{3!} = \frac{6!}{3!3!}$

$C(6, 3) = 20$

7. $C(13, 5) = \frac{P(13, 5)}{5!} = \frac{13!}{8!5!}$

$C(13, 5) = 1287$

9. Committees with no man

$C(5, 3) = 10$

All possible 3-person committee

$C(9, 3) = 84$

Committee with at least one man

$84 - 10 = 74$

11. Subset with A but not B

$C(6, 3) = 20$

Subset with B but not A

$C(6, 3) = 20$

Subset with A or B but not both

$20 + 20 = 40$

13. $C(3, 2) \times C(4, 1) \times C(5, 1)$

$3 \times 4 \times 5 = 60$

15. $n(S) = 2^3 = 8$

$n(E) = C(3, 2) = 3$

$P(E) = \frac{n(E)}{n(S)} = \frac{3}{8}$

17. $n(S) = 6 \times 6 = 36$

$n(E) = 5$

$P(E) = \frac{n(E)}{n(S)} = \frac{5}{36}$

19. $n(S) = P(5, 5) = 120$

If two people are side by side, there are three seats to fill and then double the number of ways for the order of the two side by side and times four for the different seat positions for the side by side people.

$P(3, 3) \times 2 \times 4 = 6(2)(4) = 48$

Therefore, not side by side is $120 - 48 = 72$

$n(E) = 72$

$P(E) = \frac{n(E)}{n(S)} = \frac{72}{120} = \frac{3}{5}$

21. $n(S) = 2^6 = 64$

n(no heads) = 1

n(one head) = $C(6, 1) = 6$

n(one or less heads) = $1 + 6 = 7$

n(at least two heads) = $64 - 7 = 57$

$P(E) = \frac{n(E)}{n(S)} = \frac{57}{64}$

23. $n(S) = \frac{6!}{3!1!1!1!} = 120$

$n(E) = \frac{5!}{3!1!1!} = 20$

$P(E) = \frac{n(E)}{n(S)} = \frac{20}{120} = \frac{1}{6}$

25. $n(S) = C(8,\ 4) = 70$

Committees with Alice but not Bob

$C(6,\ 3) = 20$

Committees with Bob but not Alice

$C(6,\ 3) = 20$

Committees with Alice or Bob but not both

$n(E) = 20 + 20 = 40$

$P(E) = \frac{40}{70} = \frac{4}{7}$

27. $n(S) = C(13,\ 4) = 715$

$n(\text{no women}) = C(6,\ 4) = 15$

$n(\text{at least one women}) = 715 - 15 = 700$

$P(E) = \frac{n(E)}{n(S)} = \frac{700}{715} = \frac{140}{143}$

29. $P(RB) = \frac{4}{12} \times \frac{3}{12} = \frac{1}{12}$

$P(BR) = \frac{3}{12} \times \frac{4}{12} = \frac{1}{12}$

$P(\text{one B, other R}) = \frac{1}{12} + \frac{1}{12} = \frac{1}{6}$

31. $P(\text{no red}) = \frac{4}{7} \times \frac{3}{6} = \frac{2}{7}$

$P(\text{at least one red}) = 1 - \frac{2}{7} = \frac{5}{7}$

Chapter 14 Test

1. If Ivan is in the left end seat,

$n(E_l) = 1 \times P(3,\ 3) = 6$

If Ivan is in the right end seat,

$n(E_r) = P(3,\ 3) \times 1 = 6$

$n(E) = 6 + 6 = 12$

3. $6 \times 5 \times 4 \times 2 = 240$

5. $n(E) = 5$

7. $n(S) = 6 \times 6 = 36$

$n(E') = 10$

$n(E) = 36 - 10 = 26$

9. $C(8,\ 6) = \frac{P(8,\ 6)}{6!}$

$C(8,\ 6) = \frac{8 \bullet 7 \bullet 6 \bullet 5 \bullet 4 \bullet 3}{6!} = 28$

11. $6 \times C(10,\ 2)$

6×45

270

13. $n(S) = 6 \times 6 = 36$

$n(E) = 3$

$P(E) = \frac{n(E)}{n(S)} = \frac{3}{36} = \frac{1}{12}$

15. $n(5) = 2^6 = 64$

$n(E) = C(6,\ 5) = 6$

$P(E) = \frac{n(E)}{n(S)} = \frac{6}{64} = \frac{3}{32}$

17. $n(S) = 6 \times 5 \times 4 = 120$

$n(E) = 5 \times 5 \times 4 = 100$

$P(E) = \frac{n(E)}{n(S)} = \frac{100}{120} = \frac{5}{6}$

19. $n(S) = P(6,\ 6) = 720$

$n(E) = 4 \times 4 \times 3 \times 2 \times 1 \times 3 = 288$

$P(E) = \frac{n(E)}{n(S)} = \frac{288}{720} = \frac{2}{5}$

21. $n(S) = C(8,\ 2) = 28$

$n(E') = C(5,\ 2) = 10$

$n(E) = 28 - 10 = 18$

$P(E) = \frac{n(E)}{n(S)} = \frac{18}{28} = \frac{9}{14}$

23. $P(WG) = \frac{7}{19} \bullet \frac{12}{19} = \frac{84}{361}$

$P(GW) = \frac{12}{19} \bullet \frac{7}{19} = \frac{84}{361}$

$P(\text{one G, other W}) = \frac{84}{361} + \frac{84}{361} = \frac{168}{361}$

25. $P(\text{no red}) = \frac{7}{13} \bullet \frac{6}{12} = \frac{7}{26}$

$P(\text{at least one red}) = 1 - \frac{7}{26} = \frac{19}{26}$